Oxford Series in Ecology and Evolution
Edited by Jennifer A. Dunne, H. Charles J. Godfray, and Ben Sheldon

The Comparative Method in Evolutionary Biology
Paul H. Harvey and Mark D. Pagel

The Cause of Molecular Evolution
John H. Gillespie

Dunnock Behaviour and Social Evolution
N. B. Davies

Natural Selection: Domains, Levels, and Challenges
George C. Williams

Behaviour and Social Evolution of Wasps: The Communal Aggregation Hypothesis
Yosiaki Itô

Life History Invariants: Some Explorations of Symmetry in Evolutionary Ecology
Eric L. Charnov

Quantitative Ecology and the Brown Trout
J. M. Elliott

Sexual Selection and the Barn Swallow
Anders Pape Møller

Ecology and Evolution in Anoxic Worlds
Tom Fenchel and Bland J. Finlay

Anolis Lizards of the Caribbean: Ecology, Evolution, and Plate Tectonics
Jonathan Roughgarden

From Individual Behaviour to Population Ecology
William J. Sutherland

Evolution of Social Insect Colonies: Sex Allocation and Kin Selection
Ross H. Crozier and Pekka Pamilo

Biological Invasions: Theory and Practice
Nanako Shigesada and Kohkichi Kawasaki

Cooperation Among Animals: An Evolutionary Perspective
Lee Alan Dugatkin

Natural Hybridization and Evolution
Michael L. Arnold

The Evolution of Sibling Rivalry
Douglas W. Mock and Geoffrey A. Parker

Asymmetry, Developmental Stability, and Evolution
Anders Pape Møller and John P. Swaddle

Metapopulation Ecology
Ilkka Hanski

Dynamic State Variable Models in Ecology: Methods and Applications
Colin W. Clark and Marc Mangel

The Origin, Expansion, and Demise of Plant Species
Donald A. Levin

The Spatial and Temporal Dynamics of Host–Parasitoid Interactions
Michael P. Hassell

The Ecology of Adaptive Radiation
Dolph Schluter

Parasites and the Behavior of Animals
Janice Moore

Evolutionary Ecology of Birds
Peter Bennett and Ian Owens

The Role of Chromosomal Change in Plant Evolution
Donald A. Levin

Living in Groups
Jens Krause and Graeme D. Ruxton

Stochastic Population Dynamics in Ecology and Conservation
Russell Lande, Steiner Engen, and Bernt-Erik Sæther

The Structure and Dynamics of Geographic Ranges
Kevin J. Gaston

Animal Signals
John Maynard Smith and David Harper

Evolutionary Ecology: The Trinidadian Guppy
Anne E. Magurran

Infectious Diseases in Primates: Behavior, Ecology, and Evolution
Charles L. Nunn and Sonia Altizer

Computational Molecular Evolution
Ziheng Yang

The Evolution and Emergence of RNA Viruses
Edward C. Holmes

Aboveground–Belowground Linkages: Biotic Interactions, Ecosystem Processes, and Global Change
Richard D. Bardgett and David A. Wardle

Principles of Social Evolution
Andrew F. G. Bourke

Maximum Entropy and Ecology: A Theory of Abundance, Distribution, and Energetics
John Harte

Ecological Speciation
Patrik Nosil

Energetic Food Webs: An Analysis of Real and Model Ecosystems
John C. Moore and Peter C. de Ruiter

Evolutionary Biomechanics: Selection, Phylogeny, and Constraint
Graham K. Taylor and Adrian L. R. Thomas

Quantitative Ecology and Evolutionary Biology: Integrating Models with Data
Otso Ovaskainen, Henrik de Knegt, and Maria del Mar Delgado

Mitonuclear Ecology
Geoffrey E. Hill

Mitonuclear Ecology

GEOFFREY E. HILL

Professor, Department of Biological Sciences, Auburn University, USA

OXFORD
UNIVERSITY PRESS

Great Clarendon Street, Oxford, OX2 6DP,
United Kingdom

Oxford University Press is a department of the University of Oxford.
It furthers the University's objective of excellence in research, scholarship,
and education by publishing worldwide. Oxford is a registered trade mark of
Oxford University Press in the UK and in certain other countries

First Edition published in 2019

Published in the United States of America by Oxford University Press
198 Madison Avenue, New York, NY 10016, United States of America

British Library Cataloguing in Publication Data
Data available

Library of Congress Control Number: 2018964761

ISBN 978–0–19–881825–0 (hbk.)
ISBN 978–0–19–881826–7 (pbk.)

DOI: 10.1093/oso/9780198818250.001.0001

Preface

Birdwatching brought me to mitonuclear ecology. As a pre-teen, I was introduced to birding by a high school biology teacher, and I quickly became obsessed with seeing all the birds in my *Golden Field Guide to the Birds of North America*. I certainly didn't appreciate it at the time, but a field guide to the birds of a region, through beautiful color illustrations, presents some of the most significant unexplained patterns in the natural world. All birds are neatly binned into species; the field guide suggests no ambiguity. Each species has a discrete range. Some birds thrive exclusively in warm climates. Other birds that seem identical in size, shape, and life history thrive in cold climates. Some birds are colorful; some are not. Like people, all birds come in two sexes. In some birds, the sexes look alike; in others, they do not. Explaining these patterns so starkly revealed in the pages of a field guide became the academic passion of my life. But a decade into the new millennium, despite 30 years of effort, I felt that I had made frustratingly little progress toward answering any of the basic questions that continued to confront me each time I opened a field guide. Then, I read *Power, Sex, Suicide: Mitochondria and the Meaning of Life* by Nick Lane, and for the first time I started to think about the major patterns of bird diversity from the perspective of the evolution of eukaryotes and the coevolution and coadaptation of mitochondrial (mt) and nuclear (N) genomes. It was as if I had turned on a light in a dark room.

In this book, I summarize emerging new theory and empirical observations that reinterpret key features of eukaryotic life in light of the necessity of coadapation of co-functioning mt and N genes. I call this emerging line of research mitonuclear ecology. It is the integration of studies of the coevolution and coadaptation of mt and N gene products into investigations of the evolution and ecology of whole organisms in natural environments. My purpose in creating this book is to present the new ideas and empirical observations that underlie mitonuclear ecology as an important new research focus. I propose that evolutionary ecology is poised to make a substantial leap forward in understanding fundamental features of complex organisms as evolutionary ecologists fully consider the implications of coadaptation and coevolution of mt and N genes to enable core life processes.

The origins of mitonuclear ecology lay in the biochemistry and cell biology laboratories that, in the mid-twentieth century, were in a different universe than lab groups focused on the evolution of whole organisms. In 1961, Peter Mitchell made one of the great intellectual leaps in the history of science when he proposed that chemiosmosis was the mechanism for oxidative phosphorylation and aerobic respiration in the mitochondria of eukaryotes (Mitchell, 1961) (Box 1.1). Mitchell's brilliant discovery was made at about the same time as the discovery that the mitochondrion carried its own genome (Nass and Nass, 1963; Schatz et al., 1964). Within a few years following these breakthroughs, cell biologists deduced that some components of the electron transport system were encoded by mt genes, while other components were encoded by N genes (Borst and Grivell, 1973). These were monumental discoveries by reductionistic biologists, but they went essentially unnoticed by organismal biologists focused on understanding evolution of plants and animals in natural environments. As a graduate student in the evolutionary biology program at the University of Michigan in the 1980s, I was taught nothing about oxidative phosphorylation, chemiosmosis, Peter Mitchell, or the function of mt genes. I was informed that mt genes were housekeeping genes, that variation in mt genotype was functionally neutral, and that the only reason to pay attention to mt DNA was for phylogenetic reconstructions. In retrospect, it was the biggest failure of my education in biology.

In the spring of 2012, I saw a reference to the book *Power, Sex, Suicide: Mitochondria and the Meaning of Life* by Nick Lane (2005) and on a whim I ordered a copy. I had never heard of Lane, and the topic of the book—mitochondria and the origins of complex life—was far from my focal interests as a behavioral ecologist studying plumage coloration in birds (Hill, 2002). By the end of the first chapter, I was captivated. It is not an exaggeration to say that Lane's book fundamentally changed my academic life and the direction of my career as a biologist. What I found in Lane's book was a narrative that shook the foundation of my view of biology—at once showing me the relevance of subcellular biology and forcing me to rethink fundamental aspects of evolutionary theory. I shared the book with my colleague, Jim Johnson, who is a biochemist well acquainted with mitochondrial processes, and Jim too was captivated. Jim was reading ahead of me and he sent me a note suggesting that I should page ahead to the section on the genes that code for the electron transport system. Jim wrote to me: "Lane makes the argument that the *compatibility* between the *DNA in the mitochondria...and the nucleus* is critical for overall respiratory efficiency." Jim's note was my first introduction to the concept of mitonuclear compatibility. Jim and I were already thinking and writing about the role of mitochondrial function in the production of ornamental red pigments in birds (Hill and Johnson, 2012; see also Chapter 7). After reading Lane's book we began to contemplate the role of mitonuclear interaction in condition-dependent sexual signaling. Soon thereafter, we set to work on outlining a theory of sexual selection whereby ornaments signal mitonuclear compatibility (Hill and Johnson, 2013).

Conceptualizing how mates could be sorted for mitonuclear compatibility by assessment of ornaments was one of the greatest intellectual challenges of my career, and, in considering how mate choice might maintain mitonuclear coadaptation, I came to realize that mitonuclear interactions lie at the heart of not only sexual selection but also the process of speciation and the nature of species. The role of mitonuclear coevolution in the process of adaptation fell into place and soon I was viewing essentially all of the major features of complex life through mitonuclear lenses. This book is an attempt to organize new thinking (emerging in my own head as well as within the subdiscipline of evolutionary ecology) about how mitonuclear interactions shape complex life.

The theme of this book is *mito*nuclear coadaptation and coevolution and explicitly not broader *cyto*nulear interactions. I scarcely mention plastid genomes in these chapters. I think there can be merit to making theoretical arguments as universal as possible (and hence I resisted my inclination to focus this book only on *animal* mitonuclear ecology), but there comes a point at which too broad of a focus is no focus at all. It will be challenging enough for field ecologists to wade through my opening chapters focused on genomic architecture, cell biology, and respiratory pathways. It would be counter-productive to follow these chapters with additional chapters on photosynthesis and a second set of genomic interactions between chloroplast and N genomes. Mitonuclear interactions are the common threads that bind all eukaryotes. There are, to be sure, many parallels in mitonuclear and chloronuclear genomic interactions, but I leave the development of plastid nuclear ecology for other evolutionary biologists.

I make no attempt to be balanced in my presentation of ideas and hypotheses. The purpose of this book is to succinctly outline the basic principles of mitonuclear ecology and to advocate for the importance of this approach to evolutionary questions. Topics such as the evolution of sex, the nature of species, and the process of sexual selection all have histories of investigations stretching back decades and a literature of hundreds if not thousands of papers. A balanced and comprehensive review of any of these topics would be a book-length project in itself, and such reviews are already available. I draw on previous literature primarily to set the stage for the new hypotheses emerging from consideration of the coadaptation of mt and N genomes.

My target audience is organismal biologists with limited training in cellular respiration and cell biology. With an audience of biologists, I don't take the space to explain basic concepts like transcription, translation, natural selection, genetic drift, and so forth, but I do try to be careful in what I take for granted regarding previous knowledge of biochemistry and cell biology. It is my hope that not only organismal biologist interested in cellular-level processes but also cell biologists interested in macro-evolution will find this book interesting. It is in the integration of genomics and biochemistry with ecology and organismal biology that the power of mitonuclear ecology lies. We need insights from both the top down and the bottom up. I would be exceedingly pleased if undergraduate and graduate students found the book to be a useful compass, pointing, at least in a general sense, in the direction that new research in evolutionary ecology might most fruitfully proceed.

Acknowledgments

Much of this book was written while I was on a sabbatical visit to Monash University in Melbourne, Australia, for the first part of 2018, hosted by my friend and colleague Damian Dowling. The Dowling lab has emerged as one of the most innovative research teams in the world, pushing forward the boundaries of understanding of mitonuclear ecology. Damian and his lab group could not have been more welcoming and encouraging as I struggled to write some of the most challenging chapters of the book. Damian and the Dowling lab read every chapter in early draft stages and provided hugely valuable input. While in Australia, I also had extensive discussions and received critical feedback on my book from other colleagues, particularly Paul Sunnucks, Leo Joseph, and Chris Greening. Chris Greening is a brilliant biochemist with interests that range from the physical interaction of atoms to how climate might shape the ranges of Australian birds, and he did his best to coach me through the complexities of oxidative phosphorylation. Leo Joseph, curator of birds at the Commonwealth Scientific and Industrial Research Organisation (CSIRO) and an international expert on species boundaries in birds, critiqued the chapter on speciation. Paul Sunnucks showed me the potential impact that mitonuclear thinking can have on studies of phylogeography in the field as we trapped eastern yellow robins on his study sites. The faculty and students at Monash served as vital sounding boards for my ideas, and I am very grateful for the time and effort they invested in making my book better. In Australia I also went on a lecture tour, presenting my ideas at universities throughout Victoria, New South Wales, and the Capital District. The opportunity to vet my ideas in front of hundreds of biologists from diverse backgrounds helped me gauge how the ideas in my book were likely to be received. Thanks to all of my Australian colleagues who vetted my ideas. And finally, my most important first-draft readers were the students and postdocs in my own lab and in the lab of my wife and close colleague, Wendy Hood, at Auburn University. The Hill–Hood lab groups were key in helping organize and construct chapters, and Wendy was the great motivator for getting the project done. Other colleagues including Daniel Sloan, Justin Havird, Ron Burton, Ryan Weaver, Zhiyuan Ge, Nick Justyn, Matt Powers, Yufeng Zhang, Halie Taylor, Noel Park, Kyle Heine, Chloe Josefson, Kristjan Niitepold, Tori Andreasen, Rebecca Koch, Andrea Pozzi, Rebecca Vaught, Winston Yee, Vanessa Higham, Sean Layh, Magdalena Nystrand, Ilaria Venturelli, Ekta, Yoshana Fonseka, Venkatesh Radha, and Tara-Lyn Carter provided key feedback on various chapters. Sasha Pavlova and Hernan Morales provided unpublished data and figures.

The idea for writing this book emerged initially from a weekly reading/discussion group centered on Nick Lane's book *Power, Sex, Suicide: Mitochondria and the Meaning of Life* that I organized in the Department of Biological Sciences at Auburn University in the fall of 2012. That discussion group was formalized into a graduate class, "Mitonuclear Ecology," that I taught at Auburn in 2014 and 2016. Notes from my Mitonuclear Ecology class formed the foundation of this book. The students, postdocs, and faculty who participated in those classes played key roles in shaping my thinking about the coevolution and coadaptation of mt and N genes, and to the participants in those discussion groups I owe the impetus for creating this book.

Contents

1

The genomic architecture of eukaryotes

Life is a rejection of entropy. It succeeds by harnessing and focusing energy toward the maintenance of order and stability. The greater the complexity of the organism, the more energy that is required to hold back the surging tide of chaos. Hence, the story of life on Earth is fundamentally a story of location, extraction, and production of energy. The origin of complex life was enabled by a radical redesign of simple prokaryotic cells that created unprecedented opportunities for production of energy. The linchpin of this evolutionary innovation, which changed the nature of life on Earth, was the mitochondrion.

The ascension to complexity was not an inevitable endpoint in the evolution of life. Eukaryotic life emerged from exclusively prokaryotic world in which tiny life forms extracted energy from the environment with fantastic efficiency but gained little in size or overall complexity through 2 billion years of evolution (DeLong et al., 2010). With modest capacities for energy extraction, Lane (2015a) speculated that this exclusively prokaryotic world of small and relatively simple organisms may have persisted indefinitely. There seemed to be no escape from the perpetual race to replicate. But the improbable evolution of a new type of organism, which carried two genomes, provided the radical restructuring of cellular design that jarred life out of its perpetual race for more efficient energy extraction and faster replication speed. The delegation of life processes to two genomes was an inescapable necessity in the evolution of complex life, and it is in the improbable fusion of prokaryotic life to create eukaryotic life that our story begins.

Eukaryotic evolution

Eukaryotes are chimeras

There is now compelling evidence that the origin of eukaryotes, the origin of mitochondria, and the origin of complex life were one and the same event (Lane, 2015b; Martin et al., 2015). About 2 billion years ago, two prokaryotic lineages established an intimate symbiotic relationship, evolving over an unknown but presumably long period to become ever more interdependent. One of these partners was a prokaryote belonging to the life domain archaea and the other was a prokaryote belonging to the life domain bacteria, and in a pivotal event in the history of life on Earth, these two

Mitonuclear Ecology. Geoffrey E. Hill, Oxford University Press (2019). © Geoffrey E. Hill 2019.
DOI: 10.1093/oso/9780198818250.001.0001

prokaryotic cells fused to form an entirely new organism that would evolve into a novel life domain called eukaryota, the eukaryotes (Williams et al., 2013; Poole and Gribaldo, 2014). I use the term "fused" with purpose because this was not a simple exchange of genes. Rather, this was a joining of what had been two independent organisms into a wholly new type of organism in which both founders retained some degree of independent identity. This unlikely fusion was the starting point for the evolution of complex life on Earth, but it was far from a finished process at the moment when the archaeon and bacterium joined. The 2 billion years since this chimeric fusion has been one long negotiation between the united partners regarding how to partition cellular duties, who is responsible for what genes, and, above all, how to coordinate genomic products to enable system function. Like friends sharing an apartment, the relationship requires a foundation of cooperation, but conflict is never far away.

Martin and Müller (1998) proposed that the likely circumstance that gave rise to the archaeon/bacterial chimera was an intimate symbiosis between the two cell types. This theory, which has become known as the hydrogen hypothesis, proposes that the host cell (the cell that subsumed its partner cell) was an archaeon that derived energy by ingesting hydrogen and carbon dioxide and using these materials to produce methane and energy. This methane producer, or methanogen, required access to free hydrogen, which is scarce in most Earth environments. Fortuitously, hydrogen was the waste product of a second cell type, a bacterium that derived at least some of its energy from anaerobic metabolism of organic material to produce hydrogen and carbon dioxide. The waste of one partner was the energy source for the other and so they were naturally drawn together into a tight symbiosis. Eventually, the bacterium literally moved inside the archaeon and became an internal organelle.

What followed was a series of cataclysmic events within the new chimeric cell as the components of two very different prokaryotic organisms reorganized into a single life entity. The fascinating details of this restructuring of the proto-eukaryotic cell into a true eukaryotic cell are told in brilliant prose in Nick Lane's books *Power, Sex, Suicide: Mitochondria and the Meaning of Life* and *The Vital Question: Energy, Evolution, and the Origins of Complex Life* (Lane, 2005, 2015a), as well as in dozens of articles in technical journals where data and ideas are presented one piece at a time. For the purposes of this book, the essential information is that the bacterium became the mitochondrion of eukaryotic cells and the archaeon became the "host cell" such that its genetic material evolved into the nuclear (N) genome (Sagan, 1967; Gray, 2012). There is strong evidence that this archaeon/bacterium chimeric fusion occurred only once and that all eukaryotes—animals, plants, fungi, protozoa, slime mold, brown algae—share a common ancestor (Williams, 2014; Derelle et al., 2015). In stating that all eukaryotes evolved from a common ancestor and hence that the chimeric fusion of an archaeon and a bacterium occurred only once, I am in no way suggesting that life succeeded in its initial run at this unlikely fusion and restructuring of cells. Given all of the challenges associated with this new architecture, which I review in the first five chapters of this book, one can imagine that the experiment started and failed an incalculable number of times before a thousand fortuitous events fell in sequence and the union succeeded. Keep in mind there is more than a billion years between the rise of bacteria and archaea and the appearance of the first eukaryote, and a billion years provides ample time for extremely improbable events to occur.

From the initial fusion of archaeon and bacterial cells to form a proto-eukaryote, the archaeon genome existed as one (or two in diploids) copy per cell and the bacterial genome existed as multiple copies. In discussing the origin of eukaryotes from the initial bacterium/archaeon fusion, I will focus on the genomes of the two organisms and in so doing I will set the stage for the central theme of the book: coadaptation of mitochondrial (mt) and N genomes. Because mt genes code exclusively for protein subunits of the electron transport system (ETS) or for components of the transcriptional, translational, and replicative machinery needed to produce ETS subunits, it is essential to start with a brief review of how eukaryotic cells produce ATP via aerobic respiration.

OXPHOS and the electron transport system

Most eukaryotes acquire the energy needed for life processes at least partly through aerobic respiration. The first two phases of cellular respiration, glycolysis and the citric acid cycle, occur in the cytoplasm of the cell and in the mitochondrial matrix, respectively (Figure 1.1). These initiating steps in the process of cellular respiration take food

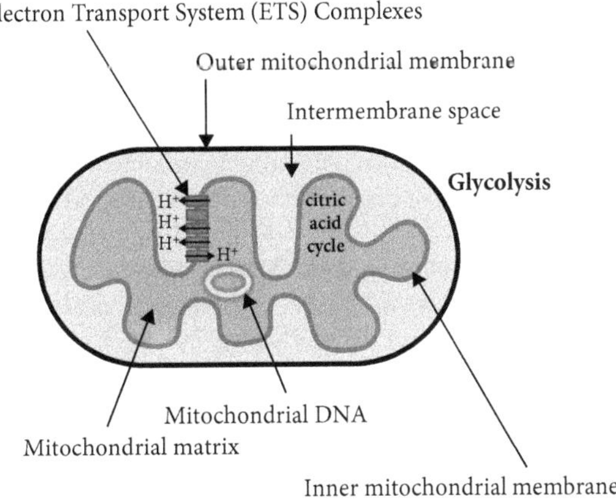

Figure 1.1 A simplified overview of key features of mitochondrial and cellular respiration. The inner mitochondrial membrane divides the cell into the intermembrane space, which lies between the inner and outer mitochondrial membranes, and the mitochondrial matrix, which is the chamber inside the inner mitochondrial membrane. Mitochondrial DNA exists within the mitochondrial matrix. The complexes of the electron transport system are embedded in the inner mitochondrial membrane, forming a critical conduit between the intermembrane space and the mitochondrial matrix. Glycolysis, the first stage of cellular respiration, occurs outside of mitochondria in the cytosol. The second stage of aerobic respiration, the citric acid cycle, occurs in the mitochondrial matrix. OXPHOS is enabled by the electron transport system.

molecules like glucose and break them down to release a small amount of ATP and a substantial amount of the energy-rich coenzymes NADH or $FADH_2$. Despite the appearance that most of the molecular decomposition of food molecules is finished by the end of the citric acid cycle when NADH and $FADH_2$ are produced, about 90 percent of the energy available in a glucose molecule is still retained in NADH or $FADH_2$. This huge remaining energy pool is captured via the process of oxidative phosphorylation (OXPHOS), which is enabled by a series of integrated protein complexes called the electron transport system (ETS) (Box 1.1). The first four complexes of the ETS are

Box 1.1 Ox Phos Wars

By the middle of the twentieth century, cell biologists had deduced the basic chemistry of cellular respiration and identified ATP as the key molecule used by organisms to store energy in a usable form. Biochemistry, like all chemistry, is founded on stoichiometry—balancing the precursors and products in reactions. In basic chemistry, atoms are never created or destroyed; stoichiometry works strictly with integers. When a scientist balances a chemistry equation, he or she is precluded from ending with, say, 73 percent of a hydrogen atom. Atoms are recombined as whole units. So, it was perplexing to say the least that the stoichiometry of cellular respiration never produced integers—it invariably produced fractions of integers. The persistent failure of the reactions driven by the ETS to stoichiometrically balance in experiments indicated that something very fundamental was missing from the theory of cellular respiration.

The paradox of equations that would not balance lasted for more than 20 years until Peter Mitchell solved the problem by hypothesizing that production of ATP involved the creation of an electrochemical gradient across a membrane (Mitchell, 1961). This process of converting food energy into a proton-motive force and then using a membrane potential as the energy source for ATP production lifted the restriction of stoichiometric precision. Protons could be released from the gradient at variable rates enabling ATP to be created in non-stoichiometric ratios.

Like many ideas that represent huge and non-intuitive advances in fields of study, Mitchell's chemiosmosis theory was not embraced quickly or enthusiastically by most of his colleagues. As a matter of fact, arguments about the mechanism by which oxidative phosphorylation generated ATP were so vitriolic that the debate became known as the Ox Phos Wars (Prebble, 2002). Mitchell was subjected to substantial ridicule in the years following the publication of his brilliant hypothesis, and it was not widely accepted as correct for years after it was initially published. In the end, Mitchell received the Nobel Prize in chemistry for his discovery, so when recognition came it came with the highest honor in science.

Studies of mitonuclear ecology build from the insights of Peter Mitchell. The chemiosmosis hypothesis of cellular respiration helped refine understanding of the ETS—the large protein complexes that use energy from electrons to pump protons across a membrane and to create a chemiosmotic energy potential (Slater, 2003). It created the foundation on which the significance of mitonuclear coadaptation and coevolution could be understood. The discovery of chemiosmosis as the basis for energy production in living organisms required a tremendous intellectual leap and great courage by Mitchell, and it stands as one of the most significant discoveries in the history of science.

often referred to as the electron transport chain or respiratory chain, and I will use these terms periodically in this book to refer to complexes that transfer electrons.

The ETS is a set of large and complex proteins that receive high-energy electrons from either NADH or $FADH_2$ and use the energy of these electrons to pump protons across the inner mitochondrial membrane, from the mitochondrial matrix (mitochondrial core) to the intermembrane space. Pumping protons across the membrane creates a net positive charge in the intermembrane space relative to the matrix (Figure 1.2). The membrane potential is the difference in electrical charge created by an excess of protons on one side of the inner mitochondrial membrane. Maintaining this membrane potential is critical for cell survival. In a healthy, properly functioning cell, the gain of protons from active pumping is counterbalanced by discharge of protons back across the membrane. Discharge of protons is typically through ATP synthase (Complex V), which phosphorylates ADP to produce ATP from the energy recaptured from protons as they move back across the membrane (Figure 1.2). Thus, food energy is used to "crank" ATP synthase (the molecule actually pivots on a molecular axis) to generate ATP. It is hard not to use the analogy of the energy potential of water held behind a dam that turns a wheel as it flows to a point below the dam, but the physical mechanisms involved with the flow of water driven by gravity and the flow of protons across an electrochemical gradient are distinct.

In its textbook description, the ETS is composed of five protein complexes designated as: Complex I, II, III, IV, and V (Figure 1.2). More technically, Complex I is

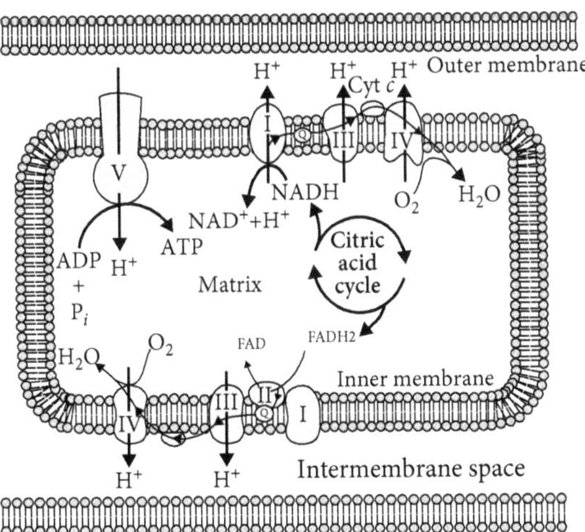

Figure 1.2 A stylized illustration of key elements of the electron transport system. Depicted is a cross section of a mitochondrion indicating the relative position of components. The top right cluster with Complexes I, III, and IV illustrates the primary path of electrons from NADH to Complex I. The bottom left cluster with Complexes II, III, and IV illustrates an alternative route of electrons from FADH2 to Complex II. Both routes end with Complex V converting the proton motive force to ATP. Illustration by Fvasconcellos at Wikimedia Commons.

NADH dehydrogenase; Complex II is succinate dehydrogenase; Complex III is coenzyme Q–cytochrome *c* reductase; Complex IV is cytochrome *c* oxidase; and Complex V is ATP synthase. Complex V is sometimes not included as part of the ETS because it does not transport electrons. Because ATP synthase is the terminal enzyme in the transfer of energy from electrons to ATP, I include ATP synthase when I refer to the ETS. In addition to the large protein complexes that span the inner mt membrane, there is a single-unit protein, cytochrome *c*, that transports electrons from Complex III to IV and a vitamin-like quinone, coenzyme Q, that transports electrons between Complexes I or II and Complex III.

The I to IV numbering of the components of the ETS suggests a linear series of actors, but this is not the case. There is core electron input from NADH in Complex I, which is the starting point for most energy production. In addition, there are side electron inputs from three flavoprotein complexes, namely Complex II, electron-transferring flavoprotein dehydrogenase, and glycerol 3-phosphate dehydrogenase (Frerman and Goodman, 2001; Ishizaki et al., 2005). The electrons from these two unequal sources converge in Complex III for most eukaryotes (Figure 1.2). But even the necessity of Complexes III and IV cannot be stated as a universal generality because in plants and some other eukaryotes, an alternative oxidase (AOX) enables the reduction of oxygen without Complex III or IV (Saha et al., 2016). Thus, for most eukaryotes and in all of the discussion in this book, there are four *core* complexes of the ETS system—Complex I, III, IV, and V—plus three side inputs, including Complex II.

The large components of the ETS are called "complexes" because they are composed of multiple, independently coded protein subunits. For instance, the mammalian Complex I is assembled from about forty-six protein subunits (Hirst, 2013). Each of these protein subunits is a three-dimensional structure that is complicated both in shape and in the distribution of reactive sites and charges. To create a functional ETS complex, each subunit must have a precise three-dimensional configuration that enables it to join and interact correctly with its neighboring subunits (Figure 1.3). The distribution of reactive sites and electrical charge across the structure of any given subunit must be complementary to all neighboring subunits. There must also be complementarity between cytochrome *c* and both Complex III and Complex IV for efficient transfer of electrons. Mutations that lead to changes in a component of an ETS complex or cytochrome *c* can compromise the functionality of a complex and in turn interfere with the functioning of the entire ETS. Such subunit incompatibility is the topic of Chapter 2 and will be the focus of much of this book.

Massive genomic restructuring

Nature favors economy over excess (Wolf and Koonin, 2013). When two organisms exist in a tight symbiosis, it is common for genes whose function is duplicated between the two partners to be lost in one of the symbionts (Moran, 2003, Moran et al., 2008; McCutcheon and Moran, 2012). For instance, the bacterium *Carsonella*

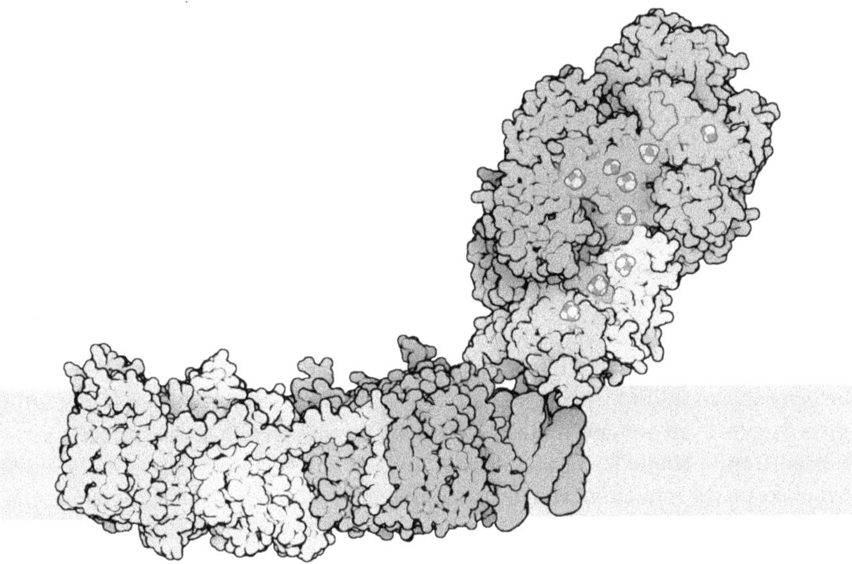

Figure 1.3 A three-dimensional reconstruction of a bacterial Complex I molecule with fourteen protein subunits, each shaded a different tone. The position of the membrane in which the molecule is embedded is shown as a rectangle. This model is presented to emphasize the intricate three-dimensional fit among subunits that is necessary to produce a functional complex. The human Complex I molecule has forty-six subunits—more than three times the structural complexity of the illustrated bacterial complex. Image from the Research Collaboratory for Structural Bioinformatics Protein Data Bank (RCSB PDB) "Molecule of the Month": Inspiring a Molecular View of Biology (Goodsell et al., 2015).

ruddii is an ancient symbiont of the sap-feeding insect *Pachypsylla venusta* and it has lost so many genes through eons of coevolution with its host that it is arguably now an organelle (Sloan et al., 2014a). *Carsonella ruddii* has lost not only genes whose functions are duplicated by host genes, but has also transferred some genes that are essential for vital biosynthetic pathways (Sloan et al., 2014a). The dynamics of genome reduction in obligate symbionts provide insights into the evolution of the N and mt genomes in early eukaryote evolution (Moran and Wernegreen, 2000; Andersson et al., 2003; Rand et al., 2004).

Once the primordial eukaryote emerged from the fusion of two formerly independent cells, genes that existed as two versions that duplicated the same function in the two entities were reduced to one version in the unified organism (Timmis et al., 2004). In most cases, the version of the gene that was retained was located in the N genome because, from the beginning, genes in the nucleus existed as one copy (technically two copies in diploids) per cell, while genes in the mt genome existed as many copies in the multiple mitochondria inside the cell. It was fundamentally more efficient to maintain most genes in the unique central genome of the cell rather than as multiple copies in mt genomes, and the nucleus quickly became the information center of the eukaryotic cell (Adams and Palmer, 2003). Moreover, as the eukaryotic genomes evolved, the N genome became sexual and evolved the capacity for recombination (see Chapter 5 for justification of the assumption that mitochondria evolved before sexual reproduction). The N genome also typically had lower mutation rates compared with the mt genome to which it was matched (Lynch, 1997) and, unlike the mt genome, the N genome avoided Hill–Robertson effects via recombination (discussed in detail in Chapter 5). Thus, the nucleus became the better depository for genomic information (Allen and Raven, 1996; Blanchard and Lynch, 2000). The gene copy that was retained in the nucleus was frequently of archaeon origin, because such genes were already located in the N genome, but sometimes the bacterial version of the gene was moved to the nucleus and the archaeon version was lost (Gray, 2012). Interestingly, even though there is now broad consensus that the mitochondrion evolved from a single bacterial ancestor, only about 14–16 percent of the proteins that make up the mitochondrion of a modern eukaryote seem to be of bacterial origin (Gabaldón and Huynen, 2004). Components of the bacterial partner have been modified, added to, and replaced in the course of evolution from free-living cell to mitochondrion (Gray, 2015).

The culling of duplicate genes from the mt genome resulted in a substantial reduction of mt genome size (Gray, 2012) (Figure 1.4), but shrinking of the mt genome was not limited to the removal of duplicate genes. Many unique mt genes that were critical for mitochondria-specific function were also transferred to the nucleus (Adams and Palmer, 2003). This transfer of functional mt genes to the nucleus necessitated that, constantly through the life of every eukaryotic organism, numerous proteins synthesized outside the mitochondrion had to be transported into the mitochondrion (Bolender et al., 2008; Calvo and Mootha, 2010). Very rapidly in the evolution of eukaryotes, nearly all of the genes that had originally comprised the bacterial genome were deleted or moved to the nucleus (Figure 1.4). Most modern species of alpha-proteobacteria, which is proposed to be the class of bacteria from which mitochondria evolved (Lang et al., 1999), have genomes with about 2000–4000 genes (Koonin and Wolf, 2008); in contrast, the mt genomes of bilaterian animals have thirty-seven genes (see Box 1.2 for a discussion of variation in mt genome size). Thus, about 99 percent of the genes that were present in the bacterial partner that evolved into the mitochondrion in the early evolution of eukaryotes were lost or moved to the nucleus (Figure 1.4). This process of gene transfer was so nearly complete that an obvious question is: why weren't *all* genes moved out of the mitochondria (Kleine et al., 2009)? Why retain that last vestige of mt genes? This entire book is about the enormous consequences of the retention of those few dozen mt genes for the evolution of complex life, so it is critical to consider why such gene retention was not a failure of an evolutionary purge, but rather an absolutely necessary feature of the genomic structure of complex life forms that derive energy from aerobic respiration.

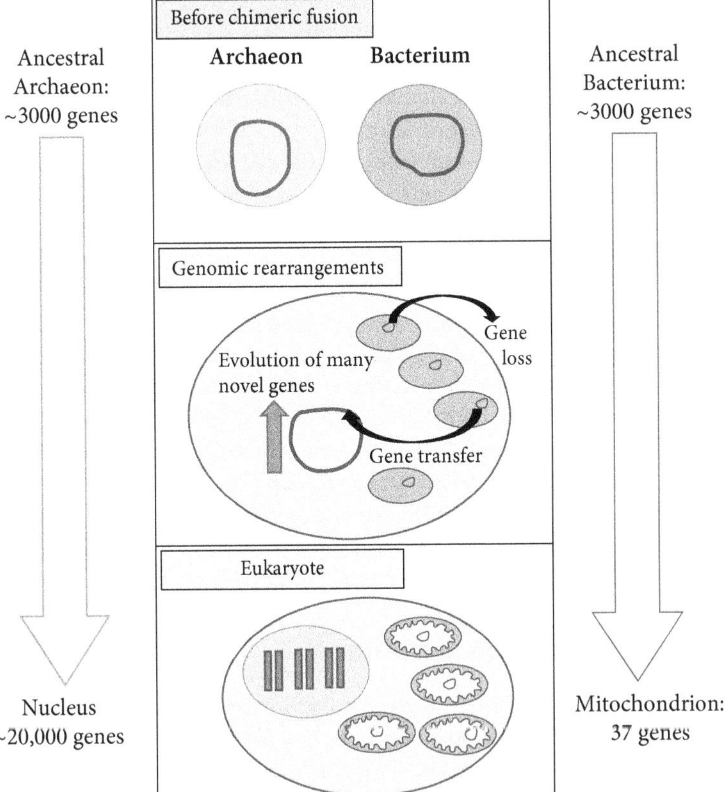

Figure 1.4 A general summary of key events in the genetic restructuring in the early evolution of eukaryotes. The three panels cover an evolutionary period of unknown duration beginning about 2 billion years ago. In the evolution of eukaryotes, many genes in the original bacterial genome were transferred to the nucleus or deleted. The number of genes in the former archaeon genome increased by more than an order of magnitude.

The mitochondrial genome

The retention of a mt genome

In the formulation of early hypotheses for why a small number of genes were retained in mitochondria, a basic assumption was that an optimal endpoint was zero genes in the mitochondrion (Björkholm et al., 2017). Researchers therefore proposed that loss of mt genes was an ongoing process that hadn't had time to reach completion, and they sought to understand the constraints on the process of gene transfer. Recent papers considering evolutionary constraints on the complete loss of mt genes have focused on difficulties related to protein transport. Because mt-encoded proteins are highly hydrophobic, they may be difficult to move across mt membranes and hence difficult to get into mitochondria if they are encoded by N genes (Claros et al., 1995; Björkholm et al., 2015). This mechanistic constraint is proposed as the reason that genes were retained in the mitochondrion. However, there are examples of hydrophobic proteins being moved into organelles and particularly chloroplasts (Allen, 2003). Other hypotheses for how constraints can explain retention of a mt genome have also been proposed (summarized in Allen, 2003).

Even though hypotheses focused on the constraints of moving hydrophobic proteins from the mt genome to the N genome remain viable explanations for the current genomic architecture of eukaryotes, I will proceed from the argument that retention of mt genes is not the result of a failed purge, but rather that in the evolution of eukaryotes there was selection to retain genes in the mitochondria. This view is supported by the observation that all eukaryotes that produce energy via aerobic respiration retain a common core set of mt genes (Boore, 1999; Björkholm et al., 2015) (Box 1.2). In addition, the loss of mt genes *was* completed in eukaryotic lineages that lost the need for a functional ETS (Smith and Keeling, 2015). Thus, in all organisms that retained aerobic respiration, core components of the mt genome were retained; when aerobic respiration was lost, so were genes remaining in mitochondria.

Co-location for redox regulation (CORR)

A key assumption in this book is that the mitochondria of most eukaryotes retained genes because genes in the mitochondria provide critical functions that enable complex life. The key to this assumption is the current leading hypothesis to explain the retention of a small genome in mitochondria: the co-location for redox regulation (CORR) hypothesis (Allen, 1993, 2003, 2017). This hypothesis states explicitly that mt genes are critical to complex life forms that derive energy from OXPHOS. The central tenet of the CORR hypothesis is that individual mitochondria must retain transcriptional and translational control of core ETS proteins so that individual mitochondria can rapidly respond to the current conditions of their inner mitochondrial membranes (Sirey and Ponting, 2016). This hypothesis focuses on the massive levels of energy needed to power complex life, with many trillions of mitochondria in billions of individual cells each creating an electrical potential across inner mitochondrial membranes that is equivalent to the energy in a bolt of lightning (Lane, 2006; Lane and Martin, 2010). The CORR hypothesis proposes that retention of mt genes was necessary to maintain control of such a massive energy generation system.

Efficient OXPHOS via a highly functional ETS is the key to eukaryotic fitness. For the ETS to work efficiently in enabling OXPHOS and capturing energy from the membrane potential, there must be precisely the correct numbers and ratios of ETS complexes to meet the demands of energy production (Lane, 2005). If the ratio of complexes is incorrect or if there are insufficient total ETS complexes, then electrons at one complex in the chain are frequently left with no empty positions in the next complex in the chain as they move toward Complex IV where the electrons are finally passed to oxygen to produce water. These stalled electrons dissipate their energy by

Box 1.2 Variation in mitochondrial gene content

Mitochondrial DNA codes for three general classes of genes: ribosomal RNA, transfer RNA, and protein-coding. Although there is variation among major clades of eukaryotes, and especially among single-celled eukaryotes, the gene content of mt genomes is remarkably conserved among all aerobic eukaryotes. Mitochondrial genomes always include genes that code for proteins that carry out core function of ETS complexes but never enough genes for complete ETS function. In all aerobic eukaryotes, mt DNA codes for at least some part of the mechanism for transcription and translation of ETS proteins. Even if none of the proteins needed for transcription is mt-encoded, as is true for many eukaryotes, the nucleotide sequence of regulatory sites in the mt DNA must be compatible with N-encoded transcriptional machinery (Ellison and Burton, 2010). The universal contribution to translation by the mt DNA of all aerobic eukaryotes is the small and large ribosomal RNAs (rRNAs). Some, but not all, eukaryotes also code for ribosomal proteins. Most eukaryotes also code for transfer RNAs (tRNAs), but these are lost completely in a few protozoa, which rely on tRNAs from the nucleus (Burger et al., 2003; Castellana et al., 2011) (Figure Box 1.2).

Among bilaterian animals, which will be the focus of a disproportionately large part of this book, there is little variation in the gene content of mt DNA. There are thirty-seven genes that include two ribosomal RNAs, a full complement of twenty-two tRNAs, and thirteen proteins that provide core subunits of Complex I, III, IV, and V of the ETS (Castellana et al., 2011).

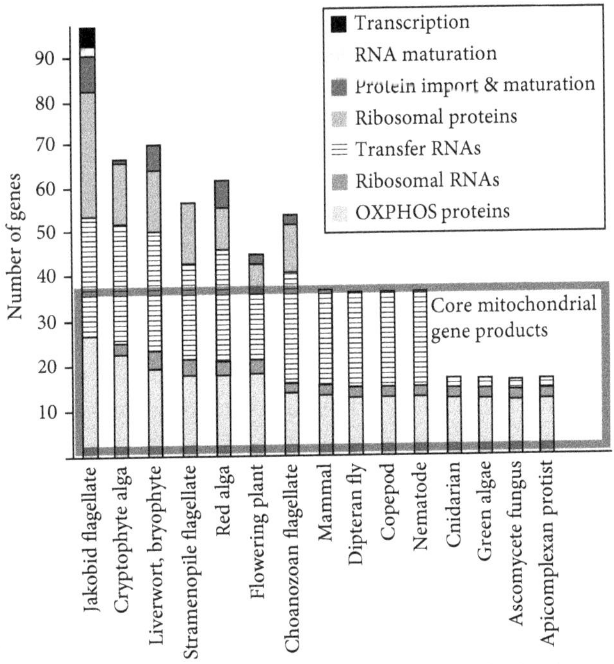

Figure Box 1.2 A graphical representation of the gene content of mitochondrial DNA from diverse eukaryotic lineages. The gray rectangle frames the core mitochondrial gene products common to all eukaryotes that derive energy via OXPHOS. Figure adapted from Burger et al. (2012).

side reactions with oxygen, resulting in the double whammy of less ATP generated and more cellular damage from free radicals (Lane, 2011c). If things go very wrong with the inner mitochondrial membrane potential, then cell death is triggered and the entire operation is shut down and disassembled (Ly et al., 2003). Only when the number and ratio of ETS complexes matches the current need of the system do electrons flow at an optimal rate producing energy efficiently and releasing few free radicals in side reactions (Allen, 2003; Lane, 2014). The CORR hypothesis proposes that the OXPHOS via the ETS can only be properly controlled if each mitochondrion has the capacity to regulate production of ETS subunits, and such control is only possible if each mitochondrion has its own set of genes for ETS components.

The challenge for eukaryotic organisms is that the conditions within one mitochondrion in a cell are not necessarily the same as conditions within other mitochondria within the same cell. There are too many small and ever-changing variables that affect the immediate conditions within a given mitochondrion. This means that it is not possible to properly regulate the ETS and maintain efficient OXHPOS within numerous individual mitochondria by regulating gene expression entirely within the N genome (Lane, 2005, 2015a; Lane and Martin, 2010). Attempting to control respiration in hundreds or thousands of mitochondria within a cell from a central command center would be like trying to control the boilers in hundreds of schools within a large school district from a central administrative office. Each boiler has to maintain an appropriate steam pressure for efficient function but each boiler is subject to untold variables as thermostats are changed, windows are opened or closed, outside temperatures change, and so forth. If boilers were checked once per day by administrators in the central office, individual systems would be perpetually maladjusted. On the other hand, if each boiler room was staffed by a technician who watched the dials and adjusted the system according to immediate conditions in that particular school, then each individual system would run much more efficiently and the district as a whole would function well.

According to the CORR hypothesis the mitochondria in all eukaryotes that engage in OXPHOS retained a small set of genes to enable within-mitochondrion transcriptional or translational regulation of proteins in the ETS complexes in response to the immediate local conditions (Allen, 2003). Without such rapid local responsiveness, membrane potential within an individual mitochondrion can drop, leading to cessation of ATP production, release of free radicals and other molecules that ultimately signal cell death. The CORR hypothesis proposes that all aerobic eukaryotes retained a core set of mt genes as a fundamental necessity of mitochondria-based respiration (Lane and Martin, 2010).

An observant reader might note that only a minority of ETS subunits are encoded by mt genes. For instance, there are approximately ninety-four subunits making up the five complexes of bilaterian animal ETS and only thirteen of these subunits are contributed by the mt genome. Thus, the great majority of proteins that create the ETS are encoded in the nucleus. Given this arrangement, how can transcriptional or translational regulation of thirteen out of ninety-four subunits enable mitochondria to control overall production of ETS complexes? Many details of the mechanisms of complex assembly remain to be worked out (Guaras and Enriquez, 2017), but it is almost surely no coincidence that the subunits that are coded for by mt DNA form the proton-translocating center of the complexes that localize *within* the inner mt membrane (Allen, 2015). Lane (2005, 2014) hypothesized that N-encoded respiratory subunits assemble around core mitochondria-encoded subunits once they are targeted to the membrane, and this hypothesis is gaining empirical support (Formosa et al., 2017). The production of N-encoded ETS subunits is dictated by the redox state of the entire cell—the average state of all mitochondria in the cell. But the specific rate of assembly of ETS complexes is dictated by mt genes that are responsive to the immediate conditions within each mitochondrion within the cell. According to this idea, N genes follow the lead of mt genes in assembling complexes. Control of mt genes therefore does provide control of complex assembly.

This proposed scenario for the assembly of ETS complexes makes the point for why it is so difficult for evolution to purge the genes for key membrane-anchored ETS subunits from the mt genome. These protein products are the enablers of locale transcriptional and translational regulation. Moreover, the signaling network that guides responsiveness to the state of the mitochondrion acts through transcription, translation, and replication of mt genes. Thus, it was essential to retain in the mt genome the genes for membrane-embedded ETS proteins as well as key components of the transcription, translation, and replication systems.

To summarize the logic for why two genomes were so important to the evolution of complex life: complex life forms require large genomes that code for huge quantities of diverse proteins. Sufficient energy to power such complexity is made possible by a cell design with a single large centralized genome and many independent mitochondria that act as power stations to support the enormous energy demands of the production of proteins encoded by the N genome (Lane and Martin, 2010). Such a cellular design absolutely necessitated the retention of a mt genome for autonomous control of the ETS by each mitochondrion to maintain the stability of the respiratory process (Lane and Martin, 2010; Lane, 2011c). The train of deductions proceeds as follows: complexity requires a large genome; a large genome requires a lot of energy; massive energy production is only possible via the combined output of many mitochondria; and control of such a power system necessitates independent genomes in each mitochondrion. As I will argue throughout this book, the necessity of coadaptation between the N and mt genomes led to the evolution of sex, two sexes, speciation, sexual selection, and adaptation.

The endpoint of gene transfer

As a result of the retention of core components of the ETS and the transcriptional and translational mechanisms needed to produce mt-encoded proteins, the phenotype of mitochondria is a product of two genomes (Figure 1.5). All of the gene products of the mt genomes—thirty-seven genes in most bilaterian animals and as many as sixty-six genes in some jakobid protozoa—stay at home and create mitochondrial form and function. Mitochondria are hugely complex organelles that complete a long list of functions that are essential for eukaryotic life (Prasai, 2017). A few dozen genes is not nearly enough to code for a functional mitochondrion; only about 2 percent of the proteins that make up a mitochondrion are encoded in the mt genome (Gray, 2015).

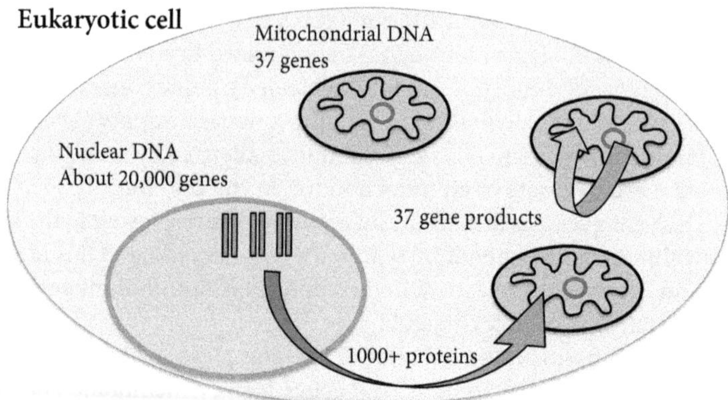

Figure 1.5 The phenotype of mitochondria is a product of both nuclear and mitochondrial gene products. For bilaterian animals, as illustrated, the phenotype emerges from more than 1000 proteins encoded on the nuclear genome and about thirty-seven gene products encoded on the mitochondrial genome. This genomic architecture means that every gene product of the mitochondrial genome functions in intimate association with the products of the nuclear genome.

Most of the proteins that form a functional mitochondrion are encoded in the nucleus, synthesized in the cytoplasm, and then imported into the mitochondrion (Bolender et al., 2008). *Mitochondrial function emerges from both N and mt gene products that must work properly together.*

In mammals, there are about 1100–1400 N-mt genes (N genes that code for products that function in the mitochondrion) (Pagliarini et al., 2008; Calvo and Mootha, 2010). These N-mt genes encode subunits of the ETS, proteins involved in the assembly of the OXPHOS complexes, enzymes that enable synthesis of cofactors, and key components that enable replication, transcription, translation, and repair of mt genes (Pearce et al., 2013; Lotz et al., 2014). A complete inventory of all of the proteins that comprise a mitochondrion, even in the best-studied eukaryotes like humans, has proven challenging for researchers (Meisinger et al., 2008). For the purposes of this book, the exact size of the mitochondrial proteome is not particularly important. The key point is that most of the phenotype of a mitochondrion emerges from the N genotype. The relatively small component of the ETS that is encoded by mt genes, however, is absolutely essential.

Given the critical necessity of compatibility in shape and function of the protein subunits of the ETS for the fitness of eukaryotes, it is a stunning paradox that the interacting subunits of the ETS are encoded by two different genomes. On first consideration, it would seem that the ETS would be the last place where mixed sets of mt and N gene products would be employed (Lane, 2015b). And yet, all eukaryotes that engage in OXPHOS depend on the co-function of products of both N and mt genes. For the remainder of this book, co-functioning products of mt and N genes will be the starting point for discussions about the evolutionary and ecological implications of mitonuclear coadaptation.

Characteristics of mitochondrial genomes

Mitochondrial genomes are different than N genomes in several basic features that have far-reaching implications for mitonuclear ecology. As noted already, mt genomes are small. For instance, in vertebrates, which have extreme size disparity between N and mt genomes, and using humans as a specific example, there are approximately 3 billion base pairs in the N genome, but only about 16,569 base pairs in the mt genome; the human mt genome holds less than one hundredth of 1 percent of the genetic code (Ballard and Whitlock, 2004). Despite variations in gene number among eukaryotes, the core function of mt genes is remarkably conserved across eukaryotes (Burger et al., 2003) (Box 1.2). This conservation of core function supports the hypothesis that genes in the mitochondria are essential to allow on-site protein production to maintain efficient OXPHOS.

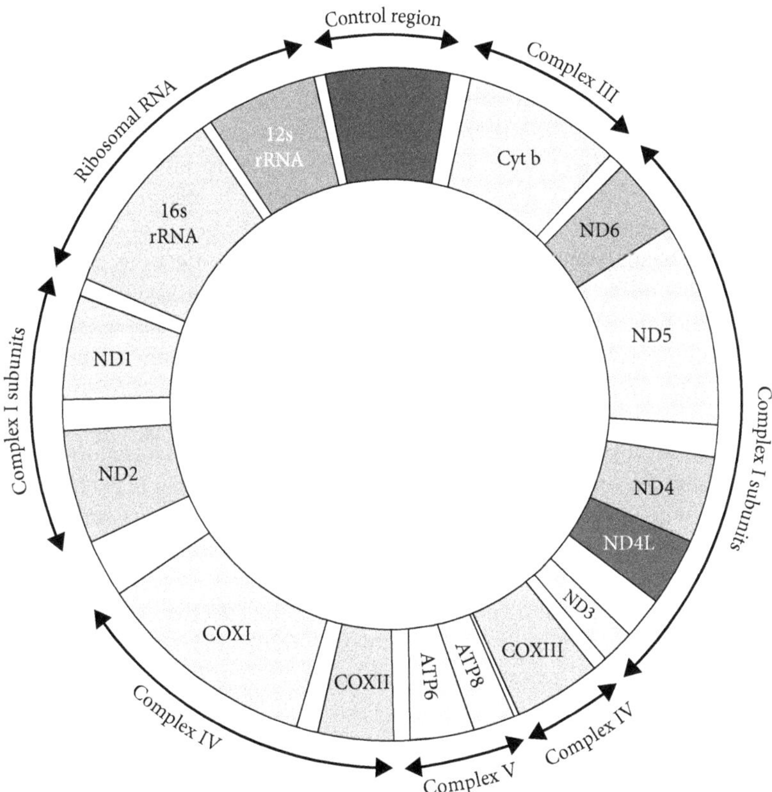

Figure 1.6 The positions of protein-coding genes, ribosomal RNA, and the control region in the human mt DNA. The gaps between gene labels are primarily tRNAs. There are no introns in mammalian RNA.

Because the mt genomes of most eukaryotes rarely or never engage in recombination, all of the genes on a mt genome behave like one linkage group (Ballard and Whitlock, 2004). Linkage of mt genes has important implications for the transmission and evolution mt genes (see Chapter 3). The previous description holds for the mt genomes of bilaterian animals and many other eukaryotes that have mt genome as double-stranded DNA that is circular (Figure 1.6). It seems also to hold for eukaryotes that have mt DNA as a linear chromosome capped by telomeres (Nosek et al., 1998; Burger et al., 2003). However, the mt DNA of some eukaryotic lineages is fragmented into multiple chromosomes (Cameron et al., 2011; Wu et al., 2015), and such division of a mt genome into multiple chromosomes could enable mt genes in these groups to behave more like N genes and to not follow the rules of genes linked on a single chromosome. The effects on evolution of having mt genes on multiple chromosomes, however, are almost completely unknown (Wu et al., 2015). The discussions that I present in this book, which are based on the assumption of mt genes acting like a single linkage group, will have to be modified going forward as we gain better understanding of inheritance of mt genes in diverse groups of eukaryotes.

The mt DNA of bilaterian animals has a high rate of mutation relative to N DNA (Figure 1.7). This is a common pattern among eukaryotes—perhaps the most common pattern—but it does not hold across all eukaryotes. In some eukaryotes, mt DNA often has a substantially lower rate of mutation than N DNA (Figure 1.7), but estimating mutation rates of mt DNA in any species is not trivial (Rand, 2008). The mt genes of bilaterian animals also do not commonly engage in recombination (Birky, 2001; Hagström et al., 2014). In contrast, the mt genes of plants and fungi routinely engage in recombination (Barr et al., 2005). Thus, the mt genomes of many eukaryotes, including all bilaterian animals, where much of this book focuses, behave as an asexual, haploid genetic component in otherwise diploid and sexual organism. This is the characterization of mitochondria that I will frequently use in my discussions of mitonuclear coevolution.

A common assumption regarding the mt genome is that it is haploid. This assumption is paradoxical given that there are typically hundreds to thousands of copies of the mt genome per cell, but the number of mt genome copies that really matters is the number of mt genotypes per gamete. If that number is one, then the assumption that mt genomes are haploid is reasonable. I will take up the issue of germline transmission of mt DNA in Chapter 6.

If mt genomes behave as haploid genomes, then the effective population size (i.e. the size of the population of genes that influences future generations) of mt genes is only half that of diploid N genes (Hudson and Turelli, 2003; Ballard and Whitlock, 2004). Moreover, in the majority of taxa in which there is exclusively maternal transmission of mitochondria (see Chapter 5), the copies of mitochondria in males are a genetic dead-end. With exclusively maternal transmission of mitochondria, the effective population size is potentially halved again. As a result, the effective population size of mitochondria genes in eukaryotes with exclusively maternal transmission is assumed to be only one-fourth that of N genes (Rand, 2001; Ballard and Whitlock, 2004). A small effective population size of mt genes has important implications for

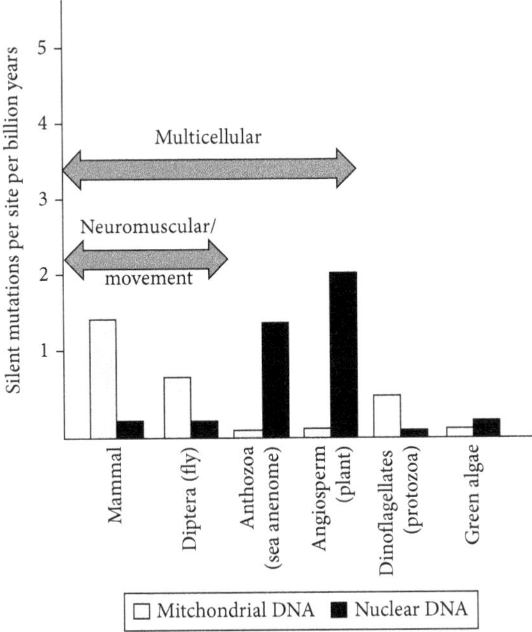

Figure 1.7 Mutation rates of N and mt DNA. Shown are approximate rates of synonymous (silent) nucleotide substitutions for some lineages of eukaryotes including lineages that are multicellular and that have movement via a neuro-motor system. Values are from Smith and Keeling (2015), Lynch (1997), and Hellberg (2006).

genome evolution, and I'll come back to this topic in Chapter 7 when I consider speciation.

Because genes in the mt genome form a single linkage group, selection must act on entire mt genomes rather than on specific genes. In other words, in taxa with no recombination of mt genes, selection responds to the average fitness of all genes in the mt genome rather than to the fitness of individual genes. This sort of tight linkage of mt genes can lead to mutational erosion over time, which is the basic premise for Chapter 3 on compensatory mechanisms and Chapter 5 on the evolution of sexual reproduction.

Selection on mt genes as a single linkage group also can lead to selective sweeps (Meiklejohn et al., 2007). Such sweeps occur when there is strong selection in a population for a particular allele at a specific locus. In response to strong selection, a locus can rapidly go to fixation such that all individuals in the population have the favored gene. If the favored locus is tightly linked to other loci, as is typically the case among mt loci, then whatever alleles happen to associated with the allele under strong selection, whether they are good or bad, will also go to fixation (Maynard Smith and

Haigh, 1974). As a consequence, fixation of one particularly good gene also leads to the fixation of all the genes that are linked to it on the chromosome, even if some of those linked genes are not good. Selective sweeps can have important implications for compensatory coevolution, speciation, and adaptation, and I will return to this topic in Chapters 3, 6, and 8.

Regardless of genome size, the mutation rate of mt genes, the potential for recombination, and whether or not inheritance of mt genes is strictly maternal, mitonuclear coadaptation remains a necessity for all eukaryotes that gain energy through OXPHOS, which is essentially all eukaryotes. Variable characteristics of mitochondria change the dynamics of the interactions of mt and N genes and can potentially change the role of mitonuclear coevolution in core processes like speciation and adaptation, and these considerations will be the focus of future sections of this book.

Classes of genes and abbreviations

Through the pages of this book, I will present many discussions of the interactions of mt and N genes. It is therefore useful to explicitly define different classes of genes involved in interactions and to introduce shorthand for the key elements that I will refer to repeatedly in this book. N genes have different levels of interaction with mt genes, and I will refer to genes that are N-encoded but whose products are targeted to the mitochondria as N-mt genes. There are between 1100 and 1400 N-mt genes in humans (Pagliarini et al., 2008; Calvo and Mootha, 2010), and the products of only a small fraction of these N-mt genes—about 180 genes in animals—engage in close functional interaction with the products of mt genes (Burton and Barreto, 2012). Thus, it is important to remember that not all products of N-mt genes have significant functional interactions with products of mt genes and hence that only a small fraction of all N-mt genes are expected to be coadapted with mt genes.

Summary

All complex life on Earth is descended from a chimera born of the fusion of an archaeon and a bacterium. From the outset, the bacterial partner became the energy production and synthesis center, and the archaeon partner became the information center. The information center existed as one entity and nearly all genes were transferred to it. The powerhouse existed as many replicants and nearly all genes were transferred from it. But complete loss of autonomy by the mitochondria was unworkable. The production of ATP via OXPHOS entailed powerful forces of energy flow that could not be properly managed from a centralized command center. Each mitochondrion *had* to retain local control of production of ETS components; a small mt genome was fundamentally unavoidable. Importantly, transcriptional and translational regulation of the ETS was not ceded entirely to the mitochondrion—only a minority fraction of the components of the ETS are under direct control of the mt genome. The portion of the ETS that is under mitochondrial control, however, is invariably the core catalytic centers of ETS function. This small concession of autonomy by the nucleus to the mitochondrion had enormous consequences for the sort of complex organism that arose from the mitochondria-powered cell line as will be thoroughly explored in the chapters of this book.

2

Forms and consequences of incompatibility

Symbiosis is a term often applied to the relationship between a mitochondrion and the rest of a cell, but describing mitonuclear interactions as symbioses is like describing the branches of government as a partnership. The judiciary, executive, and legislative branches create no government in isolation. They are not partners: they are essential parts to a whole. In the same way, there is no eukaryotic organism capable of OXPHOS without mitochondria. Complex life is possible only through a massive flow of energy, and the source of that energy is chemiosmosis via a fantastically complex yet efficient electron transport system (ETS) in mitochondria (Lane and Martin, 2010).

It seems paradoxical that OXPHOS, the core cellular process that sustains complex life, is encoded by two independently replicating sets of DNA. Dividing the blueprint for cellular respiration between two genomes creates a perpetual evolutionary challenge in the maintenance of compatibility. To make the core respiratory mechanism of eukaryotes even more improbable, mt DNA lacks mechanisms for recombination and typically has a high mutation rate, ensuring that changes will accrue. This is life careening down a perilous mountain road with two different drivers simultaneously attempting to steer as obstacles are perpetually thrown into the path. Without constant coordination between the two interdependent genomes, a bad end is inevitable. In the following sections, I consider in physiological and molecular detail the sorts of bad endings that await individuals with poorly coadapted mt and N genotypes. I first review the molecular arenas in which the products of mt and N genomes must co-function and then consider the range of mitonuclear incompatibilities that have been observed or hypothesized to occur. I end by reviewing the evidence for specific incompatibilities between mitochondrial and nuclear elements and the effects that poor coadaptation can have on cellular and organismal function.

Oxidative phosphorylation via the electron transport system

We live in a macro-scale world in which the movement of objects is dictated by gravity. It is hard for an organismal biologist like myself, with limited training in physics, to think in terms of an atomic world in which the movement of electrons and protons

Mitonuclear Ecology. Geoffrey E. Hill, Oxford University Press (2019). © Geoffrey E. Hill 2019.
DOI: 10.1093/oso/9780198818250.001.0001

has nothing to do with gravity but rather is driven by quantum energy states. Nevertheless, at least to an elementary degree, we have to come to grips with quantum energy flow if we want to understand the basics of ATP production in the ETS.

Each step in the electron transport chain moves electrons to a more electronegative acceptor, and with each transition along the ETS, electrons give up energy that is used to pump protons from the matrix side of the inner mitochondrial membrane into the intermembrane space (Rich and Marechal, 2010). Oxygen is the most electronegative electron receiver in this chain and serves as the terminal acceptor. In forming water, oxygen (O_2) removes electrons from the ETS, and this outflow of electrons through Complex IV enables the continued inflow of electrons into Complexes I or II (Figure 2.1).

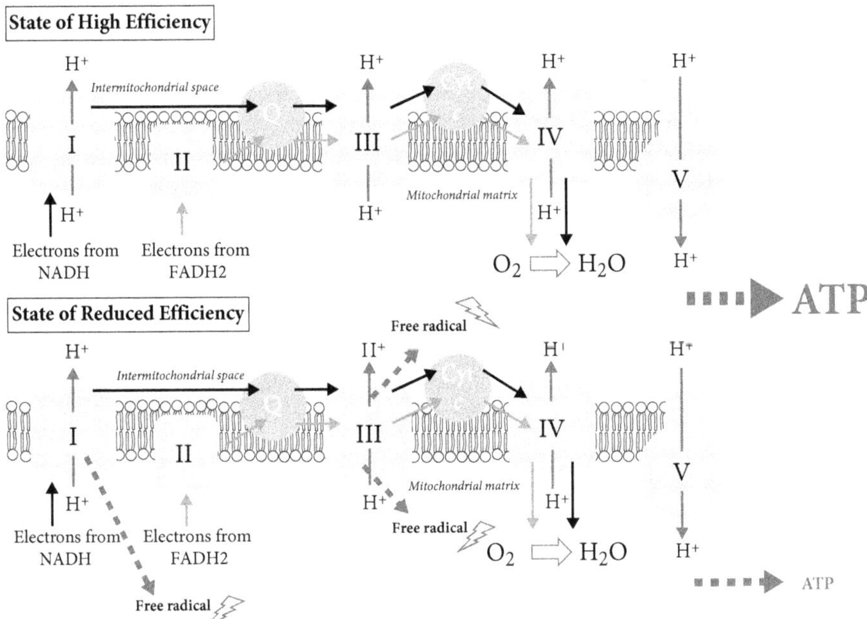

Figure 2.1 Simplified illustration of the flow of electrons in the electron transport system (ETS) when the system is running well (top) and when electron flow is impeded (bottom). The products of the citric acid cycle, NADH or FADH2, deliver electrons to Complex I and II, respectively, which serve as alternative pathways into the ETS. Electrons are then passed via ubiquinone (Q) to Complex III and then via cytochrome *c* to Complex IV. The final electron receiver is molecular oxygen, which is reduced to water. The energy stored in the proton-motive force across the inner mitochondrial membrane is converted into energy stored in ATP molecules by Complex V and protons are moved back across the membrane. When the flow of electrons is stalled, side reactions with oxygen, primarily at Complexes I and III, produce reactive oxygen species (ROS). ROS from Complex I are released into the intermembrane space; ROS from Complex III are released into the matrix. Stalled electron flow also reduces the membrane potential and causes a loss of ATP production.

For efficient function of the ETS, each step in the process must receive and pass on electrons at a rate proportional to the entry of electrons into the system. If any part of the system is impeded—for instance if there is insufficient oxygen at the terminus of the chain—then electrons stall along the chain and energy is dissipated, typically through side reactions with oxygen (Figure 2.1). Loss of high-energy electrons from the ETS not only wastes energy that could otherwise be used to produce ATP, but the capture of electrons by molecular oxygen via side reactions results in the production of superoxide, an unstable and highly reactive molecule. Superoxide and related oxygen-based free radicals are collectively called reactive oxygen species (ROS), and because these free radicals provide real-time information about the state of the ETS and inner mitochondrial membrane, they serve as critical signaling molecules in the maintenance of mitochondrial homeostasis (Shadel and Horvath, 2015). Free radicals are an inherent and necessary product of aerobic respiration. They are also highly reactive molecules that can damage DNA, proteins, and membranes as they react with whatever cellular structure they encounter, including components of the ETS (Lane, 2011c). Elevated levels of free radicals are a major negative outcome of mitonuclear incompatibilities and a poorly functioning ETS, creating a condition called "oxidative stress" (Sies, 2013).

The pumping of protons by Complexes I, III, and IV creates an electrochemical gradient—a membrane potential—across the inner mitochondrial membrane. The mitochondrial membrane potential is the reservoir of energy created by proton pumping in the same way that the force of water on the wall of a dam is the reservoir of energy in a hydroelectric system. The energy from the mitochondrial membrane potential is released when protons move back across the membrane. In a tightly coupled system, the movement of protons from the inner space back to the matrix is through Complex V, known more technically as ATP synthase, and the energy that is released is used to phosphorylate ADP molecules into ATP molecules (Brand and Nicholls, 2011). The energy in the phosphorus bond of ATP is readily available to biological systems such that ATP is the energy currency of living systems (Dzeja and Terzic, 2003). Alternatively, protons can move across the inner mitochondrial membrane in a controlled manner through uncoupling proteins (Criscuolo et al., 2005) with no ATP produced and energy dissipated primarily as heat. There can also be "leakage," with protons moving across the inner mitochondrial membrane in an uncontrolled manner without association with either Complex V or uncoupling proteins (Murphy, 1989; Jastroch et al.; 2010, Divakaruni and Brand, 2011).

The flow of electrons through the ETS is a stepping down of electrons across quantum levels. Just as physical objects fall, flow, roll, and slide to positions closer to the gravitational center of the Earth whenever they are released from whatever physical barrier is holding them in place, electrons transition to a lower energy state so long as no energy barriers impede the transition. Descriptions of "better" or "worse" ETS and mitochondrial function, which is ultimately the central concern of mitonuclear ecology, comes down to impediments to optimal electron flow. And "optimal" electron flow is not the fastest possible electron flow; it is the flow of electrons that maximizes the evolutionary fitness of the organism, which will be determined by the energy needs of the organism and the availability of electron donors and receivers.

An electron at the catalytic center of an ETS complex will be drawn to the next ETS position only if that next position enables it to exist at a lower energy state. The analogy is downhill in response to gravity. Water molecules never spontaneously move uphill and electrons never spontaneously move to higher energy states. So long as the next position in the electron transport chain is available (not already occupied by an electron) and has a lower energy state, the electron will spontaneously move to it and give up energy as it does.

"Flow" or "move" are actually inaccurate verbs to describe how electrons pass from one ETS complex to the next because electrons exist in a quantum world. Their transmission along the respiratory chain involves "transitioning" from one redox center in an ETS complex to the next. Electrons are drawn along the ETS by a process called quantum tunneling, whereby a given electron essentially instantaneously disappears from one redox center and reappears in the next (Devault, 1980; Lane, 2015a). Whether or not a given electron makes this quantum leap is determined by (1) the differences in energy state between its current position and its potential future position, (2) the proximity of the next ETS position, (3) the availability of unoccupied centers in the next ETS subunit in the chain, and ultimately (4) the availability of oxygen as an acceptor at the end of the chain (Lane, 2015a). These are not independent variables. For instance, if the flow of electrons is slowed initially by a gap between catalytic centers resulting from poorly fitting subunits, then the occupancy of catalytic centers will also be affected. In this way, the interactions of mt and N gene products play a key role in determining the rate at which electrons flow down the respiratory chain. In turn, the flow of electrons determines the pumping of protons and the membrane potential that is created as protons are accumulated on one side of the membrane. The membrane potential dictates the rate at which ATP is produced from ADP (Figure 1.2).

The spacing of catalytic centers, which is a function of the protein subunits that encase the centers, must be absolutely precise. Lane (2015) summarized the necessity for structural precision of ETS complexes, stating that there is a ten-fold reduction in the rate of electron flow for every angstrom increase in the distance between two catalytic centers. An angstrom is approximately the width of a hydrogen atom. Moreover, there is theoretical limit for the reach of quantum tunneling at about 14 angstroms and some of the spacing of some catalytic centers in Complex I are near that limit (Devault, 1980). Because the path through the ETS that is followed by electrons is created by protein subunits that are assembled to form the ETS complexes, there is a fundamental necessity that the subunits making up a complex fit together without gaps. The substitution of a single amino acid in one of the subunits of an ETS complex can shift the distance between catalytic centers by an angstrom or more, which could significantly affect the flow of electrons and hence core respiratory efficiency. Conversely, a single amino acid change in a complementary subunit can reverse such an effect (Pierron et al., 2012). In addition, the fit between subunits isn't just important for electron transfer itself. Proper subunit fit is also essential for transmitting conformational changes caused by electron transfer to the proton-translocating subunits (Fiedorczuk et al., 2016). These conformational changes are critical for all subunits to function, but they are especially central to the function of Complexes I, III, and V (Matyushov, 2013; Sazanov, 2015; Suzuki et al., 2016). The ETS complexes are highly dynamic machines, the intricacies of which aren't completely captured even in three-dimensional structural snapshots. There are numerous dynamic and complex interactions among subunits that require fine-level coadaptation between the products of the mt and N genomes.

Efficient OXHPOS requires that the inner mitochondrial membrane potential be maintained within narrow parameters (Brand and Nicholls, 2011). If membrane potential drops too low, then there is less energy to drive ATPase and the production of ATP slows (Dimroth et al., 2000). Conversely, if the membrane potential gets too high, proton pumping is inhibited resulting in a reduced flow of electrons and a greater probability of proton leakage and free radical production (Divakaruni and Brand, 2011). For each deviation from optimum function, there is a potential response by the mitochondrion. When membrane potential begins to drop, it is a sign that protons are being used faster than they can be pumped across the inner mitochondrial membrane. Mitochondria can respond to a declining membrane potential by reducing the flow of protons through uncoupling proteins as well as by pumping more protons with more ETS subunits (Aon et al., 2010). Thus, a low membrane potential should stimulate up-regulation of transcription and translation of ETS subunits, a condition that Lane (2011c) calls reactive biogenesis. More subunits pump more protons raising the membrane potential. Conversely, if the problem is a membrane potential that is too high, the immediate solution is to release protons through uncoupling mechanisms. Over a longer period, however, the system can down-regulate ETS subunits to pump fewer protons (Zhu et al., 2013; Palikaras et al., 2015).

Responses to changing membrane potential rely on the monitoring of local redox conditions by mitochondria and the capacity to respond to the local conditions by up- and down-regulation of subunit production (Lane, 2005). In turn, the capacity for up- and down-regulation of the ETS complexes depends on the function of the transcriptional and translational mechanisms that produce the mitochondrial components of the ETS (Asin-Cayuela and Gustafsson, 2007). Thus, the flow of electrons, the pumping of protons, and the production of ATP are dependent not only on the proper function of the protein complexes that make up the ETS, but also on the proper function of the biochemical machinery that produces the mt protein subunits for the ETS complexes.

Arenas of mitonuclear interaction

The fundamental principle that I extol in this book is that coadaptation of mt and N-mt genes to enable cellular respiration and core energy production is essential for aerobic eukaryotes. The implication is that if there is poor compatibility of the products of mt and N-mt genes, then specific dysfunctions will occur in mitochondria that impair cellular respiration with direct negative effects on fitness. For an effective OXPHOS via the ETS, the products of mt and N-mt genes must coordinate function to enable three critical processes: (1) protein–protein interactions in assembling and forming the complexes of the ETS (Phillips et al., 2010), (2) protein–RNA and protein–DNA interactions in the transcription and translation of mt genes that directly or indirectly affect the ETS (Taanman, 1999; D'Souza and Minczuk, 2018), and (3) protein–DNA interactions in the replication of mt genomes (Clayton, 2000a), which ultimately dictates the number of mitochondria and total respiratory capacity (Figure 2.2). In each of these arenas, mitonuclear interactions must be precise. Deviation in the fit of protein subunits by as little as the width of an atom can disrupt the flow of electrons through the ETS (Lane, 2015a). A single base-pair substitution in a tRNA can interfere with its recognition by a N protein and significantly slow developmental rates (Adrion et al., 2016). Small changes in the nucleotide sequence in the promoter region of the mt genome can inhibit transcription and replication of mt DNA (Ellison and Burton, 2010).

Evolutionary biologists tend to focus on protein–protein interactions when discussing mitonuclear compatibility (Kwong et al., 2012; Osada and Akashi, 2012; Hill and Johnson, 2013; Zhang and Broughton, 2013; Havird et al., 2015a). Indeed, the definition of a synonymous nucleotide substitution is a nucleotide replacement that does not change the structure of a protein. Historically, synonymous mutations were thought of as changes that did not affect organism function (Kimura, 1977). However,

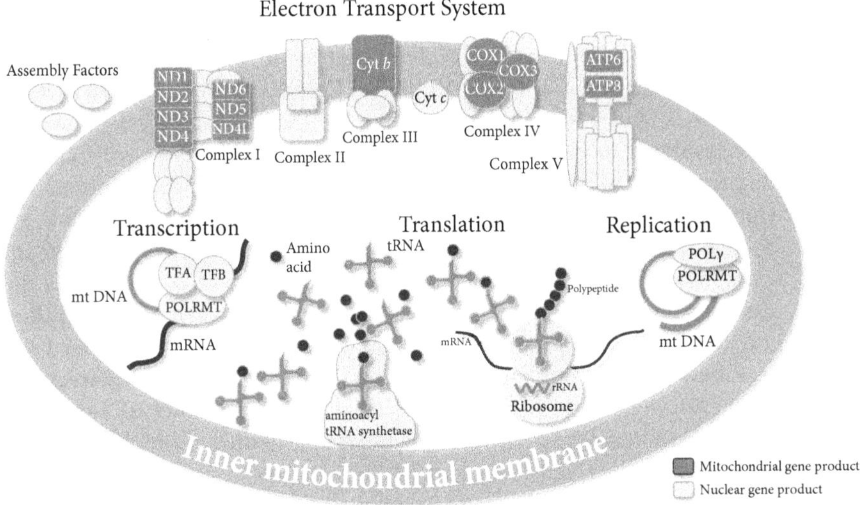

Figure 2.2 A schematic illustration of the functional interactions of mitochondrial and nuclear gene products inside an animal mitochondrion. The specific interactions will differ among different eukaryotes, but in all eukaryotes there are at least some protein–protein, protein–RNA, and protein–DNA interactions. In animal mitochondria, the product of the nuclear genome (light shading) must co-function with the products of the mitochondrial genome (dark shading) in Complexes I, III, IV, and V of the electron transport system, in the assembly of complexes, and in the transcription, translation, and replication of mt genes.

change in the nucleotide sequence for the non-translated RNA or DNA can have enormous functional impacts—occasionally positive but often debilitating—on organism function (Sauna and Kimchi-Sarfaty, 2011). If there is a loss of co-function of mt DNA/RNA and nuclear proteins that enable the production of ETS subunits, then mitochondria cannot be optimally responsive to local redox conditions via transcriptional regulation of complex proteins. The effects of compromised replication, transcription, and translation of mt gene products can have severe effects on mitochondrial function and ultimately organism fitness (Wallace et al., 2015). Thus, I will make an effort to equally emphasize RNA–protein, DNA–protein, and protein–protein interactions in discussion of mitonuclear coadaptation and coevolution.

Protein–protein interactions

Protein–protein combinations are the most well-known arenas of interaction between products of the mt genome and the N genome (Pierron et al., 2012; Hirst, 2013). All aerobic eukaryotes have the same five complexes in the ETS, but they vary somewhat in the relative contribution of mt and N gene products (Gray et al., 1998; Gray, 2012). Among bilaterian animals, the mitonuclear composition of ETS complexes is highly conserved and, because many studies of mitonuclear compatibility involve bilaterian animals, I will use the mitochondrial and nuclear composition of the animal ETS (Figures 2.2 and 2.3) as the baseline for explanations throughout this book. In mammals, Complex I, the largest of the ETC complexes and the receiver of electrons from NADH, is assembled from seven mt proteins and about thirty-seven N proteins. (Note that the number of subunits in mammalian Complex I is often stated as forty-four as in Figure 2.3 but a forty-fifth and even a forty-sixth N-encoded subunit are sometimes considered to be a part of the human Complex I (Sanchez-Caballero et al., 2016; Formosa et al., 2017)). Complex I is the point of entry for electrons from NADH. Complex II is unique among the five ETS complexes in animals in that it is entirely N-encoded, composed of four N-encoded subunits. Hence, unlike other animal ETS complexes, Complex II is not subject to mitonuclear incompatibilities and, for this reason, Complex II is an important reference enzyme when inferences about mitonuclear incompatibilities are invoked. Complex II, along with two other flavoproteins, is a second entry point for electrons into the ETS when electrons are passed from $FADH_2$ (Watmough and Frerman, 2010) (Figure 2.1). Complex III has one mt and ten N subunits. Complex IV has three mt and ten N subunits. Complex V has two mt and twelve N subunits (Figures 2.2 and 2.3).

Interactions between mt and N protein subunits of the ETS certainly concern the fit of folded proteins as assessed in static three-dimensional models, but ETS proteins also interact in dynamic manners that are not easily assessed in conventional protein models but that might play critical roles in subunit interactions (Matyushov, 2013). Protein–protein interactions among subunits of the ETS are typically the focus of discussions of potential mitonuclear incompatibilities because protein–protein interactions are the best-characterized and most intuitively comprehensible form of mitonuclear interaction (Lane, 2011c). Amino acid substitution in the proteins that form the ETS has been hypothesized to affect OXPHOS function in at least three different ways.

The first proposed mechanism by which changes in amino acid sequence of either N or mt-encoded subunits of the ETS complexes could affect respiratory function is through disruption of the precision of the static three-dimensional fit or dynamic interactions of subunits, thereby directly affecting enzyme function by altering the flow of electrons or the pumping of protons (Lane, 2011c, 2015a). The substitution of a single amino acid in one subunit of a complex can significantly disrupt function (Gershoni et al., 2014; Rak et al., 2016). Amino acid substitutions can also affect how N-encoded cytochrome *c* engages with mt-encoded subunits in both Complex III and IV in the transfer of electrons between these complexes (Willett and Burton, 2001; Solmaz and Hunte, 2008). The hypothesis that changes to the amino acid composition of ETS protein subunits can lead to poor co-functioning of mt and N components that can negatively affect OXPHOS has been called the Residue Contact Hypothesis (Sackton et al., 2003).

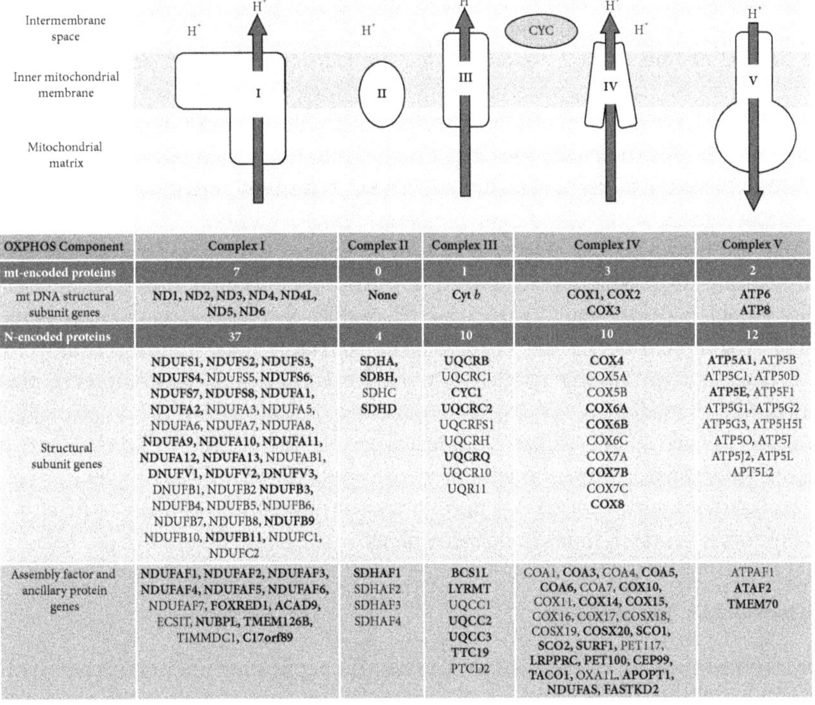

OXPHOS Component	Complex I	Complex II	Complex III	Complex IV	Complex V
mt-encoded proteins	7	0	1	3	2
mt DNA structural subunit genes	**ND1, ND2, ND3, ND4, ND4L, ND5, ND6**	**None**	**Cyt *b***	**COX1, COX2 COX3**	**ATP6 ATP8**
N-encoded proteins	37	4	10	10	12
Structural subunit genes	**NDUFS1, NDUFS2, NDUFS3, NDUFS4,** NDUFS5, **NDUFS6, NDUFS7, NDUFS8, NDUFA1, NDUFA2,** NDUFA3, NDUFA5, NDUFA6, NDUFA7, NDUFA8, **NDUFA9, NDUFA10, NDUFA11, NDUFA12, NDUFA13,** NDUFAB1, **DNUFV1, NDUFV2, DNUFV3,** DNUFB1, NDUFB2 **NDUFB3,** NDUFB4, NDUFB5, NDUFB6, NDUFB7, NDUFB8, **NDUFB9** NDUFB10, **NDUFB11,** NDUFC1, NDUFC2	**SDHA, SDBH,** SDHC **SDHD**	**UQCRB** UQCRC1 **CYC1 UQCRC2** UQCRFS1 UQCRH **UQCRQ** UQCR10 UQR11	**COX4** COX5A COX5B **COX6A COX6B** COX6C COX7A **COX7B** COX7C **COX8**	**ATP5A1,** ATP5B ATP5C1, ATP50D **ATP5E,** ATP5F1 ATP5G1, ATP5G2 ATP5G3, ATP5H5I ATP5O, ATP5J ATP5J2, ATP5L APT5L2
Assembly factor and ancillary protein genes	**NDUFAF1, NDUFAF2, NDUFAF3, NDUFAF4, NDUFAF5, NDUFAF6,** NDUFAF7, **FOXRED1, ACAD9,** ECSIT, **NUBPL, TMEM126B,** TIMMDC1, **C17orf89**	**SDHAF1** SDHAF2 SDHAF3 SDHAF4	**BCS1L LYRMT** UQCC1 **UQCC2 UQCC3 TTC19** PTCD2	COA1, **COA3,** COA4, **COA5, COA6,** COA7, **COX10,** COX11, **COX14, COX15,** COX16, COX17, COSX18, COSX19, **COSX20, SCO1, SCO2, SURF1,** PET117, **LRPPRC, PET100, CEP99, TACO1,** OXA1L, **APOPT1, NDUFAS, FASTKD2**	ATPAF1 **ATAF2 TMEM70**

Figure 2.3 The proteins that enable electron transport system function in humans sorted by origin and function. Bold font indicates proteins for which mutations are the causative factor in a human disease. Redrawn from Alston et al. (2017) in which citations for clinical studies are provided.

A second proposed mechanism by which amino acid substitutions could affect OXPHOS is through disruption of processes ancillary to the actual functioning of ETS complexes. There is a growing list of N-encoded ancillary proteins that are critical for proper respiratory chain function (bottom list in Figure 2.3) (Ghezzi and Zeviani, 2018). These ancillary proteins include assembly factors, elongation and termination factors, and translational activators that all potentially involve mitonuclear interactions (Mata, 2016; Formosa et al., 2017). Assembly proteins are a particularly interesting set of N gene products that are likely to have close interactions with mt gene products. The complexes of ETS are composed of multiple distinct protein subunits, and assembly factors control the construction of ETS complexes from protein subunits (Fernández-Vizarra et al., 2009; Mimaki et al., 2012) One could imagine that even when amino acid substitutions do not impact the basic function of an ETS complex, if they slow down or inhibit complex assembly, the effects on respiration can be significant (Lazarou et al., 2009). Among ETS complexes, Complex I has the greatest number of subunits and the most complicated assembly, and much of the literature on assembly proteins focuses on Complex I (Nouws et al., 2012; Sanchez-Caballero et al., 2016; Formosa et al., 2017). Assembly proteins are all N-encoded, but they have to function with mt-encoded ETS protein subunits in creating ETS complexes. Complex assembly and the function of other ancillary proteins are rarely discussed in the context of mitonuclear coadaptation, but more than three dozen such ancillary proteins that are not components of the respiratory complexes are known causes of human mitochondrial dysfunction (Alston et al., 2017; Lleonart et al., 2017) (Figure 2.3), and they are likely to play an important role in mitonuclear compatibility.

A final potential arena of interaction between N and mt proteins is the assembly of respiratory chain super-complexes, which are linkages among Complexes I, III, and IV, that are proposed to enhance ETS functionality (Genova and Lenaz, 2013; Letts and Sazanov, 2017; Milenkovic et al., 2017). Another commonly used name for ETS super-complexes involving Complexes I, III, and IV is "respirasomes." Super-complex assembly is a phenomena that is yet to be considered within the context of mitonuclear coadaptation among species, but the formation of super-complexes plays a critical role in mediating the effects of mitochondrial mutations in somatic cell lines (D'Aurelio et al., 2006). Super-complex assembly genes are N-encoded and they undoubtedly must co-function with mt gene products of the ETS complexes. The role of respirasomes/super-complexes in mitonuclear interactions is likely to be an active area of future research in mitonuclear ecology.

Protein–DNA interactions

Mitochondria rely on N-encoded proteins for the replication and transcription of the mt genome (Table 2.1). The implication of this fundamental dependency is that tight mitonuclear coadaptation is essential. In both the replication and transcription of DNA, N-encoded proteins must respond appropriately to specific promoter nucleotide sequences and to origin-of-replication motifs. These non-coding regions of the mt genome are subject to rapid evolution (Ellison and Burton, 2008a, 2010), which

Table 2.1 An approximate list of gene products of mammalian mitochondrial and nuclear genomes that enable transcript, translation, and replication of mt DNA and are subject to mito-nuclear coadaptation. The processes are the subject of active research and new mechanisms are being discovered.

Process	Gene Type	Origin	Gene Tally	Gene Products
Transcription	Polymerase and transcriptional factors	N	~5	POLRMT, TFAM, TFB1M, TFB2M, Abf2
Transcription	Initiation sites	mt	~3	HSP1,* HSP2,* LSP*
Transcription	Termination factor		1	mTERF
Translation	tRNA	mt	22	mt-tRNA^{Ala}, mt-tRNA^{Arg}, mt-tRNA^{Asn}, mt-tRNA^{Asp}, mt-tRNA^{Cys}, mt-tRNA^{Gln}, mt-tRNA^{Glu}, mt-tRNA^{Gly}, mt-tRNA^{His}, mt-tRNA^{Ile}, mt-tRNA$^{Leu(UUR)}$, mt-tRNA^{Lys}, mt-tRNA$^{Leu(CUN)}$, mt-tRNA^{Met}, mt-tRNA^{Phe}, mt-tRNA^{Pro}, mt-tRNA$^{Ser(UCN)}$, mt-tRNA^{Thr}, mt-tRNA$^{Ser(AGY)}$, mt-tRNA^{Trp}, mt-tRNA^{Tyr}, mt-tRNA^{Val},
Translation	Aminoacyl-tRNA synthetase	N	20	AARS, AARS2, DARS, DARS2, EPRS, FARS2, FARSA, FARSB, GARS, HARS, HARS2, KARS, NARS, NARS2, PARS2, SARS, SARS2, TARS, TARS2
Translation	Processing endonucleases	N	~2	RNase P, tRNAse Z
Translation	rRNA	mt	2	12S, 16S
Translation	Ribosomal proteins	N	~42	MRPL1, MRPL2, MRPL3, MRPL4, MRPL5, MRPL6, MRPL7, MRPL8, MRPL9, MRPL10, MRPL11, MRPL12, MRPL13, MRPL14, MRPL15, MRPL16, MRPL17, MRPL18, MRPL19, MRPL20, MRPL21, MRPL22, MRPL23, MRPL24, MRPL25, MRPL26, MRPL27, MRPL28, MRPL29, MRPL30, MRPL31, MRPL32, MRPL33, MRPL34, MRPL35, MRPL36, MRPL37, MRPL38, MRPL39, MRPL40, MRPL41, MRPL42
Replication	Transcriptional initiation sites	mt	2	P H1,* P H2*
Replication	Replication proteins	N	~4	mtSSB, Pol γ, Twinkle, POLRMT

* Not transcribed, no gene products.

Sources: Asin-Cayuela and Gustafsson, 2007; Hyvärinen et al., 2007; Sissler et al., 2017

could lead to reduced efficiency of the transcription and translation of mt genes if key N genes do not coevolve (Blumberg et al., 2014). As outlined in detail below, there is growing evidence that the interactions between N-encoded proteins and mt DNA in these gene regulatory mechanisms are key elements in mitonuclear coadaptation.

In bilaterian animals, initiation of transcription of mitochondria depends on three N proteins: mitochondrial RNA polymerase (POLRMT), mitochondrial transcription factor A (TFAM), and mitochondrial transcription factor B2 (TFB2M) (Gaspari et al., 2004; Bonawitz et al., 2006; Falkenberg et al., 2007; Gustafsson et al., 2016). In addition, mitochondrial transcription elongation factor (TEFM) and mitochondrial transcription termination factor 1 (MTERF1) are important associated N-encoded proteins, but these components of transcription may interact primarily with other N-encoded genes and probably play little role in mitonuclear coadaptation. Specific protein residues in POLRMT recognize specific nucleotide sequences in the promoter sequence, and this recognition determines strength of binding and level of transcription (Gustafsson et al., 2016). Thus, the correct co-function of N-encoded POLRMT and the promoter region encoded on the mt DNA is critical for proper regulation of the ETS (Gaspari et al., 2004; Ellison and Burton, 2008a, 2010). A number of human diseases, collectively called mt DNA depletion syndromes, are a consequence of dysfunction of mt transcription processes (Shadel, 2004; Durham et al., 2005) and the functional interactions of POLRMT and mt promoter regions appear to be an important arena for mitonuclear coadaptation (Ellison and Burton, 2010). Finally, mt genomes require N-encoded proteins for replication, including the same POLRMT gene required for mt transcription, and in this replication process N-encoded proteins have to properly function with mt DNA (Gaspari et al., 2004; Falkenberg et al., 2007). Disruption of the replication of the mt genome would obviously have serious consequences for the respiratory capacity of an organism.

Protein–RNA interactions

The third arena for the critical interaction of the products of mt and N genes is in the translation of mt-encoded proteins, and this type of mitonuclear interaction involves protein–RNA interactions (Konovalova and Tyynismaa, 2013; Meiklejohn et al., 2013). For mt genes, as for N genes, one of the key translators of a nucleotide sequence into an amino acid sequence is the aminoacyl-tRNA synthetases (mtARSs) (Ling et al., 2009) (Figure 2.4). Relative to N DNA, mt DNA employs a variant genetic code, so it requires translational components that are distinct from the components that translate N DNA. In bilaterian animals, there are twenty different mtARS proteins that are transported from the cytoplasm into the mitochondrial matrix in animals, each a complement to a different transfer RNA (tRNA) encoded by mt genes, with two redundancies (Table 2.1). These mtARS proteins are the key translators of genes encoded in mRNA to proteins because each of the twenty different mtARSs binds with one (or at most two) specific amino acids as well as exclusively with one or at most two mt-encoded tRNAs (Meiklejohn et al., 2013; Adrion et al., 2016). Using ATP for energy, each mtARS-type loads a specific amino acid onto its complementary tRNA; hence, these enzymes are the translators of a specific triplet codon to a specific amino acid (Figure 2.4).

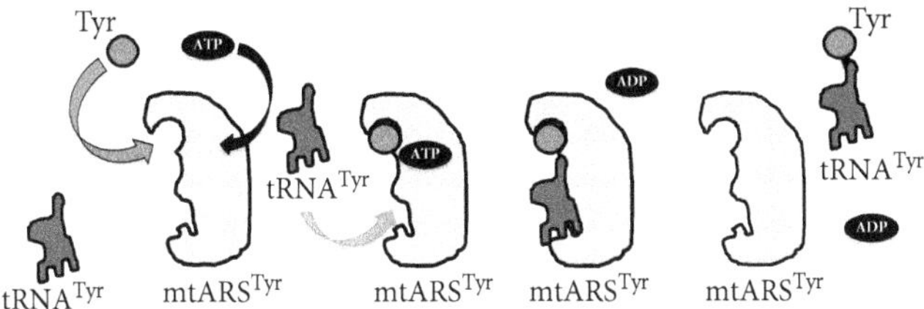

Figure 2.4 The loading of transfer RNAs (tRNA) with the correct amino acids requires tight coadaptation of aminoacyl tRNA synthetase (mtARS), which is nuclear-encoded, and tRNA, which is mitochondrial-encoded. In this example the tyrosine (Tyr) mtARS is loading Tyr onto a Tyr tRNA.

It is not hard to grasp that high degrees of coordination and coadaptation are essential for a system based on code recognition. The mt-encoded tRNAs must present an immediately recognizable RNA sequence to the N-encoded mtARS, or the two molecules will bond slowly or with inaccuracies. Indeed, the signaling that enables tRNAs to be loaded with correct amino acids has been termed "the second genetic code" (Beuning and Musier-Forsyth, 1999). In a properly functioning mitochondrial system, each type of tRNA carries one specific amino acid. In turn, each tRNA recognizes one or a few triplet base pairs in a DNA sequences, and hence it delivers the correct amino acid corresponding to the triplet codon. The capacity to accurately and rapidly read the genetic code is vital to development, growth, and maintenance of an organism (Konovalova and Tyynismaa, 2013; Sissler et al., 2017). A tRNA that sometimes carries the wrong amino acid, or that misreads or only slowly reads the nucleotide sequence, disrupts or corrupts the process of protein translation with well-documented negative consequences for the fitness of the organism. About half of the mt DNA mutations causing diseases in humans occur in tRNA genes (Moreno-Loshuertos et al., 2011), and mutations in all twenty-two mt tRNA have been linked to inherited diseases with the nervous system and neuro-muscular systems most commonly impacted but also with effects on cardiovascular, endocrine, and urinary systems (figure 2 in Sissler et al., 2017). Mutations in mtARS are also responsible for Mendelian-inherited human diseases, and mutations in nearly all mtARS have been linked to a specific disease (Sissler et al., 2017).

Post-transcriptional tRNA modifying factors are an additional type of N-encoded protein that must co-function with tRNA to enable proper translation. As their name implies, these N-encoded proteins direct essential modifications of the raw tRNA transcripts that enable them to fold and function correctly (Lai et al., 2010; Suzuki and Suzuki, 2014). Mutations in these modifying factors create yet another form of mitonuclear incompatibility and are linked to human respiratory diseases (Suzuki and Suzuki, 2014; Saoura et al., 2018). No studies of mitonuclear coadaptation have yet focused on post-transcriptional tRNA modifying factors, but because such post-transcriptional modifications involve the interaction of mt and N gene products, they will need to be taken into account in assessment of the mechanisms for mitonuclear dysfunction.

Protein–RNA mitonuclear interactions also play a key role in the function of ribosomes, which coordinate the core process of translation of mRNA (Pearce et al., 2013). Ribosomes are huge cellular structures daunting in their molecular complexity. In the simplest terms, ribosomes line up the correct tRNA according to nucleotide sequence in a messenger RNA (mRNA) so that polypeptides with the correct sequence of amino acids are produced; thus, ribosomes must correctly recognize the genetic signal of both tRNAs and mRNAs encoded by mt genes. There exist unique classes of ribosomes for prokaryotes, the cytoplasm of eukaryotes, mitochondria, and chloroplasts (Hillis and Dixon, 1991; Melnikov et al., 2012). All ribosomes are composed of both ribosomal RNA (rRNA) and proteins, and mitochondrial ribosomes (mitoribosomes) are constructed with protein encoded in the N genome and ribosomal RNA (rRNA) encoded in the mt genome (Scheper et al., 2007; Rotig, 2011; Pearce et al., 2013). Thus, the mitonuclear interactions between proteins and RNA, both within mitoribosomes and between mitoribosomes and external RNA, are intimate and extensive.

Until recently, details of the structure of any mitoribosomes were largely unknown. Recent breakthroughs in micro-structural biology (Kühlbrandt, 2014) resulted in detailed three-dimensional structural images of an intact mammalian mitoribosome (Amunts et al., 2015). The mitoribosome of mammals is composed of eighty closely interacting proteins, thirty-six of which are unique N-mt proteins, with the remaining forty-four proteins functioning in ribosomes in both the cytoplasm and mitochondria (Amunts et al., 2015). Compared to ribosomes in the cytoplasm that translate N genes, mitoribosomes are larger with better shielding of rRNA from free radicals (Amunts et al., 2015) and with the rRNA reduced in length (Van Der Sluis et al., 2015). N protein–mt RNA interactions are critical not only for the formation of a functional ribosome, but also for post-transcriptional processing of both tRNA and rRNA (Keene, 2007). Details of the molecular interactions of rRNA and ribosomal proteins in the mitoribosome remain almost entirely unknown, but numerous human diseases are caused by mutations in ribosomal proteins (Tuppen et al., 2010; Boczonadi and Horvath, 2014; Alston et al., 2017), and there is a growing focus on RNA–protein interactions in ribosomes as key elements in mitonuclear coadaptation (Barreto and Burton, 2013a).

Anterograde and retrograde signals

In discussions of the co-function of mt and N gene products to enable OXPHOS, there is an implicit assumption of the necessity for mitochondria to signal their functional state both internally to within-mitochondrion transcriptional and translational mechanisms and externally to the N genome. Through such signaling, mitochondria can trigger compensatory and adaptive cellular responses (Muir et al., 2016). Between-genome signaling, however, is rarely mentioned in discussions of mitonuclear coadaptation and coevolution. Much more focus is directed at communication between the ETS and regulators of transcription, translation, and replication of mt DNA within the mitochondrion (Lane, 2011c, 2014).

Both signaling from nucleus to mitochondria (anterograde) and from mitochondria to nucleus (retrograde) are absolutely vital to mitochondrial function and organism fitness (Woodson and Chory, 2008; Raimundo et al., 2016). The list of signaling agents involved in retrograde signals is typically restricted to the molecules associated with OXPHOS: ATP, ROS, NADH, Acetyl-CoA—along with Ca+ and simple metabolites (Woodson and Chory, 2008). These molecules are passively released depending on the state of the respiratory chain, and a common assumption has been that retrograde signaling is passive. Anterograde signaling, on the other hand, is generally presented as being largely the realm of N-encoded proteins that regulate the transcription, translation, and replication of mt DNA (Woodson and Chory, 2008; Antonicka et al., 2013; Bogenhagen et al., 2014). The protein elements of anterograde signaling have been studied in the context of mitonuclear coadaptation, as discussed above. However, there is growing evidence that non-coding RNAs may play an important role in both retrograde and anterograde signaling (Richter-Dennerlein et al., 2015; Dong et al., 2017; Vendramin et al., 2017).

The production of non-coding RNAs with important functions has been long established for the N genome, but it was recently discovered that the mt genome also produces small non-coding RNAs that likely have function at least with respect to within-mitochondrial regulation of transcription, translation, or replication, but also possibly in retrograde signaling to the nucleus (Ro et al., 2013; Pozzi et al., 2017). The mt genome may also produce long non-coding RNAs that function in the nucleus (Dong et al., 2017). In addition to small and long non-coding RNAs, small peptides of mitochondrial origin have also recently been discovered, the best known being humanin (Hashimoto et al., 2001; Lee et al., 2013). Like small non-coding RNAs, mitochondrial-derived peptides seem to function in retrograde signaling.

The discovery of peptides and small and long non-coding RNAs from the mt genome potentially challenges conventional wisdom regarding the number of genes encoded by mt genomes; for instance, it potentially expands the number of mt gene products beyond thirty-seven in animals (Lee et al., 2013; Ro et al., 2013). If these mt-encoded RNAs and peptides do have important functions, these functions are likely to be achieved only through coordination with the products of the N genome. Mitonuclear coadaptation resulting from the interactions of peptides, non-coding mt RNAs, and N-encoded proteins is an as-yet unexplored but very exciting area for new research in mitonuclear ecology.

Evidence for mitonuclear coadaptation

The basics of the form, function, and regulation of the ETS that are presented above provide a framework for understanding the potential for coadaptation of mt and N genomes to have significant effects on mitochondrial function and organism fitness. Moreover, the huge literature on inherited human diseases related to mitochondrial function provides key insights regarding the potential for corrupted mitonuclear interactions to give rise to phenotypes with poor fitness (Picard et al., 2016; Wallace, 2016; Rahman and Rahman, 2018). To directly assess the role of mitonuclear coadaptation arising from mitonuclear coevolution, it is most instructive to assess the degrees and kinds of mitochondrial dysfunctions that result when coadapted mitonuclear genotypes from different species (or occasionally subspecies or genetically divergent populations) are mixed. In the following sections I review evidence for mitochondrial dysfunction in the experiments that have been conducted on cytoplasmic hybrid (cybrid) cell lines, cybrid organisms, and hybrid organisms.

Cybrid cell lines

A basic prediction arising from the hypothesis that mitonuclear coadaptation is fundamental to aerobic eukaryotes is that there should be a loss of mitochondrial function when divergent mt and N-mt genes are forced to function together (Rand et al., 2004). In other words, if scientists experimentally combined the mt genes from one species with the N genes from another species, the result should be phenotypes that are compromised in respiratory function. Strong support for this prediction comes from studies using mitonuclear cybrid cell lines, in which N genes from one species or population are combined with the mt genes from another species or population through the transfer of genetic elements in cells growing in culture (Wallace, 1999) (Figure 2.5). These cybrid cell lines provide key insights that generally cannot be fully replicated with hybridization experiments because of the manner in

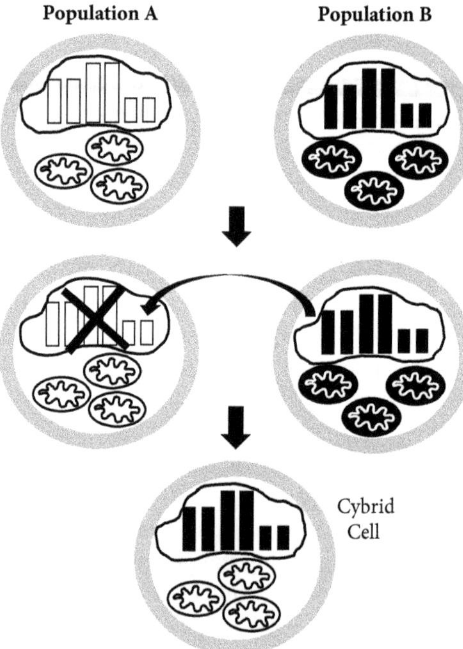

Figure 2.5 Steps in the creation of a cybrid cell line. The nucleus and its genetic material are destroyed in the host cell and replaced with the nucleus of the comparison species. Then the cybrid cell line is grown in culture.

which the cybrid cell lines are created—the nucleus of a cell from one species is destroyed and replaced by the nucleus from a second species. This mixing of novel mt and N genes without sexual reproduction eliminates any opportunity for recombination of the parental N genotypes or for selection on N genes (Figure 2.5). As cybrid cell cultures grow, the resulting phenotype is a product of the experimentally combined mt and N genes.

Experiments using cybrid cell lines are typically conducted within the context of biomedicine (Box 2.1); consequently, in one of the early cybrid studies using cell lines, a human N genome was paired with the mt genomes from a range of primates including: chimpanzees (*Pan troglodytes*), bonobos (*Pan paniscus*), gorillas (*Gorilla gorilla*), orangutans (*Pongo* sp.), African green monkeys (*Cercopithecus aethiops*), squirrel monkeys (*Saimiri sciureus*), and ring-tailed lemurs (*Lemur catta*) (Kenyon and Moraes, 1997). The mitochondrial dysfunctions arising from mitonuclear incompatibilities as these divergent genomes attempted to produce a functional ETS ranged from substantial to total system failure. When paired with human N DNA, mt DNA

Box 2.1 The impetus for studies of cybrid cell lines

The original studies of cybrid cell lines that were created using suites of primate or rodent species of variable relatedness provided unprecedented information on post-zygotic isolating mechanisms among species. The well-documented loss of function in cybrid cells due to specific mitonuclear incompatibilities has huge implications for fundamental speciation and sexual selection theory. However, these pioneering studies were not conducted by scientists with a particular interest in speciation or sexual selection; they were conducted by researchers working in biomedical labs, who were primarily interested in developing lab models for the study of mitochondrial diseases in humans. In the landmark *Proceedings of the National Academy of Sciences of the United States of America* paper on the respiratory function of cybrid cells created by combining N genes from humans with mt genomes from a range of primates, Kenyon and Moraes (1997, p. 9134) concluded "This model system may provide a new tool for approaching the pathogenesis of mitochondrial disorders associated with a severe depletion of mt DNA." The term "speciation" does not appear in this paper nor in any write-up of the cybrid studies conducted in this era. The non-evolutionary focus of the biomedical researchers who conducted these cybrid studies—and the failure of evolutionary biologists to grasp the central importance of documentation of divergent mitonuclear coadaptation between species—effectively shielded these important observations from evolutionary biologists studying speciation and sexual selection for years after they were published. It was not until about a decade into the new millennium that speciation began to be discussed in the context of mitonuclear coadaptation (Dowling et al., 2008; Lane, 2009; Gershoni et al., 2009; Chou and Leu, 2010; Burton and Barreto, 2012; but see Levin, 2003), and the cybrid studies were "discovered" by biologists interested in speciation. Once biomedicine developed lab strains of mice that carried mitochondrial mutations linked to disease, interest in between-species cybrid studies vanished in the world of biomedicine. Very slowly in recent years, manipulation by evolutionary biologists of mt and N genes in non-model systems has begun to replace the mammalian cybrid studies of the late twentieth century.

from chimpanzee, bonobo, and gorilla enabled viable cell lines and measurable respiration via OXPHOS. In contrast, when the mt DNA from orangutan, Old World monkey, New World monkey, or lemur was paired to human N DNA in cultured cells, there was no cell growth or measurable respiration (Kenyon and Moraes, 1997).

In the viable cybrids, which involved mitonuclear combinations from near-relatives of humans, respiration was conspicuously diminished (Kenyon and Moraes, 1997). A follow-up study of cybrid cell lines formed between human and chimpanzee, bonobo, or gorilla revealed a decrease in respiratory function of 20 percent, 34 percent, and 27 percent, respectively (Barrientos et al., 1998). Reduced respiration in these cybrids could be attributed specifically to poor function of Complex I. Mitochondrial Complexes II, III, IV, and V had activities in cybrid cells that were indistinguishable from parental human or non-human primate cells, and transcription and translation of ETS subunits also did not appear to be compromised in cybrids (Barrientos et al., 1998). Incompatibilities in Complex I protein subunits encoded by ape mt and human N genes were implicated as the specific source for the reduced complex function (Barrientos et al., 1998).

Follow-up studies of human–orangutan cybrids provided a glimpse into how diverse mitonuclear incompatibilities can be. Cell lines created with human N genes and orangutan mt genes were not viable, but researchers were able to generate viable cell lines by introducing specific orangutan N genes into the cultured cells, although impairment of respiratory function was still 65–80 percent even in cells with some orangutan N genes. Through these techniques, researchers found that the activity of Complex IV was reduced by 85–95 percent in human–orangutan cybrid cell cultures. Because the amino acid sequence of Complex IV proteins was conserved between human and orangutan, the impairment of Complex IV function was attributed to problems with function of Complex IV assembly proteins. The interaction of assembly proteins and complex subunits is a form of mitonuclear protein interaction that does not directly involve the co-functions of N- and mt-encoded subunits for OXPHOS. Rather, this form of mitonuclear interaction affects how efficiently subunits are combined into complete complexes (Barrientos et al., 2000).

The most significant loss of function in the human–orangutan cybrid cells, however, concerned Complex V. Both of the orangutan mt-encoded subunits of Complex V are highly divergent from other great apes, including humans. The mt-encoded orangutan APT8 shares only 67 percent of its peptides with human/chimp/gorilla, and not surprisingly Complex V had essentially no respiratory function in human–orangutan cybrids (Bayona-Bafaluy et al., 2005). Thus, dysfunction of Complex V in these cybrid lines is undoubtedly due to substantial incompatibilities between human N and orangutan mt protein subunits, the latter having many amino acid substitutions and highly divergent three-dimensional structure compared to its human counterpart. One conclusion from these studies of primate cybrids was that mitonuclear incompatibilities create total respiratory failure in primate cybrid cell lineages composed of mitonuclear components that were diverged by more than about 18 million years (the approximate date of the last common ancestor between orangutan and human as stated by these authors), but even among sister taxa such as chimp and human there was a substantial loss of respiratory efficiency when mitonuclear genes that were not coadapted were forced to function together (Barrientos et al., 1998).

These pioneering cybrid studies on primates were followed by experiments with rodents, beginning with a study in which house mouse (*Mus musculus*) and Norway rat (*Rattus norvegicus*) cybrid cell lines were created. The mouse–rat cybrid cells had 50 percent reduction in respiration compared to control lines (McKenzie and Trounce, 2000). In broader comparative studies of rodents, researchers found that respiratory dysfunction resulting from mitonuclear incompatibilities in cybrid lines was proportional to degrees of phylogenetic divergence between parental species (McKenzie et al., 2003; Enoki et al., 2014) (Figure 2.6).

The cybrid cells in these rodent studies showed no measureable impairment in transcription or translation of mitochondrial proteins, so the effects on respiration appeared to be due to protein–protein interactions in the respiratory complexes (McKenzie et al., 2003). The authors measured significant reduction in activity in Complexes I, III, and IV, with the greatest effects on Complex III. In all of these complexes there were multiple amino acid substitutions in the mt-encoded proteins between the rat and mouse. In a broad assessment of cybrid function involving multiple species of rodents with variable levels of divergence, respiratory dysfunction was again observed to be a consequence of reduced activity of Complexes I, III, and IV of the ETS, with the greatest effects on Complex III (McKenzie et al., 2003). In all of these rodent cybrid studies, there were no apparent problems with transcription or translation of ETS subunits.

Experiments with cybrid cell lines provide unique insight into mitonuclear coadaptation. They produced the first direct evidence for mitonuclear incompatibilities between divergent taxa (at least in animals) and helped to stimulate interest in studies of mitonuclear incompatibilities in the context of whole animals in different environments. However, because only monocultures of one cell type are studied, measurements of

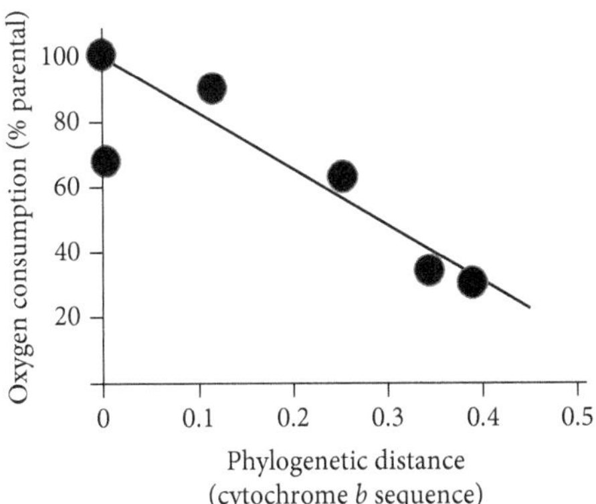

Figure 2.6 The relationship between respiratory function of cybrid cell cultures created from rodent cells, measured as percentage oxygen consumption compared with parental type, and phylogenetic distance between parental species used to construct the hybrid, as estimated from divergence in cytochrome *b* genotype. Adapted from Enoki et al. (2014).

mitonuclear incompatibilities in cybrid cell lines are necessarily restricted to relatively simple measures of cellular respiration such as oxygen consumption. It is not possible to extrapolate from studies of a single type of cell in culture to anticipate the specific effects of mitonuclear incompatibilities involving whole organisms in natural environments. For instance, with studies of cell lines cultured in the lab, there is no way to study tissue-specific, sex-specific, or age-specific effects of mitonuclear incompatibilities. In humans, inherited diseases involving the same mitonuclear interactions that are the basis for loss of respiratory function in cybrid individuals commonly manifest in effects at relatively late developmental stages and on specific organ systems (Chinnery, 2015; Alston et al., 2017). Many important negative effects of novel mitonuclear combinations might not be apparent in studies of cell cultures. To observe a full range of dysfunctions caused by mitonuclear incompatibilities, it is necessary to manipulate the mt and N gene combinations in germ lines in order to produce whole cybrid organisms.

Somatic cell nuclear transfer

Somatic cell nuclear transfer (SCNT) is a lab technique in which the nucleus from a somatic cell is transferred to an enucleated fertilized egg cell (Beyhan et al., 2007). The goal is to propagate the N DNA in a donor cytoplasmic background. SCNT is the basic cloning technique that produced Dolly the sheep in 1996 and that led to near

Box 2.2 Mitonuclear conservation biology

When students learn about the hybrid origin of eukaryotes, a pernicious tendency is to consider a eukaryote (and more specifically oneself) to be more archaeon than bacterial. According to this line of reasoning, the archaeon cell that gave rise to the nucleus is our ancestor, while the bacterial cell that was engulfed is a foreign entity within us. By a strict gene count—about 20,000 to 37—we *are* mostly a product of the nucleus. But a eukaryote is a chimeric outcome of the fusion of two prokaryotes—not capture of one prokaryote by another. By the nature of our origin, we are as much bacterial as we are archaeon.

The debate about whether the mt genome has anything relevant to add to the defining characteristics of a species becomes more than a philosophical diversion when conservation biologists seek to save a species for which only one or a few individuals exist. For more than a decade, conservation biologists have been working to refine SCNT in which the nuclear material from an endangered species is transferred to and replaces the nucleus of the fertilized egg of a donor common species. This technique is founded on the principle that the N genome holds the essence of the species and that allowing the mt genes to go extinct has no significant effect on the persistence of the endangered species. The key assumption is that saving the N genes of an endangered species is all that is required to save that species. Mitochondrial DNA is viewed as important only in so far as it enables cellular function; any mt genotype that enables function is as good as any other mt genotype. Through the chapters of this book, I will present a long counterpoint to the idea that N genes are all that really matter for the form and function of an individual. The capstone will be a species concept in which a species is defined by mt genes at least as much as by N genes.

hysteria in the late twentieth century when cloning humans seemed inevitable (Einsiedel, 2000). It was also the foundation for the plot of many novels and movies. When applied within a species, SCNT can theoretically enable the creation of multiple individuals with identical N genotypes (hence the army of identical warriors in *Star Wars Clone Wars*). When SCNT has been used to combine a N genome of one species with a mt genome of another species, the technique has been employed as a potential means to propagate endangered species (Lanza et al., 2000) (Box 2.2).

Almost universally, attempts at between-species SCNT have yielded inviable individuals with respiratory dysfunction that is proportional to genetic distance between donor and recipient (Beyhan et al., 2007). Moreover, most problems with viability in SCNT hybrids arise specifically due to mitonuclear incompatibilities (Gómez et al., 2003; Bowles et al., 2008). For instance, in SCNT involving domestic pig and house mouse, transcription of many mt gene protein products was significantly reduced (Amarnath et al., 2011). Interestingly and in contrast to the unambiguous evidence for mitonuclear coadaptation presented by the *in vitro* cybrid studies in rodents (McKenzie et al., 2003), a viable cybrid mouse generated with the N genes of *M. musculus* and the mt genes of *M. terricolor* showed no detectable differences in behavior or cognitive performance relative to parental types, but did show significant changes in patterns of gene expression (Cannon et al., 2011). In another study, cybrid mice that carried pure N DNA from either *M. musculus musculus* or *M. m. domesticus* showed post-implantation embryonic lethality when they carried the mt genome of the other subspecies (Ma et al., 2016). This *Mus* study demonstrates that even at the subspecies level there can be population-specific sets of coadapted mt and N genes that do not function well when combined.

The mitonuclear combinations in SCNT procedures are generally not as pure as in cybrid cell lines because some donor mt genes are invariably transferred with the donor N genome. However, because the SCNT procedure results not merely in cultures of cells but in whole organisms that grow and develop into adults, assessment of function through developmental stages and into adulthood often reveals problems that would not be evident in cell-culture studies (Hall et al., 2006; Beyhan et al., 2007) (Figure 2.7). Public interest in the generation of novel mitonuclear gene combinations through SCNT-like procedures rose precipitously when mitochondrial replacement therapy was introduced as a means to cure inherited diseases that are carried on mt genes (see Box 2.3).

SCNT are not entirely restricted to complex animals. Beale and Knowles (1976) created cybrid individuals by transferring mitochondria among three closely related species of the single-celled protozoan *Paramecium*, with variable levels of dysfunction of the cybrid *Paramecium* noted.

Hybrid backcrosses

In theory, hybrid experiments should provide weaker evidence for mitonuclear coevolution than experimental transfer of genetic material in cybrid studies because of the difficulty in hybrid crosses of disentangling the effect of nuclear–nuclear gene

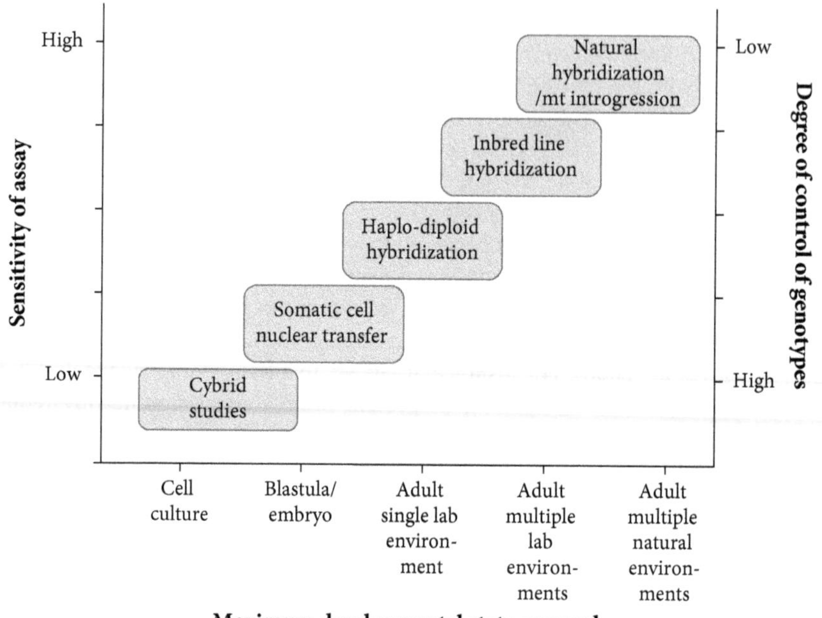

Figure 2.7 The various insights provided by different approaches to test the effects of mismatched mitonuclear genotypes. With few exceptions, researchers trade off strict control of mitonuclear genotypes with sensitivity of the approach to detect dysfunction caused by mismatched mitonuclear genotypes.

interactions from mitonuclear interactions (Figure 2.7). However, researchers have used different creative techniques in hybrid experiments to isolate the effects of mitonuclear from nuclear–nuclear gene interactions. In what has become the standard design in experiments aimed at testing for hybrid breakdown related to mitonuclear incompatibilities, individuals from inbred lines are crossed and then repeatedly backcrossed (Breeuwer and Werren, 1995; Nagao et al., 1998; Chang et al., 2015) (Figure 2.8). After multiple generations of backcrossing, the result is a population of individuals that carry the mt genes from one parental type paired with the N gene from the other parental type, just as in cybrid cell lines created through direct transfer of genetic material in a laboratory. Even in the most carefully controlled backcross experiment, however, selection during the breeding process can change gene frequencies in unpredictable ways, which can confound the results of the experiments (Burton and Barreto, 2012).

Cybrid populations from such backcrossing have been created using sister species of yeast, nematodes, fruit flies, and monkeyflowers (Sackton et al., 2003; Zeyl et al., 2005; Fishman and Willis, 2006; Lee et al., 2008; Barr and Fishman, 2010; Chang et al., 2015). These studies consistently show negative effects of pairing mt and N genomes that are not coadapted. Overall mitochondrial sequence divergences as small

Box 2.3 Mitochondrial replacement therapy

In Box 2.1 I chided the evolutionary community for not grasping the significance of mitochondrial dysfunction in cybrid experiments. It is only fair, therefore, to comment on the failure of the biomedical community to fully take account of studies by evolutionary ecologists on the functional significance of standing variation in mt genotypes within populations. These insights come mainly from detailed experimental studies of *Drosophila* (Dowling, 2014; Zhu et al., 2014), but also in a wide range of eukaryotes from monkeyflowers to killifish, individuals carry variable mt genotypes that affect how well they perform in specific environments. Moreover, the function of the mt genotype is commonly dependent on the N genotype to which it is matched. These fundamental discoveries took on new practical significance when medical researchers began experimenting with and are now applying for permits to perform mitochondrial replacement in human zygotes (Reinhardt et al., 2013).

As presented throughout this chapter, many human diseases are the outcome of mt genotype that is maternally inherited. Mitochondrial replacement therapy swaps out the mitochondria of a female parent known to carry such a genetic disorder with mitochondria from a donor from a lineage in which there is no history of mitochondrial disease (Gemmell and Wolff, 2015). In theory and in practice, this medical procedure can remove an inherited disease from a family lineage and avoid human suffering. But even as a known disease-causing mutation is removed from a personal lineage, the mixing of mt and N genes between individuals drawn potentially randomly from the population has the potential to introduce new, unanticipated mitonuclear incompatibilities. The insights from detailed studies in non-human systems provide a roadmap for best practices in mitochondrial replacement therapy that minimize the chances of stepping around one problem right into another unanticipated problem.

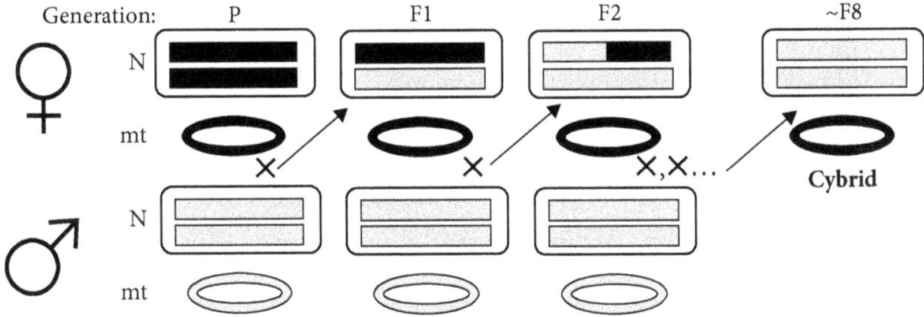

Figure 2.8 A standard experimental design using backcrossing to create individuals with mitochondrial genotype from one taxon paired to a pure nuclear genotype from another taxon taxon. Adapted from Chang et al. (2015).

as 0.1 percent can produce serious mitonuclear incompatibilities and loss of function (Meiklejohn et al., 2013; Chang et al., 2015). This raises a key issue that will be a focus in Chapter 7: overall sequence divergence is, at best, a crude index of functional divergence in coadapted mitonuclear genotypes. A single nucleotide change can, theoretically, lead to serious incompatibility and loss of fitness. Conversely, the genomes of two individuals could be divergent across thousands of loci but their genotypes could still be compatible. Incompatibilities arise from the interactions of specific genes and one key incompatibility is potentially enough to isolate populations.

Hybrid backcrossing with yeast has been particularly important to an understanding of the potential consequences of mitonuclear incompatibilities. As a case in point, the best examples of disrupted translation of mt genes arising from mitonuclear incompatibilities have been documented in hybrid backcrosses among baker's yeast (*Saccharomyces cerevisiae*) and the sister species *S. bayanus* and *S. paradoxus*. These single-celled eukaryotes interbreed freely but hybrid sexual lines have very low fitness (Chou and Leu, 2010). Lee et al. (2008) used careful genetic screening of hybrid lines from *S. cerevisiae* × *S. bayanus* crosses to link a specific N-mt gene, AEP2, with loss of transcription of mt genes. In yeast, AEP2 is a N-encoded accessory protein that initiates translation of the yeast mt gene OLT1 to produce a subunit of the ATP synthase complex. The OLT1 sequence is substantially diverged between *S. cerevisiae* and *S. bayanus* such that *S. bayanus* AEP2 cannot recognize the diverged pattern in the *S. cerevisiae* OLT1 sequence; hence, translation of a key ETS protein is not initiated in hybrids.

Further research in this yeast system resulted in the only experimental demonstration to date that loss of respiratory function in a cybrid organism is a direct result of mitonuclear interactions related to ribosome function. In a cybrid yeast that was created with the nuclear background of *S. bayanus* but with *S. cerevisiae* mt genes, cell growth was halted (Jhuang et al., 2017) (the cessation of growth was actually only observed in non-fermentable media because yeast have the capacity to shift to anaerobic respiration when OXPHOS fails). Through detailed analysis, Jhuang et al. (2017) showed that mitochondrial dysfunction in their cybrid yeast was caused by an incompatibility between mt-encoded rRNA and a N gene with a pentatricopeptide repeat (PPR) domain that binds RNA and helps stabilize it (Hénault and Landry, 2017). The dysfunction apparently arises because N gene product does not bind the mt RNA correctly leading to a breakdown of ribosomal function (Jhuang et al., 2017). Mitonuclear incompatibilities were also documented in yeast hybrids that arose because of inability to properly excise introns in COX1 mRNAs. In this case, the N-encoded splicing gene from *S. bayanus* did not recognize the code for splicing one of the two introns in the *S. paradoxus* mRNA and the result was inviable hybrid offspring (Chou and Leu, 2010). Although these examples from yeast involve mitonuclear interactions between RNA and protein that affect ribosome function, they do not demonstrate mitochondrial dysfunction related specifically to interaction between proteins and RNA in the mitoribosome. The lack of evidence for mitonuclear incompatibilities and dysfunction from interactions within the mitoribosome is undoubtedly due to the complexities of the form and function of mitoribosomes.

Hybrid studies with fruit flies identified incompatibilities in the interactions of mt-encoded tRNAs and N-encoded proteins in the translation of mt genes. Two factors determine the fidelity with which the gene sequence carried on a mRNA is translated into an amino acid sequence: (1) the loading of each type of tRNA with the cognate amino acid (Sissler et al., 2017) and (2) the accurate selection of tRNAs by ribosomes according to the triplet codon (Ling et al., 2009). To date, the best documented example of compromised protein translation arising from mitonuclear incompatibilities concerns a hybrid line of *Drosophila* with the mt genome from *D. simulans* and N genome from a lab population of *D. melanogaster* which carries a SNP that results in an amino acid substitution in mtARS (Meiklejohn et al., 2013) (Figure 2.9). Cybrids that carry a *D. simulans* mt genotype and N genome of *D. melanogaster* with the mutant mtARS showed decreased OXPHOS activity specifically for those complexes that include mt-encoded (and translated) subunits (Hoekstra et al., 2013; Meiklejohn et al., 2013). The reduced OXPHOS activity in the cybrid flies resulted from reduced protein synthesis due to poor interaction between a nucleotide substitution in the tRNA anti-codon arm that differentiates *D. melanogaster* and *D. simulans* and a single amino acid substitution in the binding pocket of the mtARS in the lab strain of *D. melanogaster* (Figure 2.9). Interestingly, the effects of this tRNA–mtARS interaction

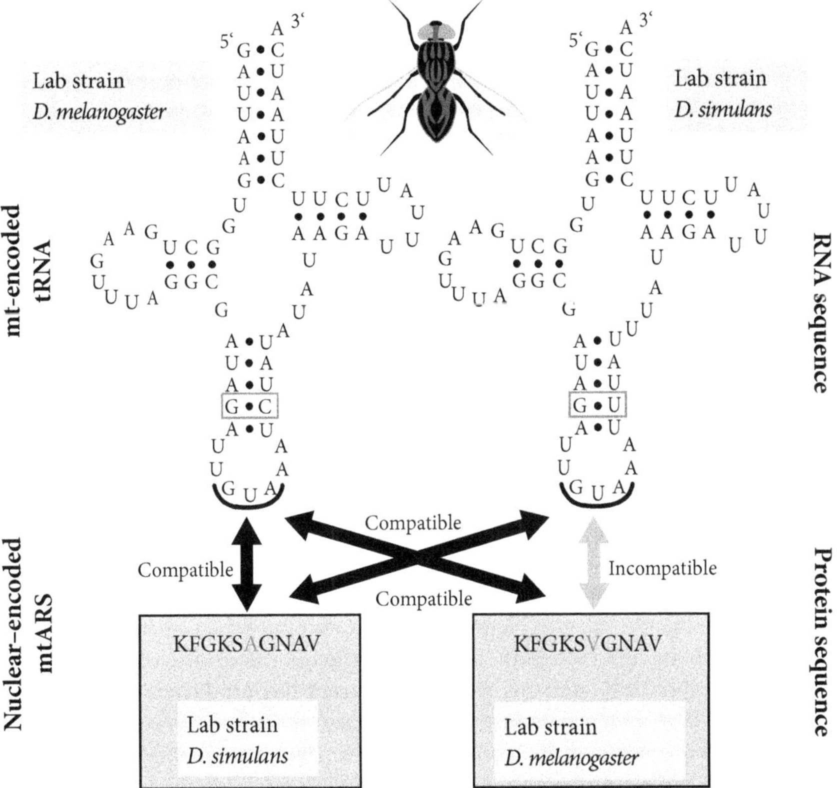

Figure 2.9 OXPHOS dysfunction arising from incompatibilities between N-encoded mt aminoacyl tRNA synthase (mtARS) and mt-encoded tRNA. A single base substitution in tRNA (within the gray frame) that does not change the anti-codon matched to a single amino acid change (shaded letter in box) in the binding pocket causes a severe developmental dysfunction. Adapted from Hoekstra et al. (2013).

were temperature-specific. At room temperatures, the negative fitness effects were significant. At cooler temperatures, however, the cybrid flies showed essentially normal patterns of development (Hoekstra et al., 2013).

In many cybrid and hybrid experiments that have created novel combinations of mt and N genomes, reduced OXPHOS is observed but the specific incompatibility cannot be identified (Burton and Barreto, 2012). Given that there are twenty distinct protein–RNA interactions involving tRNA/mtARS, each of which entails interaction of specific recognition sites between the nuclear and mitochondria components, it seems very likely that tRNA–mtARS interactions are a common source of mitonuclear incompatibilities. This should be an area of active mitonuclear research in coming years.

There is an extensive literature on cytonuclear incompatibilities arising from hybrid crosses in plants, but plants have two organelles that each carry a genome: mitochondria and chloroplasts. Rarely is it possible to isolate the effects of mitonuclear interactions from the effects of chloroplast–nuclear interactions (Greiner and Bock, 2013), but see Barr and Fishman (2010). When the effects of one organelle genome on hybrid fitness are found, it is almost always the effects of the chloroplast genome that are revealed (Greiner et al., 2011). For instance, Moison et al. (2010) documented significant within-species variation in the N, mt, and chloroplast genomes within the flowering plant species *Arabidopsis thaliana*, on par with the genomic variation within *Tigriopus* copepods. Furthermore, in crosses between genetically divergent populations, they demonstrated that the cytoplasm donor had a significant effect on the germination rates. The authors were not able to isolate mitonuclear from cytonuclear effects. Most of the literature on mitonuclear interactions in plants concerns cytoplasmic male sterility, and I will take up this topic in detail in the next chapter on cooperation and conflict.

Hybrid crosses: Classic studies with Tigriopus *copepods*

In some investigations that are focused on understanding hybrid incompatibilities related to mitonuclear gene interactions, hybrids are continually crossed with each other rather than backcrossed to parental lines. Single-generation maternal and paternal backcrosses can then be used to demonstrate mitonuclear effects independent of nuclear–nuclear effects (Figure 2.10). Research published by the laboratory of Ronald Burton on the tide-pool copepod *Tigriopus californicus* employs such a hybridization approach, and this research program has produced by far the most complete and compelling demonstration of hybrid dysfunction resulting from the interactions of the products of mt and N genomes. *Tigriopus* copepods are very small (about the size of a grain of sand) and are restricted to small splash pools above rocky shorelines that are typically disconnected from the ocean. There is surprising genetic structure among copepods from rock outcroppings only a few dozen kilometers apart or between mainland and near-shore island rocks. Divergences in mitochondrial sequences among populations within this species along the coast of California can be as high as 22 percent (Willett and Ladner, 2009).

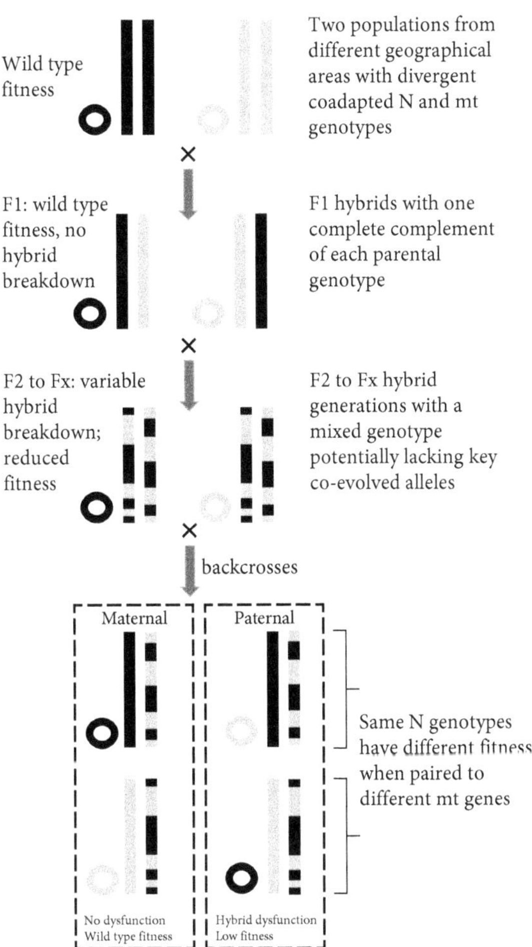

Figure 2.10 Overview of the experimental design and gross outcomes of hybrid experiments conducted on *Tigriopus californicus*. Restoration of full respiratory function via maternal but not paternal backcrosses implicates mitonuclear coadaptation in hybrid dysfunction. Adapted from Burton and Barreto (2012).

When Burton and colleagues crossed individuals from genetically divergent populations of *T. californicus* copepods, they found no negative fitness effects in F1 offspring. Indeed, F1 hybrids typically had slightly higher fitness than either parental types indicating a positive effect of novel combinations of N genes, a phenomenon known as heterosis (Edmands, 2008; Ellison and Burton, 2008b) (see Fig 6.6). In F2 and subsequent hybrid generations, however, the average fitness of hybrids was significantly lower than that of adults (Burton et al., 2006). Not all hybrid individuals showed reduced fitness—some did quite well, but most did poorly (the potentially most interesting individuals with the most impactful mitonuclear incompatibilities failed to develop and were lost from the study, which is an unavoidable shortcoming of this approach). Overall, F2 hybrids show significantly greater variation in fitness than parental types (Barreto and Burton, 2013b). The negative effects on fitness in F2 and subsequent generations arise specifically from dysfunction of aerobic respiration, which is associated with increased release of free radicals (Barreto and Burton, 2013b) and decreased production of ATP (Rawson and Burton, 2002; Ellison and Burton, 2006).

Two lines of evidence demonstrate definitively that respiratory dysfunction in hybrid copepods arises due to incompatibilities in coadapted mitonuclear genes. First, ETS Complexes I, III, IV, and V are all compromised in hybrid copepods, but there are no negative effects on Complex II (Ellison and Burton, 2006). As presented in detail above, Complexes I, III, IV, and V are all composed of interacting mt and N gene products. Complex II is the only complex that is composed entirely of N gene products. The negative effects on hybrid copepods are restricted to enzymes with both mitochondrial and nuclear components. Second, all of the effects of hybrid dysfunction are alleviated if the hybrid line is backcrossed to the maternal line (Burton and Barreto, 2012; Burton et al., 2013) (Figure 2.10). Hybrid dysfunction is not alleviated through backcrossing with the paternal line. The only explanation for fitness rescue via maternal backcross is that the female provides one complete set of N genes to co-function with her mt genes, hence restoring full organism function (Figure 2.10).

Burton and colleagues used the copepod system to document specific forms of mitonuclear incompatibilities leading to hybrid dysfunction. They found high levels of amino acid substitutions in mt-encoded COX1 and COXII as well as in N-encoded cytochrome *c* (CYC) among divergent populations (Rawson et al., 2000). CYC facilitates the transfer of electrons from Complex III to Complex IV, and protein–protein interactions between CYC and COX genes are necessary for proper respiratory function. In hybrid crosses between populations in which individuals carried a common mt type but variable N genotype, respiratory function (Edmands and Burton, 1999) as well as population fitness (Willett and Burton, 2001) were highly correlated with CYC genotype, implicating these CYC–COX protein interactions in mitonuclear hybrid breakdown. Definitive demonstration of divergent coadaptation of these CYC/COX proteins was made by showing significantly higher activity of Complex IV when it is paired with same-population CYC proteins versus CYC from different populations (Rawson and Burton, 2002).

In follow-up studies, Burton and colleagues tested the hypothesis that mitochondrial dysfunction in hybrid copepods can result from poor co-function of N-encoded transcriptional machinery and the promoter region on the mt DNA. They focused on the interaction of mtRPOL and the promoter region of mt DNA. In their hybrid crosses, they found that specific combinations of mtRPOL proteins and mt DNA promoter sequences showed reduced levels of mt gene transcription in response to stressors (Ellison and Burton, 2008a). They then sequenced the entire mt genome from three divergent populations of *T. californicus* copepods, and found numerous nucleotide substitutions in the control region, which contains the mt DNA transcriptional promoter (Burton et al., 2007). The works by Burton and colleagues on copepods remain the only studies on interactions between N proteins and mt transcriptional promoter sites as a basis for mitonuclear coadaptation, but this may turn out to be among the most common sources of mitonuclear incompatibilities between populations (Burton et al., 2013).

A subsequent study by the Burton lab group on the relationship between mitochondrial transcriptional response and mt DNA copy number provides evidence that incompatibilities between mtRPOL and the promoter regions of mt DNA affect not just transcription of mt DNA but also replication (Ellison and Burton, 2010). Because the same mtRPOL enzyme functions in the initiation of both the transcription of mt genes and the replication of the mt genome, disruption of transcription can lead to disruption of replication (Clayton, 2000b). Among the poorly functioning hybrid copepod lines, Ellison and Burton (2010) found a negative relationship between mt transcriptional response and mt DNA copy number. These observations suggest that mitonuclear incompatibilities between N-encoded mtRPOL and mt promoter regions may disrupt a balance between transcription of mt genes and replication of the mt genome.

Other research focused on the interactions of N-encoded proteins with mt nucleotide sequence employ a "a pure *in vitro* system" wherein reactions related to mt DNA replication are observed with no other cellular components. Gaspari et al. (2004) employed such a pure *in vitro* system when they created combinations of mouse or human TFAM/TFB2M/mtRPOL with mouse or human mt DNA promoter regions. Within-species mt DNA replication was well supported in the *in vitro* system, but between-species replication was not. The authors attributed the failure of the hybrid systems to the failure of the mtRPOL to recognize the heterospecific promoter sites. Work on mitonuclear effects on transcription and replication of mt genes should be a focus of future research.

A hybrid-cross approach was also used to study mitonuclear coadaptation in parasitic wasps in the genus *Nasonia* (Nagao et al., 1998; Ellison et al., 2008). These wasps are haplo-diploid, with females diploid and males haploid. The haploid state of males is convenient for studies of mitonuclear coadaptation because there exists only a single nuclear allele at each locus to interaction with mt genes. Just as in *Tigriopus* copepods, matching the N genome of one *Nasonia* species to the mt genome of a sister species created dysfunction of ETS Complexes I, III, and IV, but not Complex II, and protein–protein interactions were implicated in the hybrid dysfunction (Ellison et al., 2008). Hybrid crosses between *Drosophila* species also created ETS dysfunction (Sackton et al., 2003).

Within-species mitonuclear studies

Studies of mitonuclear coadaptation are still in their infancy, but as the basic necessity of mitonuclear coadaptation is established through studies mixing mt and N genes between species, there is growing interest in the interactions of mt and N genes within species (Dowling et al., 2008; Wolff et al., 2016a). Conplasmic lines of organisms carry a common nuclear background and a diversity of mt genomes derived from standing variation or variation among subspecies variants (Tourmente et al., 2017). A growing number of studies are using such lines for the assessment of the effects of polymorphic mt DNA (Yu et al., 2009; Scheffler et al., 2012; Yee et al., 2013; Wolff et al., 2016a). More than anything else, these studies of conplasmic lines show the potential for the evolution of mitonuclear coadaptation. Many species have standing variation in mt genotypes and these mt genotypes function with various efficiencies in different nuclear backgrounds. The outcomes of these within-species mitonuclear interactions are environment dependent, and I'll return to the topic of within-species mitochondrial variation in Chapter 9.

Summary

Mitonuclear ecology is founded on the premise that mitonuclear coadaptation is essential for eukaryotic life. At the heart of mitonuclear ecology is the prediction that combining mt and N-mt genotypes from individuals from divergent populations should create combinations of mt and N-mt that are not coadapted, leading to poor mitochondrial function and reduced fitness. There is substantial and growing evidence from eukaryotes as diverse as baker's yeast and lowland gorillas documenting that when mt genes from one population are combined with N genes from heterospecifics or diverged foreign populations, there are inevitably mitonuclear incompatibilities creating poor performance and reduced fitness. The potential arenas for such mitonuclear incompatibilities are, first, in the co-functioning of mt- and N-encoded protein subunits making up the complexes of the ETS. Both cybrid and hybrid experiments that combine divergent mt and N genotypes in yeast, nematodes, flies, and mammals have demonstrated respiratory dysfunction arising from mitonuclear incompatibilities in protein subunits. Consequential mitonuclear incompatibilities are not, however, limited to protein–protein interactions. The mt components of the ETS are transcribed and translated within the mitochondria via mechanisms that require close coordination of mt and N-mt gene products. Dysfunction in protein–RNA and protein–DNA interactions during transcription, translation, and replication of mt genes is as important as dysfunction arising from protein–protein interactions in determining the fitness effects of mitonuclear incompatibilities. Disruptions of mitochondrial transcription and translation due to mutations in both N and mt components are the cause of numerous human diseases, and recent studies on hybrid breakdown in animals and fungi implicate mitonuclear incompatibilities in these regulatory mechanisms as frequent causes of respiratory dysfunction in hybrids. As knowledge of ETS function and retrograde and anterograde signaling accelerates exponentially in the age of genomics, more potential arenas of mitonuclear co-function and coadaptation are being discovered. The necessity for mitonuclear coadaptation has been empirically validated. In the following chapters, I'll present the enormous implications of mitonuclear coadaptation for the evolution of complex life.

3

Compensatory coevolution

Organisms evolve as integrated sets of genes. Changing one component in the whole necessarily affects other components that will, in turn, change in response. These statements could have been written by any graduate student in Thomas Hunt Morgan's lab in the 1930s. The concept of coevolution of genes in a genome is as old as the concept of a gene (see, for instance, Dobzhansky, 1937b and Wright, 1942). In modern evolutionary biology, coevolution of the molecular components of biological systems is still mostly discussed with consideration of proteins encoded by the N genome (e.g. Pazos and Valencia, 2008; de Juan et al., 2013), and within this nuclear-focused coevolutionary arena, compensatory coevolution is unnecessary. Coevolution of N-encoded proteins involves natural selection eliminating deleterious variants of protein-coding genes and promoting beneficial variants that enable enhanced co-function with other proteins (Choi et al., 2005; Socolich et al., 2005). Such coevolution of N gene products is universally accepted as an inevitable consequence of natural selection. In contrast, the hypothesis that there is compensatory coevolution of mt and N genotypes, wherein N-encoded genes change in ways that compensate for deleterious changes to the mt genes, is neither well integrated into evolutionary theory nor without controversy.

In this chapter, I present an overview of the implications of and the evidence for compensatory coevolution of mt and N-mt genes. I also outline alternative hypotheses to the mitonuclear compensatory coevolution hypothesis that might explain the different rates of evolution observed for mt, N, and N-mt genes. I explore the hypothesis that the many N-encoded subunits both in eukaryotic electron transport system (ETS) enzymes and in the mitoribosomes were recruited specifically to regulate and control the core catalytic reactions undertaken by ETS complexes and mitoribosomes in the face of mutational erosion of mt genes. I also consider the evidence that mutational erosion can in some cases be reversed through introgression of entire mt genomes. Compensatory coevolution has important implications for the evolution of sex and a sequestered germ line, speciation, and adaptation. Here, I set the stage for future discussions of those topics.

Mutational erosion

The problem with non-recombining genomes

Because of their vital functions in enabling cellular respiration, mt genes are subject to strong purifying selection (Stewart et al., 2008; Kerr, 2011). Indeed, the premise that

Mitonuclear Ecology. Geoffrey E. Hill, Oxford University Press (2019). © Geoffrey E. Hill 2019.
DOI: 10.1093/oso/9780198818250.001.0001

mt genes must be subject to perpetual intense purifying selection was the basis for the assumption that only functionally neutral variation could exist in a mt genome because any other variation would have been eliminated. Importantly, however, with no recombination, mt genes form one linkage group on the mitochondrial chromosome. As a consequence, natural selection can only judge the overall fitness of an entire mt genotype; it cannot select for mt genes individually (Maynard Smith and Haigh, 1974). This means that bad genes can hide among good genes and be perpetuated (Fay and Wu, 2000; Chun and Fay, 2011; Hartfield and Otto, 2011) (Figure 3.1). Interference in the selection on one allele by selection on a second allele is called the Hill–Robertson effect (Hill and Robertson, 1966), and Hill–Robertson effects are the theoretical basis for the accumulation of deleterious alleles in non-recombining lineages (Box 3.1). In short-hand language commonly used in the evolutionary literature, mitochondria experience mutational erosion as they accumulate deleterious alleles, where erosion is in reference to loss of both function and fitness (Lynch and Blanchard, 1998; Neiman and Taylor, 2009). The term "mutational meltdown" is used to describe the point at which accumulated deleterious mutations undercut function and fitness to the point of inviability (Rand, 2008).

Recombination during sexual reproduction can break linkages between alleles, permitting natural selection to independently promote good genes and to eliminate bad genes from the population (Felsenstein, 1974). Because of recombination, most N genes of sexually reproducing eukaryotes are not subject to strong Hill–Robertson effects and do not suffer from mutational erosion. A fundamental paradox of mitochondrial evolution is that with a high mutation rate (at least in bilaterian animal taxa) and no recombination, mutational erosion of mt genes is theoretically inevitable, but an accumulation of bad genes in the mt genome is untenable because mt genes code for core respiratory function that underlies fitness (Loewe, 2006). A potential solution to this paradox is the mitonuclear compensatory coevolution hypothesis.

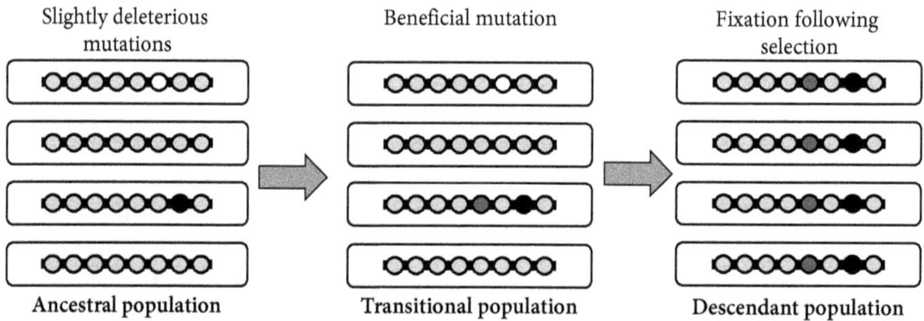

Figure 3.1 A simplified pictorial example of how Hill–Robertson effects can lead to the fixation of deleterious alleles in a population of haploid, non-recombining genomes. Dots represent loci; light gray dots are selectively neutral; black dots and white dots represent slightly deleterious alleles that are the result of spontaneous mutations. Dark gray dots represent novel beneficial alleles. Arrows indicate transitions between generations. Positive selection for the beneficial allele leads to fixation of that allele along with the deleterious allele that happened to exist on the same chromosome.

Box 3.1 Evidence for mutational erosion of mt gene products

The accumulation of deleterious mutations in the mt genome and the erosion of structural and functional integrity of mt gene products is predicted by theory (Loewe, 2006) and supported by the high nucleotide substitution rates of many mt genomes (Figure 3.1). Until recently, however, there was little direct evidence for accumulation of deleterious alleles in the mt genome. Van Der Sluis et al. (2015) posited that if deleterious mutations have accumulated in the mt genomes of eukaryotes throughout their evolutionary history, then the negative effects of such mutations should be observed in the mt sequences and mt gene products of modern eukaryotes. Conversely, equivalent negative effects should be absent from N-encoded gene products, which are predicted to be much less vulnerable to mutational erosion. The researchers then tested for such effects in both mt-encoded rRNA and mt ETS proteins.

For their assessment of deleterious mutations in rRNA in mitoribosomes, Van Der Sluis et al. (2015) were able to estimate the stability of that portion of mitoribosomes arising from RNA base-pairing. The match between complementary RNA strands had been used previously to estimate the stability of ribosomes in bacteria (Lambert and Moran, 1998). Van Der Sluis et al. found that, compared with bacteria, the rRNA of mitoribosomal had both fewer overall base pairs and a larger number of weak base pairs; they estimated that animal mt rRNA functions with 260 fewer hydrogen bonds than rRNA of bacteria. Based on this analysis, the researchers concluded that mitoribosomal structure is inherently less stable than the ancestral bacterial ribosomal structure as predicted if mitoribosomes are subject to the negative effects of accumulated deleterious mutations in the mt genome (Box Figure 3.1).

To test for evidence of mutational erosion of mt OXPHOS genes, Van Der Sluis et al. (2015) assessed structural deficiencies of OXPHOS complexes based on recently available protein models. For this analysis, they used the proportion of unprotected backbone hydrogen bonds. In previous studies, the number of unprotected backbone hydrogen bonds had been used as a measure of protein stability (Fernández and Scott, 2003). Van Der Sluis et al.

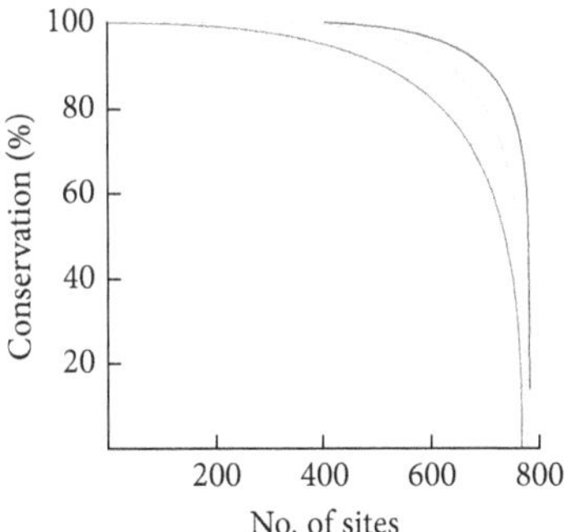

Figure Box 3.1 The relative stability of rRNA molecules as estimated based on the base pair conservation plotted against the number of base pairs examined. The higher the base pair conservation, the more stable the ribosome. Mitochondrial-encoded rRNAs (bottom curve) are less stable than bacterial (top/outer curve) or chloroplast (middle curve) rRNAs, as predicted if mt RNAs are subject to accumulation of deleterious alleles. From Van der Sluis (2015).

(Continued)

Box 3.1 *Continued*

(2015) calculated the proportion of unprotected hydrogen bonds for animal Complexes III and IV, modeled with supernumerary proteins removed. They found that animal Complexes III and IV with supernumerary proteins removed were significantly less stable than their bacterial counterparts. When Van Der Sluis et al. (2015) then re-ran the analyses with supernumerary proteins returned to the enzyme structures, Complexes III and IV were assessed to be as stable as bacterial complexes. This study both provides direct evidence for erosion of structural and functional integrity of mt gene products and demonstrates that the supernumerary proteins that have been recruited compensate for this stability lost through accumulation of deleterious mutations.

The mitonuclear compensatory coevolution hypothesis

Compensatory vs complementary coevolution

In the first two chapters of this book, I made the case for the general necessity of mitonuclear coadaptation. However, simply achieving coadaptation between co-functioning sets of mt and N gene products is not sufficient. Mitonuclear compatibility must be maintained over evolutionary time, which presents a perpetual challenge for eukaryotic organisms because neither the N genome nor the mt genome is static. If change involves adaptations in both mt and N genes, then coevolution can be envisioned as a partnership leading to ever better function and increased fitness as an innovation encoded by one genome enables further innovation in the complementary genome (Figure 3.2). Through this process of complementary coevolution, each step is a functional improvement over a previous state. Because the adaptations encoded by mt and N genes build on each other and are dependent on mutations, this process will necessarily be unique to each lineage and can lead to the evolution of population-specific mitonuclear genotypes. Moreover, the uniquely coevolved mitonuclear genotype of one lineage might be incompatible with the uniquely coevolved mitonuclear genotypes of any other lineages.

On the other hand, if evolutionary change in the mt genome involves fixation of slightly deleterious mutations via Hill–Robertson effects, mitonuclear coevolution may be directed more toward maintaining functional integrity of the OXPHOS system than toward innovation in function. In the early 2000s, evolutionary biologists proposed that for pairs of N and mt gene products that have tight functional interactions, a deleterious change in the product of one genome could be compensated for by a change in the other genome if the change in the second gene product reversed the deleterious effect in the first (Schmidt et al., 2001; Rand et al., 2004) (Figures 3.2 and 3.3). This hypothesis potentially explained how mitonuclear coadaptation was maintained in the face of mutational erosion of mt genes and why we don't see a decline in fitness of eukaryotes across evolutionary time as predicted by theory (Lande, 1994; Lynch and Blanchard, 1998). I'll refer to this as the mitonuclear compensatory coevolution hypothesis.

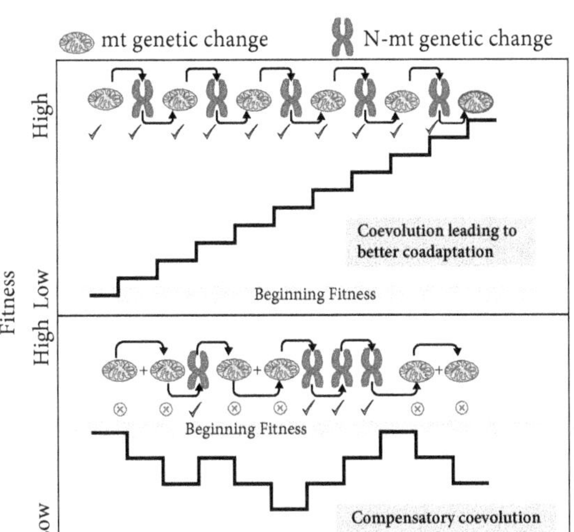

Figure 3.2 Complementary vs compensatory coevolution of mt and N-mt genes. The symbols for the mt genome or the N genome indicate the position in evolutionary time at which there are changes in these genomes. Checks are adaptations; Xs are deleterious changes. The effects of evolutionary changes on fitness are shown by the black line. In complementary coevolution (top panel), adaptive changes in one genome lead to complementary adaptive changes in the other genome and fitness perpetually rises. In compensatory coevolution (bottom panel), changes to the mt genotype cause loss of fitness that is restored by subsequent changes in the N genotype.

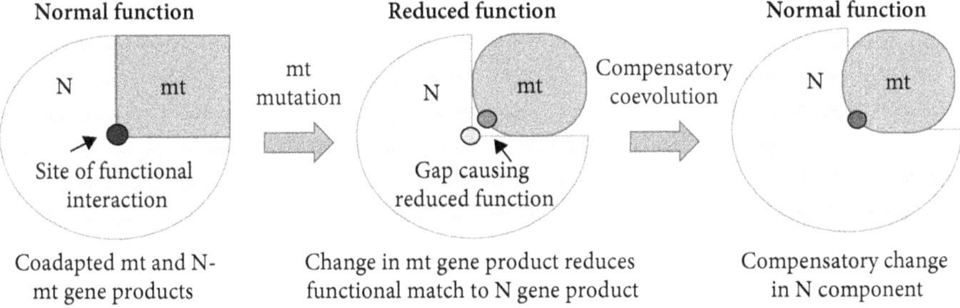

Figure 3.3 A schematic representation of the basic concept of compensatory coevolution of mt and N genes. The large light-gray shape represents the product of a N-mt gene and the smaller dark-gray shape represents the product of a mt gene. Function results through the interaction of N and mt functional sites. In this example, a mutation in the mt gene that causes poor functional interaction with a co-functioning N gene product fixes in the population. A subsequent evolutionary change in the N-mt gene restores configurational match and restores full function.

This simple dichotomy, in which mitonuclear interactions are either mutually positive or compensatory (Figure 3.2), is a useful framework for introducing the concept of compensatory coevolution, but there can be many other forms of synergistic interactions of mt and N-mt alleles. For instance, a change that is neutral in one genome might create an opportunity for a new beneficial change in the other. Moreover, the fitness of both mt and N-mt genes is necessarily dependent on the other genome. For example, a N allele with negative fitness effects might be neutralized or even become positive when paired to different mt alleles. It is the complex interplay of the two genomes creating one function that necessitates a discussion of coevolution and not just a consideration of the evolution of each genome independently.

In theory and in practice, it is generally thought that a compensatory change to a N gene would offset a deleterious change to a mt gene rather than vice versa. There are two simple reasons for this asymmetry in compensation. First, mt genes typically have more alleles that induce deleterious effects that need reversing than N genes because mt genes of most eukaryotes are much more vulnerable to mutational erosion than N genes (Rand et al., 2004). Second, through recombination, N genes are able to more rapidly evolve solutions to evolutionary problems (Havird et al., 2015a). Changes to mt genes that compensate for deleterious N-mt genes are also theoretically possible and have been demonstrated in lab experiments, as I'll present below. Compensatory coevolution and complementary coevolution involving mt and N-mt genes are not mutually exclusive processes; both should be occurring in all eukaryotic lineages, although the much greater frequency of deleterious versus beneficial mutations, especially in the mt genome, may make compensatory coevolution the primary driver of mitonuclear coevolution.

Evidence for compensatory coevolution

The most direct evidence for compensatory coevolution comes from documentation of mutual changes to interacting pairs of gene products, each of which would lead to dysfunction by itself but in tandem create no change in function. Such a documentation is not proof that one gene evolved under selection to compensate for a deleterious change in the partner gene, but it does provide at least strong circumstantial evidence that compensatory coevolution has occurred.

Some of the most compelling demonstrations of compensatory coevolution involving mt genes come from studies of mutations that cause pathogenic effects in humans. Biomedical researchers have amassed long inventories of mutations to the mt genes that cause specific pathologies in humans (Alston et al., 2017). Interestingly, the same alleles that cause pathologies in humans are sometimes the fixed genotype in non-human taxa. Fòr instance, a single nucleotide substitution on the mt-encoded $tRNA^{Asn}$ gene causes severe neuromuscular dysfunction when it occurs as a rare inherited disease in the human population (Hao and Moraes, 1997). By searching the gene sequences of other mammals, the researchers found that, surprisingly, this same human-disease-causing tRNA sequence is the fixed genotype in another species of mammal; in other words, the human-pathogenic form of the tRNA gene is the only form of the gene occurring in a population of non-human animals. Because the individuals in the species with the human-pathological tRNA sequence are healthy and fully functional, there must be something in the non-human genomes that mitigates the problems that the genotype causes in humans. Researchers therefore call the healthy form of the genotype "compensated pathogenic deviation" (Barešić and Martin, 2011).

To study this phenomenon, researchers used genomic databases or did their own sequencing in order to search for human-pathogenic genotypes that are the typical (wild-type) genotype in non-human taxa. Through this screening of genotypes, Kondrashov et al. (2002) and Barešić et al. (2010) found numerous cases of human-pathogenic genotypes for mt-encoded tRNA that were fixed in non-human animal populations. Without exception, in each of the cases, there was a second compensatory change that counteracted the problem that the variant allele caused in humans. It is important to note that these documented cases of genetic compensation are not examples of N genes compensating for mt genes or vice versa. These studies with tRNA genes involved potential cases of a change in a mt gene compensating for a deleterious mutation in the same mt gene product. However, these documented examples provide the best evidence for compensatory coevolution of co-functioning genetic elements, and they establish the plausibility of the mitonuclear compensatory coevolution hypothesis.

In the most detailed study of compensatory coevolution of mt tRNA loci, Kern and Kondrashov (2004) began with a list of eighty-six mutations to mt tRNAs that caused human pathologies. They then searched public databases for these human-pathogenic genotypes in non-human animals. They found fifty-two cases in which a genotype that caused disrupted mitochondrial function and disease in humans enabled full translational function in the non-human taxa. These compensated pathogenic deviations mostly occurred in the structural arms of the tRNA. In the human-pathogenic genotype, the correct folding of the tRNA was disrupted by a mutation. In each of the cases, where the human-pathogenic sequence occurred in non-human animals, the authors could show that there had been a corresponding change in the complementary RNA sequence of the arm that reversed the deleterious effect and enabled function (Figure 3.4). Similar studies have documented mutational changes in protein-coding genes that are deleterious in one taxon but neutral in another taxon because they have been compensated for by a second amino acid change (Kulathinal et al., 2004; Barešić and Martin, 2011).

In a follow-up study of compensatory coevolution of interacting nucleotides in tRNA molecules, Meer et al. (2010) proposed that such coevolutionary events might not only restore tRNA function but that they might also enable lineages to acquire novel genotypes that otherwise would not have been possible. The implication was that compensatory coevolution might be a mechanism for the evolution of novel adaptive genotypes. These studies of mt-mt and N-N compensatory changes dramatically demonstrated that compensatory coevolution is potentially widespread and an important part of eukaryotic evolution (Kulathinal et al., 2004; Barešić and Martin, 2011). The key question in the context of this book is whether compensation between mt and N-mt gene products can also evolve.

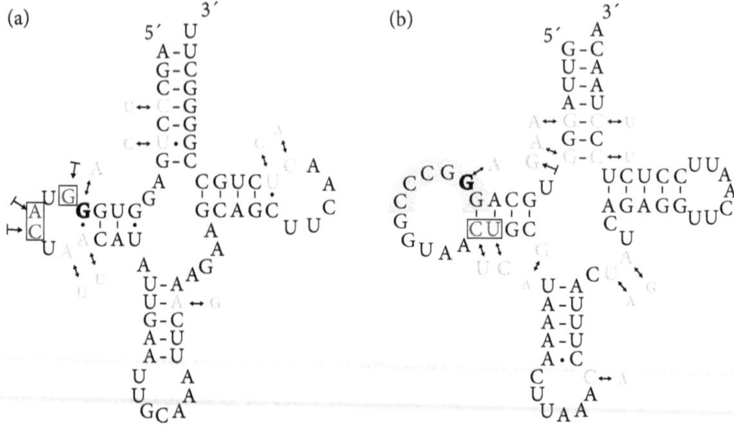

Figure 3.4 Two examples of compensatory coevolution between mt genes coding for tRNA. (a) Rabbits have a nucleotide position (bold black) on tRNA^{Cys} that causes a dysfunction in translation of mt proteins in humans. (b) Dogs have a nucleotide position (bold black) that causes dysfunction of tRNA$^{Leu\ (UUR)}$ in humans. Apparent compensatory changes (font in box) found in functional non-human animals but not human tRNAs in both cases negate the deleterious effects and restore full function. Letters at the outer ends of arrows denote the common and functional genotype of humans. The human pathogenic mutation in tRNA$^{Leu\ (UUR)}$ completes a sequence (shaded) that is prone to the formation of a dimer between two tRNA$^{Leu\ (UUR)}$ molecules; the compensatory substitutions prevent the formation of the dimer. There are many examples of alleles for tRNA that cause pathologies in humans but that are fixed and functional in non-human animals with compensatory changes negating dysfunction. Reprinted from Kern and Kondrashov (2004).

Evidence for N compensation for deleterious mt genes

The approach of using human-pathogenic mutations as the starting point in a search for examples of compensatory coevolution was applied to mitonuclear interactions by Azevedo et al. (2009). They began with forty-nine missense disease-associated mutations in the human mt genome that affected mt-encoded OXPHOS proteins. Using genetic databases, they then searched the genomes of about 200 mammalian species looking for cases where the human-pathogenic sequence was the wild-type sequence. They found three cases where the human deleterious mutation was fixed as the wild-type genotype of a species. These three cases involved mt-encoded cytochrome *c* oxidase subunit I (COX1), COXIII, and cytochrome *b* (CYTB), respectively. The deleterious mutations in the two COX genes were reversed in non-human mammalian species by a second amino acid substitution in the same mt-encoded genes and thus, as in the tRNA studies described above, these cases involved mt genes compensating for other mt genes. However, the deleterious mutation in CYTB was reversed by a change to an N-encoded Complex III protein, cytochrome c1 (Complex III subunit 4), with which CYTB closely interacts to enable unimpeded election flow (Azevedo et al., 2009).

The sequence of evolution of the deleterious and compensating alleles in the Azevedo et al. (2009) study was interesting. In the mt/N evolutionary changes affecting COXIII and CYTB function, phylogenetic reconstructions indicated that the compensatory allele appeared first, which enabled the fixation of the otherwise deleterious allele. So the solution preceded the problem. In the mt and N change-affected COXI, evidence suggested evolutionarily simultaneous change in the deleterious mt genes and the compensating N-mt gene. These potentially different routes to an evolutionary outcome that suggests compensatory co-evolution highlight the complexities of the coevolution of coadapted mt and N-mt genes.

In assessments of mitonuclear coevolution not involving alleles linked to human mitochondrial diseases, a core testable hypothesis is that there should be close functional interaction between gene pairs proposed to be involved in compensatory coevolution. This hypothesis was central to a study of the evolution of gene sequences coding for protein subunits of Complex IV among mammals, which provides another likely example of compensatory coevolution of mt and N genes (Schmidt et al., 2001). Phylogenetic reconstructions indicated there has been sequence evolution in two closely co-functioning ETS subunits in humans, mt-encoded COXIII and N-encoded COXVIIa, and that these changes were complementary: the change to COXIII caused an amino acid substitution that removed a positive charge while the change to COXVIIa substituted an amino acid that removed a negative charge. The net result was no change in Complex IV function when the two changes were expressed together. Schmidt et al. (2001) suggested that in this case a change in the mt nucleotide sequence may have compensated for a deleterious change in the N genotype, but they could not discount the possibility that the reverse had occurred. In a much more detailed comparison a decade later, Osada and Akashi (2012a) undertook a phylogenetic comparison of species of primates with a focus on the three-dimensional structure of the Complex IV subunits to test for compensatory changes among interacting sites. They found that amino acids in mt- and N-mt-encoded subunits that were in close physical proximity in the structure of Complex IV tended to evolve in a correlated manner among primate lineages. Intriguingly, and unlike what was found in the study by Azevedo et al. (2009) that I describe above, amino acid substitutions in mt-encoded proteins were often followed by amino acid substitutions in co-functioning N-encoded subunits. They concluded that this temporal order of substitutions supported true compensatory coevolution as a major factor in accelerated evolution of Complex IV proteins in primates (Osada and Akashi, 2012). These studies of Complex IV subunits in primates represent some of the best direct evidence for compensatory coevolution between mt and N genes.

Compensatory coevolution among specific interacting mt and N-mt genes has also been implicated in other studies. In a study of Complex I subunits among various mammal species, Mishmar et al. (2006) noted that a specific amino acid substitution in the N-encoded NDUFC2 was repeatedly associated with a specific amino acid substitution in mt-encoded subunit ND5 with which it closely co-functions. These patterns of coordinated change suggested that there may have been a compensatory interaction between these sites, but the authors acknowledged that the pattern could also be due to repeated fixation of the same nucleotide substitutions due to adaptive benefits. Gershoni et al. (2014) followed up with detailed studies of mt- and N-encoded proteins in human Complex I. They experimentally demonstrated N-encoded NDUFC2 is essential for stability and proper function of Complex I, further supporting the hypothesis that N-encoded proteins provided compensatory function as they coevolved with mt-encoded products. The implication was that population-specific compensatory coevolution had rescued individuals from otherwise deleterious alleles. Moreover, evidence for coevolution of mt-encoded COX1 and co-functioning N-encoded proteins is not restricted to mammals; Lee et al. (2008) showed that the N-encoded splicing protein *Mrs1* changed across lineages of yeast in concert with changes to mt-encoded COX1 in a pattern consistent with compensatory coevolution. Additional evidence for compensatory coevolution comes from a study of co-functioning mt-encoded tRNA and N-encoded aminoacyl-tRNA synthase in *Drosophila* (Meiklejohn et al., 2013). Researchers documented incompatibilities between tRNA and N-encoded aminoacyl-tRNA synthase in hybrid crosses between *D. simulans* and *D. melanogaster* (see also Figure 2.10) and concluded that the reversibility of deleterious alleles—with alleles for both the problem and the solution existing in populations—suggested that these interacting mt and N gene products might evolve through compensatory coevolution; although, as in nearly all research invoking compensatory coevolution, there are other explanations for these patterns.

Experimental evidence of compensatory coevolution

Theory predicts that compensatory coevolution will most typically involve a change in the N genome that compensates for a deleterious change in the mt genome (Rand et al., 2004; Havird et al., 2015a). It is unexpected, therefore, that the only *experimental* demonstration of compensatory coevolution between mt and N genes involves the rescue of a deleterious N allele by an evolutionary change in the mt genome. The *gas*-1 line of the nematode *C. elegans* carries a single nucleotide mutation in the protein-coding sequence for a N-encoded subunit of Complex I. Animals with this mutation have reduced ATP output and increased reactive oxygen species (ROS) production because of Complex I dysfunction (Hartman et al., 2001). In an experiment designed to provide an opportunity for the evolution of compensatory coevolutionary changes, Christy et al. (2017) propagated twenty-four replicate lineages of the *gas*-1 mutant in large populations for sixty generations. At the end of the experiment, most populations showed the low fitness that is characteristic of the founding populations of *gas*-1 nematodes, but a few lines showed fitness about the same as that of the lines with no ETS deficiencies. Genetic analysis revealed that the deleterious Complex I SNP that caused low fitness in the *gas*-1 lineage was still present, so fitness had not been recovered by a simple reversal of the *gas*-1 mutation. Rather, the researchers found that, in each of the recovered lines, there was a change in the nucleotide sequence of a mt-encoded subunit that functioned in intimate association with the affected N-encoded subunit that is defective in the *gas*-1 lineage. Interestingly, the specific change to the N gene product in each recovered line was different. When this same mitochondrial change was expressed in the fully functional *C. elegans* background, the benefits of the mitochondrial mutation were lost—animals experienced a slight loss of fitness. The conclusion was that a mutation in the mt genome had reversed the negative effects of the *gas*-1 mutation in the N genome. This nematode study is the first experimental demonstration of compensatory coevolution between mt and N-mt genes, and it involves compensation for a deleterious N allele by a sequence change in the mt genome.

Patterns of mutation and selection in mt and N genomes

Rates of evolutionary change among mt, N, and N-mt genes

Studies comparing the rates of evolutionary change among mt, N, and N-mt genes or their protein products provide some of the best evidence for coevolution of mt and N genes and at least indirect evidence for compensatory coevolution. The fundamental observation that is the starting point for nearly all comparative studies focused on compensatory coevolution is that mt genes that code for subunits of ETS complexes seem to be under much stronger stabilizing selection, also called purifying selection, than N-mt genes that code for subunits in the same complexes. Stabilizing selection is selection for the gene products that already exist, so evidence for stabilizing selection is a lack of change in nucleotides that code for the structures of proteins. Mitochondrial genes that code for ETS proteins change very slowly over time. N genes, on the other hand, show a much greater propensity to accumulate changes to the nucleotides that code for the structure of proteins (Figure 3.5).

There are different ways to estimate the rate of evolutionary change for a specific gene or a class of genes, but the most widespread approach is to compare the rate of change in the nucleotide sequence in sections of the DNA that code for the amino acid sequences of proteins (non-synonymous change) versus the rate of change in the nucleotide sequences that do not code for amino acids in proteins (synonymous change). The ratio of non-synonymous changes to synonymous changes is commonly written as d_N/d_S (or alternatively K_N/K_S). The basic logic of this ratio is that synonymous mutations are presumed not to affect fitness and to be hidden from natural selection, so they provide an indication of the background mutation rate. A high rate of synonymous mutations indicates a high mutation rate that would, presumably, also affect non-synonymous sites in the DNA sequence. If there are few changes to non-synonymous sites, this is taken as evidence that purifying selection is removing the non-synonymous mutations. Hence, low d_N/d_S is taken as evidence for strong purifying selection on a gene or class of genes. Conversely, a higher d_N/d_S is evidence for evolution in that lineage. The pattern that is consistently observed for genes coding for subunits in the ETS is that d_N/d_S of mt genes is much lower than d_N/d_S of N-mt genes (Nabholz et al., 2013; Popadin et al., 2013) (Figure 3.5).

One of the leading hypotheses to explain a higher rate of evolutionary change in N-mt genes compared with mt genes is that the high rate of evolutionary change in N-mt genes that code for ETS subunits is a result of selection for compensatory

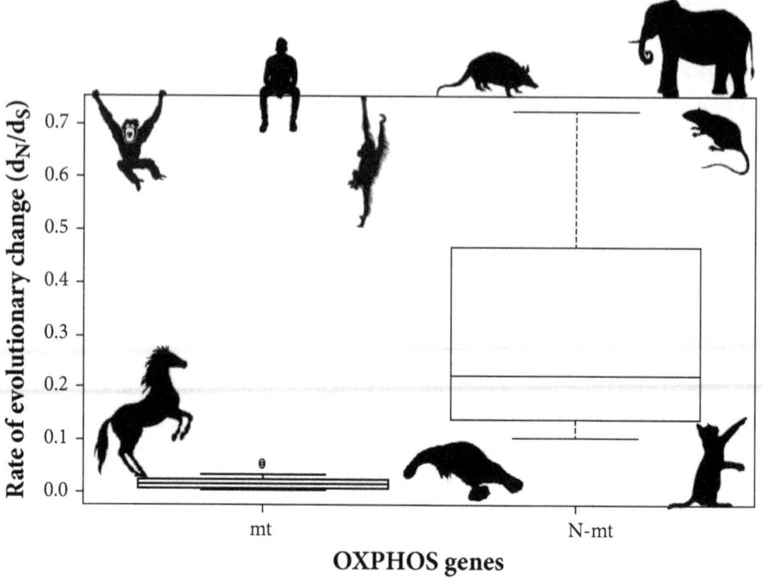

Figure 3.5 The rate of evolutionary change of subunits of the electron transport system (COX genes) that are either mt- or N-encoded in mammal species. Despite the higher mutation rates for mt DNA in these species, mt genes accumulate changes to the amino acid sequence slowly compared with the N-mt genes. One explanation for this common pattern is that N-mt genes evolve rapidly to compensate for mutational erosion of mt genes. From Popadin et al. (2012).

coevolution to correct deleterious mutations in the mt genome (Osada and Akashi, 2012; Barreto and Burton, 2013a; Meiklejohn et al., 2013). Evolutionary changes to the mt-encoded ETS subunits are scarce but not absent and the compensatory coevolution hypothesis proposes that strong selection to correct deleterious changes to mt genes would drive rapid evolution of the N-mt genes that interact with them. This hypothesis requires that compensatory coevolution leads to higher rates of evolution in N-mt genes compared with mt genes rather than "matched" rates with each non-synonymous change in a mt gene compensated exactly by one change in the complementary N-mt gene. The empirical observation of higher rates of evolution of N-mt genes suggests that multiple non-synonymous changes to N-mt genes are required to compensate for single changes to mt genes.

Compensatory coevolution between mt and N-mt genes is predicted to lead to particularly rapid mutational change in N-mt genes compared with N genes whose products are not functionally linked to mt genes. Thus, along with a prediction that functional evolutionary change will be greater in N-mt genes compared with mt genes, the compensatory coevolution hypothesis also predicts that functional evolutionary change will be greater in N-mt genes than N genes that do not co-function with mt genes. This hypothesis is tested most effectively by comparing N-mt and N genes that perform the same function inside and outside the mitochondrion, and Barreto and Burton (2013a) conducted such a comparison by examining the rates of evolution of N proteins that serve as functional components of ribosomes. These authors took advantage of the two distinct types of N-encoded ribosomal proteins in eukaryotes: cytosolic ribosomal proteins that function exclusively with other N-encoded gene products in ribosomes in the cytoplasm, and mitoribosomal proteins that must co-function with mt-encoded rRNA in ribosomes in mitochondria. They reported that, for both arthropods and yeast, the proteins in mitoribosomes (encoded by N-mt genes) had significantly elevated rates of amino acid change compared with proteins in cytosolic ribosomes (encoded by N genes; Figure 3.6).

In a subsequent study, Barreto et al. (2018) carried out whole-genome sequencing of mt and N genomes for eight populations of the copepod *Tigriopus californicus*. These copepod populations are highly divergent in N and especially mt genotype, and they would likely be classified as different species if they were, for instance, birds. Barreto et al. (2018) found evidence for strong purifying selection of N-mt genes, but also high rates of amino acid substitution in N-mt gene products that, based on protein models, have close functional interaction with mt gene products. This observation of particularly rapid evolution of N-mt genes that co-function with mt genes supported a key prediction of compensatory coevolution—a conclusion that was further supported when the pattern held after they controlled for differences in gene expression among different N genes. (I take up the significance of gene expression in relation to gene evolution in the next section.) When Barreto et al. (2018) looked in more detail at which N-mt genes were evolving, they reported that four mitoribosomal proteins and five OXPHOS proteins showed positive selection for at least one amino acid site. Again, these observations support the hypotheses that there has been compensatory coevolution in these diverging populations of copepods (Hui, 2018).

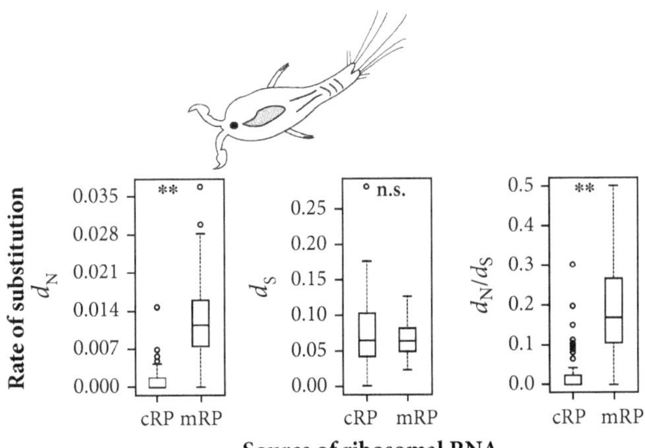

Figure 3.6 The rate of non-synonymous (d_N) and synonymous (d_S) changes to N genes coding for ribosomal proteins functioning in either the cytosol (cRP) or the mitochondrion (mRP) in the copepod *Tigriopus californicus*. d_N/d_S is frequently used as a measure of the rate of evolution. The faster rate of N-encoded genes that closely co-function with mt genes is consistent with the compensatory coevolution hypothesis. Adapted from Barreto and Burton (2012).

In another comparative study of evolution rates of cytosolic- and mitoribosomal proteins, Sloan et al. (2014b) took advantage of tremendous variation in the mutation rates of mt genes in the angiosperm genus *Silene*. These authors predicted that if mitochondrial mutation rates and selection for compensatory responses are driven by evolutionary rates in N-mt genes, then lineages with fast mt DNA evolution should show faster N-mt gene evolution compared with lineages with slowly evolving mt DNA. If differences in functional constraints are solely responsible for generally higher rates of evolution of N-mt genes, then N-mt genes should not differ in evolutionary rates between lineages with slow- versus fast-evolving mt DNA. They observed that *Silene* species with fast-evolving mt DNA exhibited increased amino acid sequence divergence in mitoribosomal proteins but not cytosolic ribosomal proteins. They concluded that high mutation rate in mt genes likely selected for more rapid changes in mitoribosomal proteins to enable compensatory coevolution.

Adrion et al. (2016) also compared cytosolic-functioning with mt-functioning proteins, but instead of ribosomal proteins that co-function with rRNA, they focused on aminoacyl-tRNA synthetases (ARS) that co-function with tRNA. They compared the rate of non-synonymous changes to mt ARS proteins both that function in the cytosol with no mt gene products and that function in mitochondria with mt tRNA molecules. They predicted that if proteins functioning with mt gene products engage in compensatory coevolution, then they should evolve faster than their functional equivalents in the cytosol with no interaction with mt genes. In examination of evolutionary patterns in mammals, birds, and flies their prediction held: mt ARS proteins evolved faster than N ARS proteins (Figure 3.7). Even though the pattern was

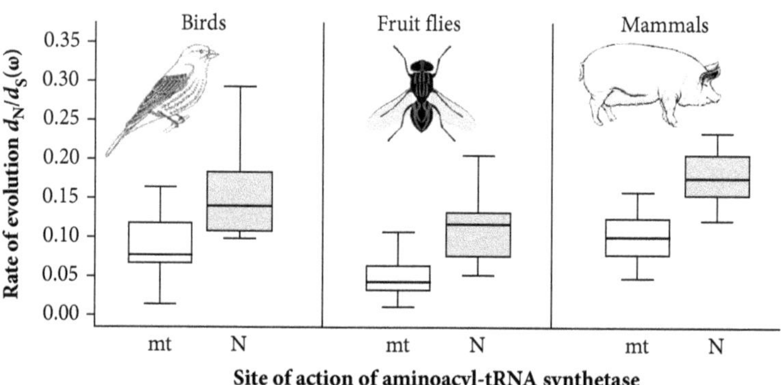

Figure 3.7 Rates of non-synonymous versus synonymous evolution of genes coding for aminoacyl-tRNA synthetases that function either in the cytosol with no interaction with mt gene products (white boxes) or in the mitochondrion in close association with mt tRNAs (gray boxes). The higher rates of evolution of proteins that co-function with mt gene products suggests that there may be compensatory coevolution. Boxes indicate medians and interquartile ranges. Adapted from Adrion et al. (2015).

consistent with compensatory coevolution, Adrion et al. (2016) pointed out that other explanations were also possible (see section on "Alternative explanations for patterns in comparative data").

A genomic study of a lineage of insects hints at the sort of novel insights that might arise from a consideration of the need for mitonuclear coadaptation and compensatory coevolution. Yan et al. (2018) found a strong correlation between the rates of evolutionary change in mt genes that code for OXPHOS proteins and N-mt genes that code for OXPHOS proteins and that make physical contact with mt-encoded proteins. They then used this strong association as the basis for an algorithm to search for other "mitochondria interacting nuclear proteins" across the entire proteome of the study taxa. Basically, they predicted that any proteins that showed the pattern of accelerated evolution that they associated with mitochondria interacting nuclear proteins could very well be previously unrecognized mitochondria interacting nuclear proteins. They found several likely candidates for previously unrecognized mitochondrial interacting nuclear proteins.

There is a worrisome bias in studies that have sought to test the compensatory coevolution hypothesis by comparing the rates of evolution in mt, N, and N-mt genes—comparative studies have mostly been conducted on animal taxa, and most commonly on mammals. Daniel Sloan and his lab group, whose work on *Silene* I mentioned above, are one of the few plant-oriented labs in the animal-dominated field of mitonuclear ecology, and they consistently bring a broader, across-eukaryote perspective to comparative studies of mitonuclear hypotheses. In a sweeping comparative study, Havird and Sloan (2016) included representatives from essentially all major eukaryotic groups. They used this broad comparative approach to assess the patterns of synonymous versus non-synonymous evolutionary changes in mt and N genes. First, they used the rate of synonymous substitutions (d_S) to estimate the relative background mutation rate for mt genes for each eukaryotic group in their analysis. (Recall that synonymous mutations are theoretically shielded from natural selection because they are proposed not to affect fitness. Thus, the rate at which they accrue should reflect mutation rates.) They found wide variation in the rate of mutation in mt genes across eukaryotes.

Havird and Sloan (2016) further predicted that if N genes were compensating for deleterious mutations in the mt genes, then there should be a higher rate of non-synonymous change (changes that affect protein function) in N-mt genes in lineages with high mt mutation rates. As predicted, they observed that the rate of non-synonymous changes in N-mt genes coding for ETS subunits increased with increasing mt mutation rate. To more definitively link the relationship between mt mutation rate and evolution of N-mt genes, Havird and Sloan (2016) then repeated the comparison of rate of non-synonymous substitutions in N genes with the relative background mutation rate of mt genes, but this time they normalized the rate of mt mutation using either N-mt genes or N genes that code for components of glycolysis, the latter having no mt component and hence involving no coevolution of mt and N genes. They found that, unlike the pattern they observed when relative mt mutation rates were normalized using N-mt genes, there was no longer a significant association between background mutation rate and rate of change of N-mt genes when normalizing with glycolysis genes. These patterns of evolutionary change are consistent with rates of gene evolution being driven by mitonuclear coevolution. The comparative study of Havird and Sloan (2016) is a particularly important investigation because it extends patterns that are well confirmed in bilaterian animals, and particularly mammals, to all eukaryotes.

Alternative explanations for patterns in comparative data

In previous paragraphs, I rather strongly advocated for compensatory coevolution as an explanation for the commonly observed high rates of evolution in N-mt genes compared with mt genes and other N genes (Figures 3.6 and 3.7). However, there are alternative explanations that may better explain at least some of the patterns in comparative studies on rates of evolution.

The primary alternative hypothesis for why N-mt genes accumulate amino acid changes in proteins faster than mt genes is that purifying selection on N-mt genes is weak compared with purifying selection on mt genes (Nabholz et al., 2013; Popadin et al., 2013; Zhang and Broughton, 2013). I will refer to this as the relaxed functional constraints hypothesis. Weaker purifying selection creates the opportunity for more rapid evolution of N-mt genes via drift (Adrion et al., 2016; Havird and Sloan, 2016). The functional explanation for proposing weaker purifying selection on N-mt genes versus mt genes is that many N-encoded ETS proteins are peripheral to the core functional proteins of the ETS complexes (Zhang and Broughton, 2013; Van Der Sluis et al., 2015), although a minority of N-encoded subunits are central to complex function. All of the mt-encoded protein subunits, in contrast, are core proteins that directly mediate proton translocation using energy released from electron transport and that are likely to be subjected to purifying selection (Pierron et al., 2012). Weaker selection on N-mt genes would enable more changes to accumulate and lead to a higher d_N/d_S.

Another factor that is proposed to affect the strength of purifying selection on a gene product is the abundance of protein produced. Genes from which proteins are produced in greater abundance are proposed to be under greater functional constraint than genes that produce proteins in lower abundance. This argument assumes that non-synonymous mutations will reduce the efficiency of protein folding and that the costs of such inefficiencies scale with the number of proteins units being produced (Nabholz et al., 2013). It is argued that mt genes produce more proteins than N-mt genes and therefore should be subject to stronger stabilizing selection (Nabholz et al., 2013) (Box 3.2).

Nabholz et al. (2013) attempted to test these alternative explanations for rates of evolution of N-mt and mt genes in a comparative study of diverse lineages of bilaterian animals. First, they used protein hydrophobicity as a rough index of the functional importance of ETS proteins. Core catalytic functions related to OXPHOS occur within the inner mitochondrial membrane (Pierron et al., 2012) and thus are primarily enabled by proteins that seek the membrane (hydrophobic) as opposed to proteins that seek an aqueous environment and will reside partly or entirely outside the

Box 3.2 Why so much mt transcript?

Recent studies have revealed that cells contain more—often approximately an order of magnitude more—messenger RNA (transcript) from mt genes that code for OXPHOS subunits than from N genes that code for OXPHOS subunits (Nabholz et al., 2013; Havird and Sloan, 2016). Because the translated proteins from N and mt genomes that form the ETS exist at equi-molar ratios, it is curious that there are so many more mt transcripts than N transcripts. One possible explanation could be that each cell contains only one N genome but it contains hundreds or thousands of mt genomes, and more replicants means more transcript produced. However, if it were simply the case that more genome copies lead to more transcription, then all mt genes should be equally highly expressed, but Havird and Sloan (2016) showed that this is not the case. For instance, some mt genes in the plant *Arabidopsis thaliana* produce thirty-six times more transcript than other mt genes in the same genome (Havird and Sloan, 2016). In addition, the OXPHOS proteins encoded for by N and mt genes exist in the abundances expected if they are all used in the construction of OXPHOS complexes—there are not lots of extra mt proteins (Calvo et al., 2016). Why, then, are there so many "extra" transcripts of mt genes?

Havird and Sloan (2016) considered that the differences in transcription might relate to something about the mt genomes themselves. Maybe the specific genes that exist in the mt genome produce much more product than do genes in the N genome and therefore have to be transcribed at a higher rate. To test this idea, they took advantage of the fact that, in plants, several genes that encode OXPHOS components can be located either in the mt genome or in the N genome. For instance, the cytochrome *c* oxidase gene Cox2 is located in the mt genome of most plants but it has been transferred to the N genome in some species of legumes. Even though it has the same function, it is expressed at a much higher level when it is a mt gene than when it is a N gene. It seems, therefore, that it is the location of the genes, and not the genes themselves, which causes mt genes to be transcribed at a higher rate.

Havird and Sloan (2016) then hypothesized that it might be necessary to hold a lot of transcript in reserve because of gross inefficiencies in the process of the transcription of mt genes. This hypothesis emerges from the observation that, while the translational efficiency of mt genes appears to be on par with translation of N genes, the transcription of mt genes is slow compared with transcription of N genes (Couvillion et al., 2016). As outlined in detail in Chapter 1, a central hypothesis for the retention of a genome in mitochondria is that mt genes are needed for rapid production of OXPHOS components in response to changing conditions within the mitochondrion (the CORR hypothesis). If the process of gene transcription is slow, however, responsiveness to changing conditions within the mitochondrion will be slow. Thus, it is proposed that creating excess transcripts is a means to alleviate the bottleneck created by slow transcription. Why mt transcription is so inefficient compared with N transcription remains to be explained (although see Ellison and Burton, 2008a), but the inefficient mitochondrial transcription hypothesis is currently the best explanation for the high number of copies of mRNA from mt genes.

membrane (hydrophilic). They found that mt-encoded OXPHOS proteins were much more hydrophobic on average than N-encoded OXPHOS proteins, which supports the hypothesis that mt-encoded proteins serve more vital roles in cellular respiration than N-encoded proteins and will be subjected to stronger purifying selection (Figure 3.8). The key assumption of this part of the Nabholz et al. (2013) study is

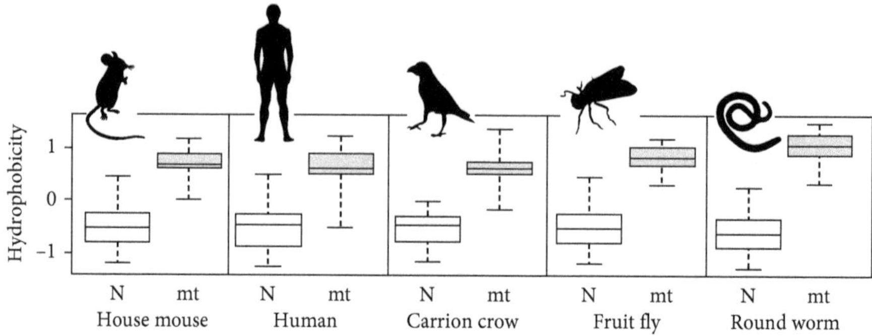

Figure 3.8 Differences in the hydrophobicity of mt-encoded and N-encoded OXPHOS proteins. Hydrophobicity, as estimated from the amino acid composition of each OXPHOS subunit, determines the degree to which a subunit will be positioned inside or outside of the inner mitochondrial membrane. Core electron transport system processes occur within the membrane so hydrophobicity is used as an index of the functional significance of subunits to OXPHOS. Across diverse animals, mt-encoded proteins are more hydrophobic than N-encoded proteins. Boxes contain quartiles and whiskers indicate extreme values. Adapted from Nabholz et al. (2013).

that the most important biochemistry is at the membrane. However, core processes related to electron transport and proton translocation occur in the hydrophilic portion in Complex I and to a lesser extent Complex V. In Complex I, all the electron transfer happens in the N-encoded core hydrophilic subunits; these subunits then interface with the mitochondrial hydrophobic subunits to trigger conformational changes that lead to proton pumping (Vinothkumar et al., 2014). As we learn more about the function of each subunit of each ETS complex, more refined predictions will be possible.

Second, Nabholz et al. (2013) compared the abundance of transcripts from mt versus N genes that code for OXPHOS proteins, assuming that transcript number provides an estimate of the abundance of proteins produced by each genome. Big differences in the transcriptional and translational machinery of mt and N genes could make this assumption invalid (Taanman, 1999; Litonin et al., 2010; Mick et al., 2011). Nevertheless, they found that across diverse animal taxa, mt genes were transcribed at much higher rates than N-mt genes (Box 3.1) (Figure 3.9). In an analysis considering all of the potential determinants of rates of evolution of mt and N-mt genes, Nabholz et al. (2013) concluded that transcript number was the best predictor of rate of protein evolution. Thus, they concluded that strength of purifying selection and not compensatory coevolution is the driver for high rate of amino acid substitution in N-mt-encoded proteins. The conclusion that transcript number and not compensatory coevolution best explains the rate of evolution of mt versus N genes was further supported by Adrion et al. (2016) in a study of rates of evolution of ARS proteins that function either in the cytosol or mitochondria; they also

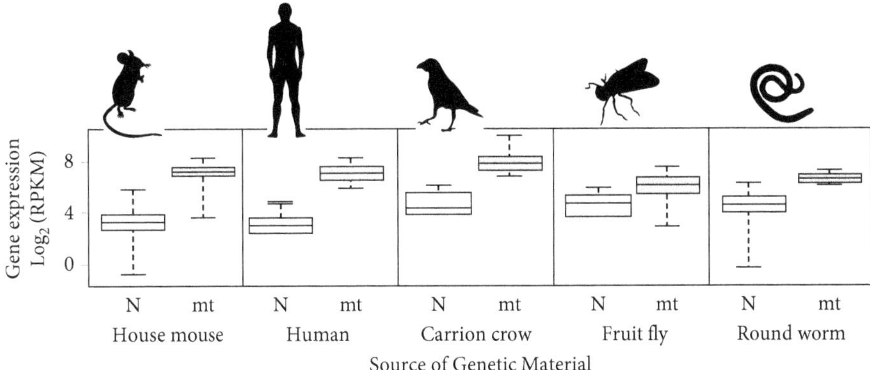

Figure 3.9 Differences in expression (measured as reads per kilobase per million mapped reads) of mt and N-mt genes coding for OXPHOS proteins. Across diverse animal taxa, mt genes are expressed at a higher level than N genes. Shown are medians and interquartile ranges. Adapted from Nabholz et al. (2013).

concluded that transcript number was the best predictor of evolution rate of these ARS proteins.

A drawback to the animal-focused studies of Nabholz et al. and Adrion et al. is that there is relatively little variation in the mutation rates of mt DNA in the animal taxa that were studied. In other words, the difference in transcript abundance between mt and N genes is confounded by the difference in mutation rate between the genomes, and it is difficult to deduce which difference is driving the higher rates of N d_N/d_S. To circumvent this problem, Havird and Sloan (2016) conducted a subsequent comparative analysis using data from a much more diverse range of eukaryotes that vary in the mutation rates of mt genomes. They found that the relationship between the rate of evolution of mt and N-mt genes varies dramatically among eukaryotes but that, across diverse eukaryotes, transcript abundance is consistently higher for mt genes than N-mt genes. They concluded that protein abundance (as estimated from transcript abundance) in particular and strength of purifying selection on N-mt genes in general are not the primary drivers of difference in rates of evolution of mt and N-mt genes. Instead, Havird and Sloan (2016) proposed that mutation rates of mt and N-mt genes were the best predictor of rates of evolution, consistent with compensatory coevolution. This conclusion was further supported in a study of species within a genus of plants that have highly divergent rates of mt evolution (Havird et al., 2015b).

Compensatory co-evolution of mt and N-mt genes and differing strengths of purifying selection created by different expression levels are not mutually exclusive hypotheses for the rates of evolution of N and mt genes. Both of these hypotheses can potentially explain some of the patterns in comparative data. A key topic in future research should be: what is the relative importance of coevolution?

Whole-gene and whole-genome mechanisms of compensatory coevolution

Compensation through protein subunits

Discussions of the mitonuclear compensatory coevolution hypothesis generally focus on deleterious point mutations that induce small dysfunctions which are counteracted by point mutations in co-functioning genes. We might call this evolutionary process allelic compensatory coevolution. But there is another level of compensatory coevolution that seems to have shaped the modern eukaryotic OXPHOS system and the mt translational machinery that enables it. Evidence suggests that during the evolution of the complexes of the ETS and the mitoribosomes of eukaryotes numerous N-encoded protein subunits were added to a core functional mechanism (Berry, 2003; Desmond et al., 2011; Pierron et al., 2012). For instance, Complex I of metazoans has forty-five protein subunits, but only fourteen of these Complex I subunits—seven mt-encoded and seven N-encoded—have functional equivalents in most bacteria (Elurbe and Huynen, 2016; Hirst, 2011) (Figure 3.10). Structural modeling studies indicate that the fourteen core protein subunits are all that are critically needed for full enzymatic function of Complex I (Hirst, 2013). The role and necessity of the remaining thirty-one "supernumerary subunits" remain incompletely resolved (Hirst, 2011; Stroud et al., 2016).

Given that the ETS of mitochondria is of bacterial origin, we can surmise that the ETS of early eukaryotes was more bacteria like, with fourteen core subunits

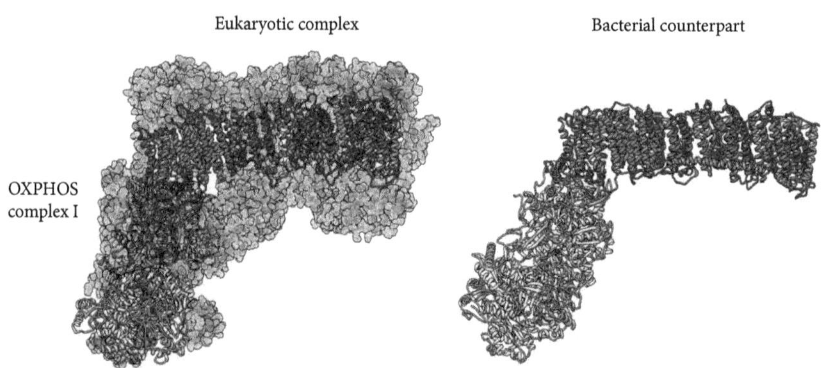

Figure 3.10 Examples of Complex I enzymes in a eukaryote (sheep, *Ovis aries*) and a prokaryote (*Thermus thermophilus*). These enzymes enable the same function in eukaryotes and prokaryotes and share a similar core structure (subunits shaded dark). In eukaryotes, numerous supernumary proteins have been recruited to the core structure (shaded light gray). Supernumerary proteins are always nuclear-encoded in eukaryotes, while core subunits are coded for by both mt and N-mt genes. The prokaryote reconstruction is shown to indicate homology with subunits in the eukaryotic enzyme. Reproduced from Sloan et al. (2018) with permission.

(Hirst, 2011). Complexes III, IV, and V were also originally smaller with fewer N-encoded subunits (Berry, 2003). N-encoded subunits appear to have been recruited to ETS complexes over time. Likewise, over their evolutionary history, mitoribosomes recruited approximately twenty-five new N-encoded protein subunits to the original ribosome structure inherited from bacteria (Desmond et al., 2011). The incorporation of additional subunits into the ETS complexes and mitoribosome increased the cellular energetic cost to produce these structures and necessitated the involvement of numerous assembly factors for complex biogenesis (Stroud et al., 2016). The question is: Why? What benefits of additional subunits balanced the costs of producing and maintaining larger ETS complexes and mitoribosomes?

One hypothesis for the evolution of supernumerary subunits is that they provide a shield that protects the enzymatic core from damaging free radicals (Hirst, 2011). However, this hypothesis is contradicted by the observation that the rates of free radical production are similar in the "shielded" ETS of eukaryotes and the "unshielded" ETS of prokaryotes (Esterházy et al., 2008). A new hypothesis, which is complementary to the free-radical shielding hypothesis, proposes that N-encoded protein subunits were recruited to core ETS complexes and mitoribosomes to provide structural stability as mitochondrial components were degraded by mutational erosion during mitochondrial evolution in eukaryotes (Van Der Sluis et al., 2015). In other words, it is proposed that recruitment of N-encoded novel ETS subunits was a form of compensatory coevolution.

According to this hypothesis, the composition of the ETS complexes and ribosomes in a primitive eukaryotic ancestor was much like the composition of these structures in bacteria. The genes for some of the protein subunits of ETS complexes and mitoribosomes were transferred from the mt genome to the N genome, but as discussed in Chapter 1, it was necessary to retain genes for core OXPHOS and ribosomal function in the mt genome. Genes on the mt genome were subject to mutational erosion, which would have caused a decline in core mitochondrial processes over time. Van Der Sluis et al. (2015) proposed that the recruitment of novel N subunits to the ETS complexes and to the mitoribosome was a form of compensatory coevolution. According to this hypothesis, the architecture of the modern eukaryotic ETS and mitoribosomes is the outcome of recruitment of N-encoded proteins specifically to counteract the negative effects of deleterious mt alleles.

Both mitoribosomes and ETS complexes of mitochondria appear to evolve through two distinct phases (Van Der Sluis et al., 2015) (Figure 3.11). A constructive phase, which occurred early in eukaryotic evolution, is proposed to involve the recruitment of novel protein subunits to reverse adverse effects of mutational erosion of mt-encoded genes. The specific mechanism by which supernumerary proteins alleviate the problems created by deleterious mutations to mt genes remains to be resolved in any detail, but Van Der Sluis et al. (2015) proposed that the negative effects of deleterious mt alleles can occur at three levels: (1) biogenesis (folding and assembly), (1) stability, or (3) activity. It is intriguing to note that when the functions of auxiliary proteins of ETS complexes or mitoribosomes are known, they invariably are involved in assembly, stability, and regulation (Berry, 2003; Smits et al., 2007; Gershoni et al., 2014). There are

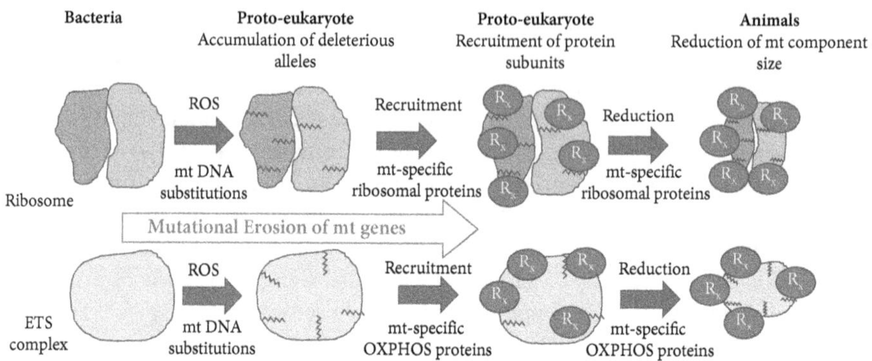

Figure 3.11 Schematic illustration of a hypothesis for the evolution of the mitoribosomes and ETS complexes composed of mt and N components. The accumulation of deleterious alleles in mitochondrial genome coding for core components (cracks in structures) led to the recruitment of nuclear proteins (balls labeled R_x). Once in place, nuclear subunits enabled reduction in the peptide length of mt OXPHOS subunits and the nucleotide sequence of mt rRNA. Adapted from van der Sluis (2015).

about seventy-five proteins involved in ETS complexes and ribosomes of mitochondria that are not found in prokaryotes. Presumably, it was necessary for these recruited protein components to be N-encoded because recombination during sexual reproduction enables N-encoded gene products both to evolve rapidly in response to mt changes and to avoid their own mutational erosion (Havird et al., 2015a).

Several observations provide support for the hypothesis that N-encoded protein subunits were recruited to ETS complexes and mitoribosomes as a form of compensatory coevolution. First, it has been demonstrated that at least some supernumerary proteins are important for enzyme stability (Angerer et al., 2011; Gershoni et al., 2014). Second, there has been no recruitment of N-encoded proteins to the mitochondrial enzymes that lack mt-encoded components. For instance, enzymes of the citric acid cycle are generally reduced in eukaryotes compared with prokaryotes—there are no cases of recruitment of novel proteins (Huynen et al., 1999). Third, when Complex II is composed exclusively of N-encoded proteins, as in metazoans and fungi, then it exists as a simple four-subunit enzyme just as in bacteria (Yankovskaya et al., 2003). When Complex II includes mt-encoded subunits as in some plants, however, it also has additional N-encoded subunits (Huang and Millar, 2013). Finally, in plants the structures required for photosynthesis are encoded by both N and chloroplast genes, potentially setting the stage for the same types of recruitment of protein subunits as in mt structures. However, chloroplast genomes accumulate deleterious mutations much more slowly than do mt genomes (Lynch, 1997) and thus there has been virtually no recruitment of supernumerary proteins to chloroplast ribosomes or photosynthetic enzymes (Van Der Sluis et al., 2015). All of these observations are consistent with the hypothesis that N-encoded proteins are recruited to compensate for problematic mt-encoded proteins.

The second evolutionary phase for eukaryotic ETS and mitoribosomes proposed by Van Der Sluis et al. (2015) is a reductive phase. This evolutionary process, which occurred most conspicuously in the evolution of metazoans, involved the gradual reduction in the length of mt RNAs and, to a lesser degree, mt-encoded ETS subunits (Van Der Sluis et al., 2015). If mutational erosion is a perpetual problem for DNA sequences located in the mt genome, then reducing sequence length would be one mechanism for reducing the negative effects of mutational erosion: shorter sequences are smaller targets for random mutational changes (Lynch et al., 2006). Alternative explanations are also possible. For instance, loss of mt RNA and mt-encoded ETS subunits could be the result of genetic drift, with unnecessary residues being lost, or could be the result of a secondary factor in a wider evolutionary force (drift or selection) to reduce mt genome size. Whether or how the addition of supernumerary components created the opportunity for reduction in the sequences for mt-encoded products is unknown.

It is interesting to speculate that N-encoded proteins were recruited to the ETS complexes and mitoribosomes to compensate for mutational erosion of mt-encoded products, but once in place they enabled novel adaptations. Azevedo et al. (2009) proposed that the evolution of compensatory mechanisms can sometimes predate the evolution of the deleterious change. According to this idea, as more and more nuclear protein subunits were recruited to both the OXPHOS and mitoribosomal complexes, a capacity for mutational compensation was created. But even beyond a heightened capacity to compensate for deleterious mutations to mt genes, the addition of N-encoded subunits may have improved the efficiency of eukaryotic complexes compared with prokaryotic complexes and thus been driven by adaptive evolution. By this hypothesis, once they evolved in response to the need for compensatory coevolution, supernumerary proteins become the target of natural selection enabling fine tuning of mitochondrial processes in response to changes in the environment and in the life history of the organism. For instance, it is speculated that the recruitment of supernumerary proteins to stabilize the assembly of Complex I led to gradual improvements in energy conservation of Complex I (Hirst, 2011).

These studies suggest that there are multifaceted reasons for the recruitment of the supernumerary subunits. One reason for their evolution is almost certainly because they increased stability/efficiency relative to the progenitor, supporting the increased energy production required for a much larger genome with far more protein products (Lane and Martin, 2010). However, compensation for mutational erosion was likely also a strong complementary force. Another key conclusion regarding supernumary elements in the ETS is that, regardless of why they were initially recruited, supernumery subunits now provide nearly all the sites for compensatory evolution (Elurbe and Huynen, 2016). The evolution of the structures of ETS complexes and ribosomes as they transitioned from biochemical machines that provided energy for individual bacteria to machines that that supported the massive energy needs of complex organisms has scarcely been studied. The work of Van Der Sluis et al. (2015) shows the potential insights that can be gained by considering mitonuclear coevolution and coadaptation in such discussions.

Mitochondrial introgression as a compensatory mechanism

A final hypothesized mechanism for genetic rescue is the introgression of an entire mt genome to replace a mt genome corrupted by mutational erosion (Sloan et al., 2017) (Figure 3.12). Genetic introgression is the movement of a genotype from one distinct population to another distinct population. Introgression can involve any specific N gene, a portion of the mt genome (a phenomenon most common in plants), or the entire mt genome (the most common form of mt introgression). Any introgression of mitochondria between diverged populations comes at the cost of disruption of coadapted sets of mt and N genes (Hill, 2016; Baris et al., 2017). The magnitude of this cost is proportional to the degree to which the genotypes have diverged and the extent of the mitonuclear incompatibilities that arise when the introgressed mt genes must co-function with the native N genes. In Chapter 2, I outlined what is known about the functional consequences of mitonuclear incompatibilities arising from mixing mt and N genes that are not coadapted. When considering introgression of mt genomes, however, the cost of mixing mt and N genes that are not coadapted must be weighed against the benefits of a potentially more fit mt genome. As a general rule, introgression of mt genomes will be favored when the loss of fitness from hybrid dysfunction is less than the loss of fitness from the mutational load of the native mitochondrion (Sloan et al., 2017) (Figure 3.12, panel B). Such introgression will typically be possible only between closely related species where divergence in mt and N-mt genotypes does not create severe incompatibilities.

A documented introgression of mt genotype from one fish species to another closely related fish species represents a potential case of a corrupted mt genome being replaced with a less corrupted, more functional mt genome (Hulsey et al., 2016; Figure 3.13). These researchers studied cichlid fish in the genus *Herichthys*. The mt genome of one *Herichthys* cichlid species is replacing the mt genome of a sister taxa without significant flow of N genes. Moreover, the recipient species has a small population size and high rate of non-synonymous mutations to its mt genome—conditions that should lead to mutational erosion of mt genomes. The implication of this example is that a mt genotype from one species is replacing the mt genotype of a sister species because the fitness gains from a mt genome with fewer fixed

Figure 3.12 Graphic illustration of the hypothesis that introgression can serve as a means of compensation for mutational erosion of mt genes. Change in fitness over evolutionary time is shown for two hypothetical species with divergent coadapted mitonuclear genotypes. Steps up and down in fitness are the result of single evolutionary events. In this example, loss of fitness is a result of fixation of a deleterious mt allele and rise in fitness is the result of compensatory coevolution. The fitness of a hybrid with N genes from one parental type and mt genes from the other is indicated. (A) Hybrid fitness is below the fitness of each parental population so natural selection prevents introgression of mt genotypes. (B) Introgression increases fitness because the fitness costs of disrupted coadaptation are less than the fitness loss from retaining the corrupted mt genotype. (C) Introgression of coadapted N-mt genes following introgression of mt genome provides greatest fitness advantage. These examples assume no environmental change over time.

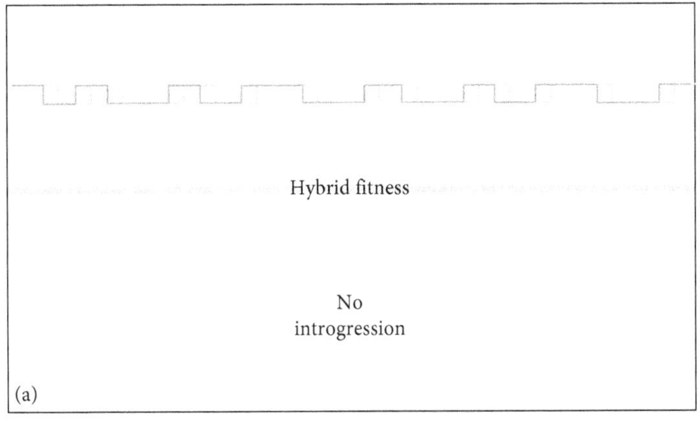
Hybrid fitness
No introgression
(a)

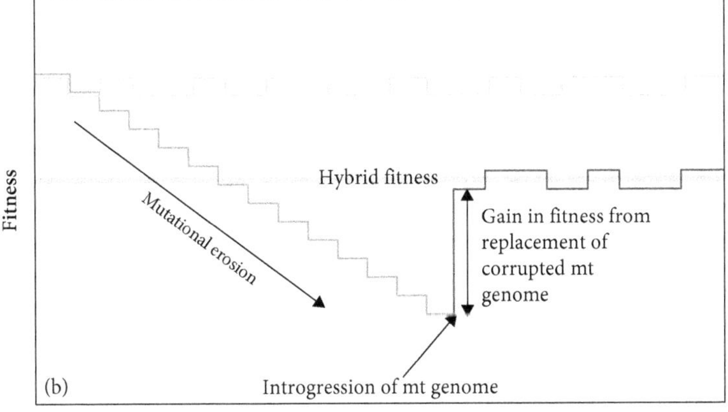
Fitness
Hybrid fitness
Mutational erosion
Gain in fitness from replacement of corrupted mt genome
(b)
Introgression of mt genome

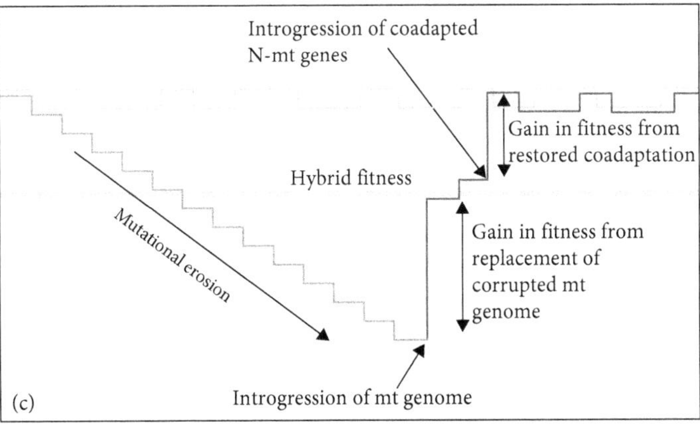
Introgression of coadapted N-mt genes
Gain in fitness from restored coadaptation
Hybrid fitness
Mutational erosion
Gain in fitness from replacement of corrupted mt genome
(c)
Introgression of mt genome
Evolutionary time

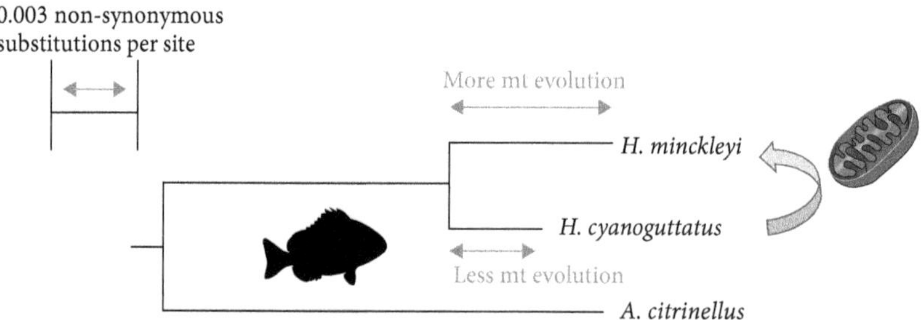

Figure 3.13 An example of introgression of the mt genome from one species to another that is consistent with the hypothesis that introgression alleviates a lineage's mutational load in the mt genome. In this example, the mt genome of a ciclid fish species in the genus *Herichthys* with small population size and high mt non-synonymous substitution rate was replaced by the mt genome of a sister species of ciclid with a larger population and lower rate of non-synonymous substitution. The branch lengths on the tree are proportional to the rate of non-synonymous evolution of mt genomes and the arrow indicates the direction of mitochondrial introgression. Adapted from a figure in Sloan et al. (2017) based on data from Hulsey et al. (2016).

deleterious mutations are greater than the fitness gains from a fully coadapted, but corrupted, mt genome.

The above considerations are made with no accommodation for influences of environment. There will be circumstances in which adaptations to the mitochondria of one species would also increase the fitness of another species (Bonnet et al., 2017). In this case, we could restate the circumstances that favor introgression of mt genomes in terms of both gain of adaptive elements and potential loss of corrupted genotypes. So long as there is a net positive benefit to the population, novel mt genotypes will be favored over current mt genotypes and introgression will occur (Pereira et al., 2014; Bonnet et al., 2017). Interestingly, in the same study of mt introgression between fish species highlighted above, there are also indications that the introgressing mt genomes may carry alleles that are better adapted to local thermal environments, further increasing the benefits of introgression (Hulsey et al., 2016).

Even if the mt genome that could potentially be replaced by introgression is corrupted by deleterious alleles and even if new mt genome brings better adapted alleles, the fitness costs of mixing mt and N-mt genotypes that are not coadapted must be overcome. If coadapted N-mt genes can be introgressed along with the mt genome (Figure 3.12, panel C), however, then a lineage potentially gets all of the benefits of a less corrupted or better adapted mt genotype without the costs of mitonuclear incompatibilities (Sloan et al., 2017). The likely scenario for such co-introgression would be initial introgression of a mt genotype with loss of fitness due to incompatibilities compensated for by gains in fitness from a better adapted or less corrupted mt genotype. Secondarily there would be strong selection for the spread of N genes that are coadapted with mt genes. Co-introgression of mt and N-mt genes is at present little more than a hypothetical concept, but in a study of two sister species of *Drosophila* Beck et al. (2015) presented evidence that the mt genome of one species had replaced the mt genome of the second species through introgression. They subsequently looked for evidence of gene flow of N-mt alleles and reported evidence that alleles for three N-encoded COX proteins, which function in close association with mt-encoded COX proteins, had also introgressed between species. Introgression of mt and N-mt genes between cold- and warm-adapted species of rabbits may provide another example (Seixas et al., 2018).

Studies of the fitness costs and benefits of introgression of mt genome as well co-introgression of mt and N-mt have huge implications for hypotheses related to mito-nuclear coadaptation and speciation. The introgression of mt genomes from one species is also central to ideas related to mitonuclear adaptation and sexual selection. The fitness costs and benefits of mitochondrial introgression should be an area of active research going forward. I will return to the topic of introgression of mitochondria between species in the chapters on speciation, sexual selection, and adaptation (Chapters 7–9).

Summary

For a lineage to retain the capacity to efficiently produce energy, N-mt and mt genes must be coadapted. The challenge is that the mt and N genomes evolve independently. Thus, coevolution of mt and N-mt genes is necessary to maintain coadaptation over evolutionary time. Moreover, the effectively haploid and non-recombining mt genome is subject to perpetual accumulation of deleterious alleles that are predicted to erode the function of core respiratory processes. Compensatory coevolution of N-mt and mt genes is a hypothesis for how eukaryotes maintain coadaptation across evolutionary time and avoid the negative effects of mutational erosion of mt genes.

The mitonuclear compensatory coevolution hypothesis takes three forms. Most widely discussed and best supported by empirical data is the hypothesis that the evolution of small changes to N gene products counteracts deleterious changes in mt gene products. Researchers have found specific examples of what appear to be such compensatory changes in N-encoded proteins. Moreover, comparative studies show that N gene products that co-function with mt gene products evolve faster than N gene products that do not interact with mt gene products. These observations are consistent with the compensatory coevolution hypothesis, but other explanations are also consistent with the patterns.

A second hypothesized form of compensatory coevolution is the recruitment of N-encoded proteins to structures with both mt and N components. Recent studies of the structures and evolutionary histories of ETS complexes and mitoribosomes suggest that these structures have recruited numerous N proteins to help stabilize and moderate function. Structures with no mt component, such as Complex II in many eukaryotes, do not have the sort of N-encoded supernumerary proteins observed in other ETS complexes and in the mitoribosomes.

A third hypothesized mechanism for escaping the accumulation of deleterious alleles in the mt genome is the capture of a less corrupted mt genome from a sister taxon through introgression. Introgression of a foreign mt genome introduces incompatibilities between N-mt and mt genes, so the benefits of escaping deleterious alleles must be more than the cost of lost coadaptation. There are currently no data that directly support this sort of adaptive introgression but no researchers have yet looked at the mutational load in introgressed versus replaced mitochondria.

Compensatory coevolution of mt and N-mt genes is a central hypothesis in mitonuclear ecology and testing, and refining this foundational idea should be the focus of research by evolutionary biologists.

4

Coevolution, co-transmission, and conflict

Imagine that you are an automobile manufacturer and that your company produces most of the components for car models on-site in your own factory. Within your factory, you have control of manufacturing processes. You demand high component quality, coordinate changes among components as new models are created, and deal promptly with performance issues that arise. Now imagine that, through a quirk of the history of your company, a set of core and critical engine components must be manufactured in another factory that is independently owned and operated. These components from the other company have to integrate precisely with components produced by your company or engine performance will suffer. Unfortunately, this other factory sometimes has issues with quality control. They have a tendency to make changes to the components that they ship without coordinating with your team. They also have an annoying history of making deals that are beneficial to themselves but that are detrimental to your company and to your effort to produce a top-quality automobile. It would be challenging to produce a competitive automobile under such circumstances. The nucleus of eukaryotic cells is confronted with similar challenges as it attempts to produce an efficient and effective system of aerobic respiration in partnership with a mt genome.

The central theme of this book is that mt genes and N-mt genes must be tightly coadapted to enable core respiratory function. Given the positive fitness effects that are achieved when the products of mt genes and N-mt genes co-function properly, one might anticipate that, once mitonuclear coadaptation is achieved through coevolution, mechanisms would evolve to preserve functional gene combinations by co-transmission of these coadapted gene sets across generations. It seems paradoxical, therefore, that mt and N genomes are replicated and transmitted across generations as physically separated units that can be inherited independently (Wallace, 2010). The complications that arise from co-functioning genes located on different genomes give rise to many of the defining characteristics of eukaryotes and are the focus of this book. Given the perpetual risk of poor mitonuclear gene combinations that comes with having the mt and N genes located on different genomes, it seems that there must be substantial compensating benefits for not co-transmitting mt and N genomes. In this chapter, I'll consider hypotheses for why mt and N-mt genes are not more tightly co-transmitted across generations.

As if there were not already sufficient chaos from having mt and N-mt genes on different genomes, the transmission of coadapted mt and N genes across generations

Mitonuclear Ecology. Geoffrey E. Hill, Oxford University Press (2019). © Geoffrey E. Hill 2019.
DOI: 10.1093/oso/9780198818250.001.0001

is further complicated by discrete mating types. In most eukaryotes, mitochondria are transmitted only through female lineages, making males a genetic dead-end for mitochondria. As a consequence, the fate of males imposes no natural selection on mt genotypes. From the perspective of a mt gene, males and females may be valued differently, and genomic conflict can arise. Having two sexes also creates an opportunity for sex chromosomes, which raises the potential for enhanced or diminished co-transmission of mt and N genes depending on position of N-mt genes on autosomes or sex chromosomes. Sex linkage of N-mt genes can also have significant effects on genomic conflict. In this chapter I will explore how co-transmission, sex linkage, and genomic conflict shape the coevolution of mt and N-mt genes. This chapter will set the stage for discussions of the evolution of sex, speciation, sexual selection, and adaptation.

Co-transmission and coevolution

The tradeoff between co-transmission and evolability

At face value, co-transmission of mt and N genes seems as though it would be so beneficial for holding together coadapted mt and N-mt genotypes that it is puzzling that so many features of the reproductive tactics and genomic architecture of eukaryotes work against co-transmission (Wade and Goodnight, 2006). Considerations of the value of holding together coadapted gene complexes are as old as the new synthesis (Dobzhansky, 1948; Wallace, 1953). The same basic arguments that are now being applied to mitonuclear interactions have long been made regarding the value of linking highly co-functional N genes (Turner, 1967; Fenster et al., 1997; Neiman and Linksvayer, 2006). The perceived dilemma is that once a high-performing, well-adapted phenotype is produced in a population, it seems like a mistake to engage in sexual reproduction and scramble the gene sets that gave rise to such a functionally sound form. Why not have mechanisms for capturing great phenotypes and transmitting them unchanged across generations? Many eukaryotes do have the capacity for asexual reproduction and thus can propagate successful genotypes across multiple generations, but nearly all eukaryotic lineages return regularly to sexual reproduction (See Chapter 5). One widely stated reason for recombination of N genotypes is that environments change and sexual reproduction and recombination enable lineages to better adapt to a changing world (Tooby, 1982; West et al., 1999). Gene combinations that are most fit in a current environment may not be the best combinations in the future. But even ignoring environmental change, it is still untenable for any lineage to try to make permanent a particular genotype across evolutionary time. Mutation inevitably introduces change to genotypes, and most mutational change is bad. Sexual reproduction and recombination enable deleterious mutations to be continuously removed from the N genome (Otto and Lenormand, 2002).

I briefly discussed Hill–Robertson effects in Chapter 3. Stated simply, a Hill–Robertson effect is selection on alleles at one locus affecting the selection on alleles at

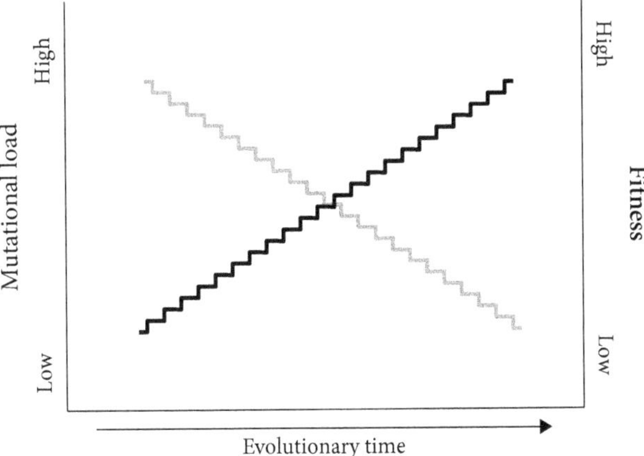

Figure 4.1 A pictorial illustration of the basic prediction of reduced fitness with increasing mutational load via Muller's ratchet across evolutionary time. The black line illustrates an increasing mutational load ratcheting upward with the fixation of each deleterious allele. The light gray line shows the predicted change in population fitness, ratcheting downward with the fixation of each deleterious allele.

other loci (Hill and Robertson, 1966; Comeron et al., 2008) (Figure 3.1). Hill–Robertson effects are a product of gene linkage, which is the tendency for two or more genes to be transmitted across generations as connected units. If selection increases the frequency of one of an allele at one locus in a linkage group, then the frequency of the other alleles on other loci in the linkage group will also change (Roze and Barton, 2006). Hill–Robertson effects are typically spoken of as an evolutionary problem, not a desired outcome, because Hill–Robertson effects contribute to Muller's ratchet and mutational erosion (Felsenstein, 1974). Muller's ratchet is the accumulation of deleterious mutations in non-recombining genomes; with the fixation of each new deleterious allele, the genome ratchets toward lower fitness (Figure 4.1). The dilemma is that, if genes are tightly linked and transmitted as a set, then selection cannot act on the fitness of a particular gene in the set—it can only select on the set as a whole (McVean and Charlesworth, 2000). As a result, bad genes can hide among and be selected along with good genes, or good genes can fail to spread because of counterbalancing effects of linked bad genes (Figure 3.1).

Consideration of gene linkage, Hill–Robertson effects, and Muller's ratchet becomes very relevant to discussions of co-transmission of mt and N-mt genes because the benefits of co-transmission of coadapted mt and N-mt genes are counterbalanced by the negative consequences of the Hill–Robertson effects that result when genes are linked (i.e. co-transmitted). It is important to note that the Hill–Robertson effect not only promotes Muller's ratchet through reduced selection against deleterious alleles, but that it can also depress the effectiveness of selection for beneficial mutations (Hill and Robertson, 1966). Making the evolution of beneficial traits related to mitonuclear coadaptation more difficult has important implications for compensatory coevolution between mt and N-mt genes (Chapter 3). The deleterious effects of a mt allele that is fixed via Hill–Robertson effects might be reversed if the mt allele is paired with a compensating N-mt allele, but the evolution of compensating N-mt genes is only possible if mt and N-mt genes are not tightly linked. Linkage favors the status quo and holds current genotypes together. Compensatory coevolution, on the other hand, requires change in one of the mitonuclear partners. Thus, independent assortment of loci is the key to effective natural selection. For N genes to evolve compensatory changes that counteract the effects of deleterious alleles in the mt genome, and hence avoid mutational meltdown in the mt genome, natural selection has to *independently* sort among different combination of mt and N genes. Without independent assortment of mt and N-mt genes, natural selection has no means to find better adapted gene combinations. If mt and N genes are linked and co-transmitted (and thus assort non-randomly), then selection can act only on the linkage groups as a whole (Hill, 2014a). These considerations lead to a basic principle in mitonuclear ecology: *co-transmission of mt and N-mt genes thwarts the coevolution of mt and N-mt genes.* Likewise, even if deleterious genes are not involved, certain combinations of mt and N alleles might lead to better function than other combinations of mt and N alleles. But again, the only way in which natural section can try out different combinations of mt and N-mt alleles is if mt and N-mt assort independently.

Independent transmission of mt and N-mt genes is therefore essential for both compensatory and adaptive coevolution of mt and N-mt genes. I propose that the low co-transmission of mt and N-mt genes that is characteristic of eukaryotes is evidence that the benefits of mitonuclear coevolution are greater than the benefits of tight co-transmission of N-mt and mt genes (Hill, 2014a). This hypothesis can be tested by assessing the accumulation of deleterious mutations in lineages with varying degrees of linkage, and therefore varying degrees of co-transmission, between mt and N-mt genes.

Sex chromosomes

In the course of this chapter I will frequently invoke sex linkage of N-mt genes, so it is important that I present a brief review of what sex chromosomes are and the different forms in which they exist among eukaryotes. The N genome of all eukaryotes is, at least for part of the life cycle, a diploid genome. There are two copies of each gene maintained as paired sets of chromosomes. Autosomes are paired sets of chromosomes that carry the same set of genes, but not necessarily the same set of alleles. Sex chromosomes are paired sets of chromosomes that are unequal in gene content and usually also unequal in size. They are called sex chromosomes because, in many eukaryotes, whether an individual received two big sex chromosomes or one big and one small sex chromosome determines the sex of an individual. Systems in which females carry two big chromosomes are called XY systems. In XY systems, females are the homogametic sex, having two X chromosomes. Males are the heterogametic sex, having one large X chromosome and one smaller Y chromosome, the latter having reduced gene content. Humans, of course, are an XY species, as are all placental mammals and many other bilaterian animals. The other common system of sex determination in animals is ZW, in which males have the two large chromosomes, designated by convention as Z chromosomes, and females have one large Z chromosome and one smaller W chromosome. Thus, females are the heterogametic sex in ZW systems. ZW sex determination is common to all birds and is also found in some other vertebrates and arthropods. Many species of eukaryotes have no sex chromosome. In these species either other mechanisms such as developmental temperature are used to determine whether an individual will be a male or a female, or all individuals in a population are both male and female (hermaphrodites).

A few features of the naming conventions of sex chromosomes may be useful to point out to non-geneticists. There need be no homology among X chromosomes. For instance, the X chromosome of a fruit fly evolved completely independently from a different autosome relative to the X chromosome of a placental mammal (Steinemann and Steinemann, 1998). X simply designates the big chromosome in a system in which females are the homogametic sex. Second, the disparity in gene content between X versus Y or Z versus W is highly variable (Rice, 1996; Charlesworth and Charlesworth, 2000). Y and W chromosomes can have a gene content that is nearly as large as that of X and Z chromosomes, or Y and W chromosomes can be much reduced compared with X or Z. In some taxa, the smaller sex chromosome is lost altogether yielding an XO sex determination system. In haplodiploid systems, males lack not only a second sex chromosome but also a second copy of all alleles at all loci. Thus, females in these systems are diploid and males are haploid, but otherwise sex determination is like in the XO system. Sex chromosomes evolve from autosomes, and in many eukaryotic lineages sex chromosomes are lost and evolve repeatedly (Charlesworth et al., 2005).

Sex linkage and co-transmission

With the evolution of sex chromosomes came the potential for sex linkage of N-mt genes that co-function with mt genes. When genes are located on sex chromosomes, they are called "sex-linked," and sex-linked genes have different patterns of inheritance than genes located on autosomes (all of the chromosomes that are not sex chromosomes). The smaller chromosome (Y or W) is carried only by one sex, so it cannot exclusively hold genes that are essential for common function of both males and females, such as genes for a functional electron transport system (ETS) and cellular respiration. Therefore, discussion of sex-linked genes in this book will concern only genes on the larger chromosome (X or Z depending on system).

In XY sex determination systems, females are XX and males are XY, so two-thirds of the genes that are X-linked are carried by females. Because the entire mt genome also is inherited through female lineages in most eukaryotes, X-linked genes and mt genes are transmitted together at a higher frequency than mt genes and N genes on autosomes (Figure 4.2). This is potentially of tremendous significance to mitonuclear coevolution and mitonuclear ecology because it means that positioning N-mt on the X chromosome is a readily available means to enhance the co-transmission of mt and N-mt genes (Drown et al., 2012; Hill, 2015b) (Figure 4.2). Positioning N-mt on the X chromosome will also necessarily impede coevolution of mt and N-mt genes.

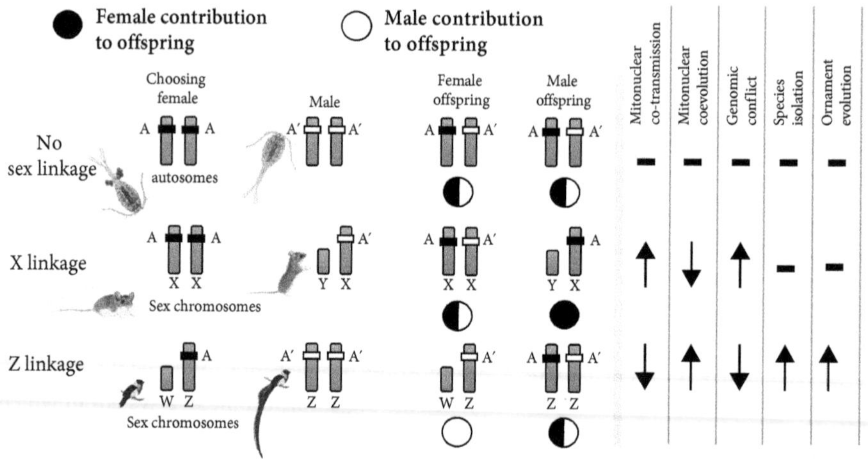

Figure 4.2 The effects of chromosomal position on patterns of inheritance of N genes. Allele A from the female is shaded black; Allele A′ from the male is shaded white. The proportion of the pie chart that is black or white indicates relative contribution of paternal and maternal alleles at that locus in the F1 generation. For autosomal genes, males and female each contribute one allele. In the XY system, females have a larger contribution at sex-linked loci. In ZW systems, males have a larger contribution at sex-linked loci. Adapted from Hill and Johnson (2013).

The effects of sex linkage are distinctly different in ZW sex determination systems, and these differences have big implications for the effects of sex linkage on mitonuclear coevolution. In the ZW system, females are ZW and males are ZZ. Two-thirds of sex-linked genes are therefore carried by males, and mt genes and sex-linked genes are co-transmitted at a frequency lower than the frequency of co-transmission between mt genes and genes on autosomes (Figure 4.3). Placing N-mt genes on the Z chromosome, therefore, is a means to avoid co-transmission of mt and N-mt genes and to promote mitonuclear coevolution (Hill and Johnson, 2013; Hill, 2014a).

Beyond effects on co-transmission, sex linkage of N-mt genes can have important consequences for the potential for mitonuclear dysfunction when individuals that carry divergent mitonuclear genotypes produce offspring. In Chapter 2, I noted that in hybrid crosses with *Tigriopus* copepods, the negative effects of mitonuclear incompatibilities were not observed until F2 and subsequent generations (Burton et al., 2006). *Tigriopus* copepods have no sex chromosomes, so there is no opportunity for sex linkage (Foley et al., 2013). Thus, in the F1 generation, a hybrid copepod receives one copy of each chromosome from the female parent along with one copy from the male parent. As a consequence, at each locus with a N-mt gene there is at least one maternally derived allele that is compatible with the maternal mt genome (Figure 4.3). Whether the compatible N-mt genes contributed through the maternal line can fully compensate for incompatible alleles contributed through the paternal line will depend on the dominance interactions of the two alleles (Turelli and

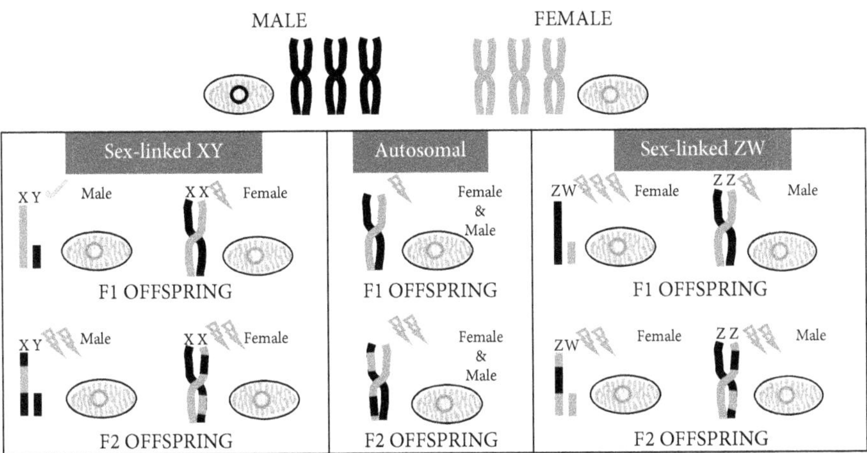

Figure 4.3 The interactions of mt (circular genome) and N (paired bars) genes depending on whether N-mt genes are Z-linked, X-linked, or autosomal. N genes that are compatible with the light gray maternal mitonuclear type are light gray. N genes that are compatible with the black paternal mitonuclear type are black. Lightning bolts indicate predicted incompatibilities if black and light gray genes are from diverged populations. A checkmark indicates no incompatibilities. One lightning bolt indicates possible incompatibilities. Two lightning bolts indicate moderate incompatibilities. Three lightning bolts indicate complete incompatibilities. Adapted from Hill (2018).

Orr, 2000; Raj et al., 2010). I'll revisit this issue in Chapter 6 when I consider the role of mitonuclear compatibility in speciation.

The situation can be different if an organism has sex chromosomes and the potential for sex linkage. In XY systems, if N-mt genes are positioned on the X chromosome, then males inherit both mt genes and X-linked genes exclusively from their mother. As a consequence, when there is sex linkage of N-mt genes in an XY system, the males will have the same mitonuclear compatibility as their mother for genes on the sex chromosomes, regardless of the compatibility of their father's N genes (Figure 4.3). In an XY system, females receive one X from the female parent and one X from the male parent, so there is at least one compatible allele at each locus as with taxa lacking sex chromosomes. The net result is that if N-mt genes are X-linked, there may be no negative effects of hybrid pairing between individuals with incompatibilities in N-mt and mt in the F1 generation. Expression of mitonuclear incompatibilities can be delayed until the F2 and later hybrid generations just as with no sex linkage (Burton et al., 2006; Hill and Johnson, 2013). This delay in the negative effects of hybrid pairing in XY sex determination systems has big implications for speciation and the genetic structure of populations (Chapter 7).

In ZW systems, female receive their single Z chromosome exclusively from their male parent. As a consequence, if N-mt genes are positioned on the Z chromosome, a female offspring has exclusively paternal N-mt genes to match to maternal mt genes, such that mitonuclear incompatibilities are fully revealed in the F1 generation for females. Males receive one Z chromosome from mom and one from dad, so male hybrid offspring are much less likely than female hybrids to show issues with compatibility (Figure 4.3).

The net effects of positioning N-mt genes on X chromosomes is that, relative to placement on autosomes, co-transmission is promoted and coevolution is hindered (Figure 4.2). Conversely, the net effect of placement of N-mt genes on Z chromosomes is that co-transmission is hindered and coevolution is promoted (Figure 4.2). In Chapter 6, I will discuss how sex linkage can affect postzygotic fitness following hybridization and hence how sex linkage can promote or hinder the process of mitonuclear speciation (Hill, 2017). Sex linkage of N-mt genes is potentially an explanation for Haldane's rule—the observation that in hybrid crosses the fitness of the heterogametic sex is typically lower than the homogametic sex (Wu et al., 1996; Hill and Johnson, 2013) (Box 4.1). And finally, sex linkage of N-mt genes affects not

Box 4.1 Haldane's rule

When organismal biologists perceive a widespread pattern in the natural world, the association can become stated so often that the observation becomes entrenched as a rule. The propensity for animals to be bigger in colder climates is commonly referred to as Bergman's rule. The tendency for organisms to become larger and more complex across evolutionary time is Cope's rule. More than a dozen such biological rules have been proposed, but perhaps no rule in biology holds as universally, with so few exceptions, or emerges from such an important underlying genetic mechanism as Haldane's rule, named for the brilliant British/Indian early-twentieth-century evolutionary biologist J. B. S. Haldane, who first formally defined the pattern (Haldane, 1922). Haldane's rule states that hybrid dysfunction is greater in the heterogametic sex—hybrid males suffer more in XY mammals and hybrid females suffer more in ZW birds. Wu et al. (1996) summarized sex-specific effects of hybridization for mammals and fruit flies (XY) as well as birds and butterflies (ZW). They found few exceptions to Haldane's rule (Table Box 4.1). A taxonomically much broader analysis of Haldane's rule, including both vertebrates and invertebrates, concluded that that about 88 percent of animal species conformed to Haldane's rule (Schilthuizen et al., 2011).

The disproportionate negative effects of hybridization on the heterogametic sex are usually attributed to the dependency on a single set of genes from the unmatched X or Z chromosome in the heterogametic sex. In hybrids, it is proposed that incompatibilities in sex-linked genes of two parental genotypes will be revealed in the heterogametic sex (Schilthuizen et al., 2011). For XY taxa, this hybrid dysfunction must come from nuclear–nuclear gene interactions because XY males obtain both their X chromosome and their mt genome from their mothers (Figure 4.4). There is no novel mixing of mt and N genes in XY male hybrid offspring.

For ZW taxa, however, mitonuclear interactions could play a key role in creating dysfunction in female offspring. In hybrid female offspring in ZW systems, the Z chromosome comes from the father while the mitochondria come from the mother. Consequently, any Z-linked N genes that differ between the hybridized species to the extent that they affect

Box 4.1 *Continued*

Table Box 4.1. The effects of hybridization on male and female hybrid offspring within various clades of animals. Data from Wu et al. (1996).

	Sex Determination	One of the Sexes Inviable		Both Sexes Viable; One Sex Sterile	
		Male Inviable	Female Inviable	Male Sterile	Female Sterile
Drosophila	XY	14	9	199	3
Mammals	XY	0	1	25	0
Birds	ZW	2	21	0	30
Butterflies	ZW	4	36	0	15

co-function with mt genes will be fully revealed in female hybrids. In male hybrids, in contrast, one Z chromosome will come from the mother and one from the father, so the effects of incompatible Z-linked genes might be masked (Hill and Johnson, 2013; Hill, 2014a). This disproportionate effect on female hybrids of mitonuclear incompatibilities would create the pattern noted by Haldane for ZW taxa. To date, the role of coadapted mt and N-mt genes in creating greater dysfunction in female hybrid offspring in ZY systems has not been tested.

only co-transmission of N-mt and mt genes and the risk of mitonuclear incompatibilities when sexually reproducing individuals carry incompatible mt and N-mt genes, it can also play a key role in genomic conflict between the mt and N genomes. As I will discuss in the next section, X linkage promotes genomic conflict while Z linkage hinders it (Figure 4.2).

Given the potential significance of the chromosomal position of N-mt genes, a few studies have begun to map the chromosomal locations of N-mt genes. In a pioneering study, Drown et al. (2012a) looked at the position of N-mt genes on autosomes versus sex chromosomes in birds and mammals (Figure 4.4). They found that the proportion of N-mt genes occurring on X chromosomes of mammals was much lower than expected by chance. In contrast, the proportion of N-mt genes on the Z chromosome in birds was not different than expected by chance (Figure 4.4). In a follow-up analysis that included a much broader range of animals including nematodes, flukes, arthropods, birds, mammals, and fish, Dean et al. (2014) found that, except for mammals and nematodes, N-mt genes occurred on both X and Z chromosomes at a frequency that was not different than random placement.

Both of these studies assessing the location of N-mt on chromosomes looked at the position of all N-mt genes. To date, no study has focused specifically on those N-mt genes with close functional interaction with mt genes. Drown et al. suggested that the under-representation of N-mt genes on X chromosomes had evolved so as to alleviate the negative effects of genomic conflict (Drown et al., 2012). Alternatively, by chance, the autosome that gave rise to the mammalian X chromosome may have had

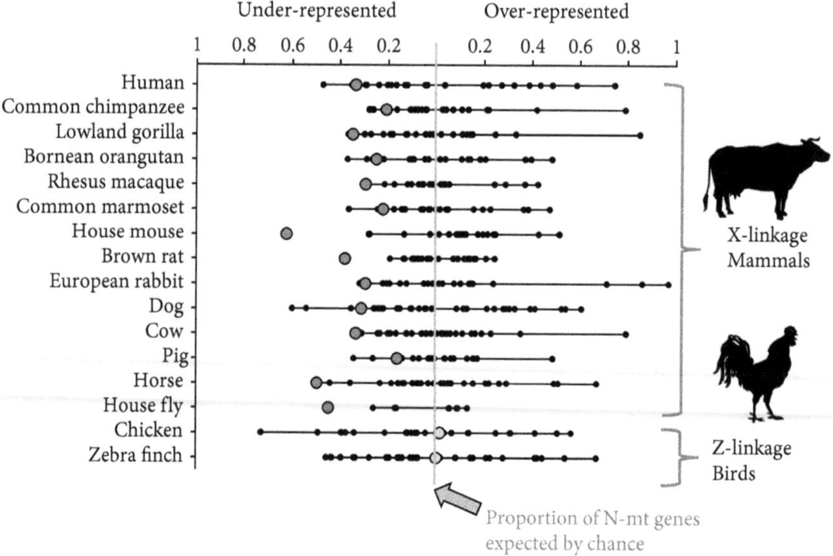

Figure 4.4 An assessment of sex linkage of N-mt genes in birds and mammals. Depicted is the ratio of observed N-mt genes versus expected N-mt genes on each chromosome across the N genome for fourteen species of mammals and two birds. Expected values are based on a random distribution of genes. Autosomes are shown as small black dots (there are tens of autosome in each genome) and sex chromosomes are shown as larger shaded circles. N-mt genes are highly under-represented on X chromosomes but present at a frequency expected by chance on the Z chromosome. Adapted from Drown et al. (2012).

fewer N-mt genes and this pattern has subsequently been inherited by all mammals. A more interesting but entirely speculative explanation is that N-mt genes have been moved from the X chromosome to avoid Hill–Robertson effects and to enable mitonuclear coevolution (Hill, 2014a). The chromosomal position of N-mt genes and particularly the position of those N-mt genes with close functional association with mt genes should be an active area of research going forward. The potential for the linkage of sets of N-mt genes is also a fascinating topic for future research.

Genomic conflict

In forging the chimera that became the common ancestor of eukaryotes, both the archaeon that gave rise to the N genome and the bacterium that gave rise to the mitochondrion gave up a degree of independence. Cooperation and co-function are undoubtedly the dominant endpoints in mitonuclear interactions, or I would not be sitting here drawing ideas from a nervous system that requires a both a fantastically complex blueprint in the N genome and a continuous and enormous flow of energy from a billion mitochondria. Cooperation and co-function may win out in all successful lineage of eukaryotes, but conflict between the mt and N genomes is also inexorably and incessantly in play (Partridge and Hurst, 1998).

The potential for conflict between mt and N genomes is largely a product of maternal transmission of mt DNA, which is by far the most common means by which mt genes are passed between generations across diverse eukaryotes (Birky, 2001). Maternal inheritance of mt genes means that males are a genetic dead-end for mt genes and that mitochondrial fitness is derived exclusively through females. Thus, traits that favor females, even if they come at the expense of males, are potentially advantageous to the mt genome (Frank and Hurst, 1996; Gemmell et al., 2004). As a consequence, there can be selection for mt genes that increase the fitness of females at a cost to males and at a net loss of fitness to the N genome (Rice 2013) (Figure 4.5). An interesting aside is that the compensatory coevolution theory for the evolution of sex (which will be presented in detail in Chapter 5) holds that the N genome evolved sexual reproduction and mating types in order to better compensate for mutational meltdown in mitochondria (Havird et al., 2015a). That being the case, mitochondria responded to the evolution of sex, which was a lifeline cast to it by the N genome, by pursuing strategies to undercut males to the detriment of the N genome. It would seem that in genomic battles as in human society no good deed goes unpunished.

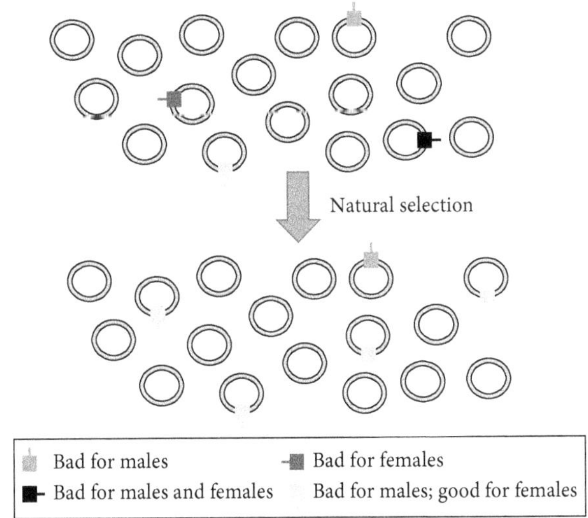

Figure 4.5 A simplified illustration of mother's curse effects on selection for mt alleles (gray circles represent mt genomes in different individuals in a population and boxes depict specific alleles at different positions within the mt genome). Beginning with a population with mutations that are either bad for males, bad for females, bad for both males and females, or bad for males and good for females, natural selection will remove any mutations that harm females whether or not they harm males. Mutations that harm males but do not help females have neutral fitness and can increase only through drift. Mutations that harm males but are beneficial to females will increase in frequency due to natural selection.

There are two perspectives from which to view genomic conflict, and they lead to different conclusions regarding the ubiquity of conflict versus cooperation (Wade and Drown, 2016). One view, which is the view that I've advocated through the first three chapters of this book, puts an emphasis on whole-organism function and hence on cooperation and coadaptation. According to this view, the necessity of coadaptation between N-mt and mt genes and the need for effective co-function between their gene products to enable cellular respiration lead inevitably to cooperative coevolution, with selection promoting favorable N and mt gene combinations and eliminating deleterious combinations (Wade and Goodnight, 2006; Brandvain and Wade, 2009). The numerous examples of intricate coadaptation of mt and N genes creating core function in eukaryotes, as presented in the first three chapters of this book, can be viewed as strong evidence in support of mitonuclear coevolution that is guided by the need for cooperation.

The alternative way to think about mitonuclear interactions is a gene-centric view (Dawkins, 1976) in which the mitochondrion is an "encapsulated slave" (Maynard Smith and Szathmary, 1998) with genetic interests that are largely at odds with the N genome (Haig, 2016). This line of thinking leads to descriptions of a "genomic battlefield" with the mt and N genomes locked in a perpetual evolutionary struggle for supremacy with casualties on all sides (Johnson, 2010). Because mitochondrial transmission is exclusively through females, any changes to the mt genes that promote female survival or fecundity will lead to higher mitochondrial fitness and should spread in a population even if such traits have negative effects on males and the N genome. The existence of such male-harming mt genes that are passed through female lineages via mitochondria was dubbed "the mother's curse" in one of the seminal papers presenting the theory behind mitonuclear genomic conflict (Gemmell et al., 2004).

There can be no doubt that genomic conflict is real. There is substantial empirical evidence that genomic conflict between mt and N genomes leads to patterns of evolution in which females benefit or are unaffected and males are harmed. In humans, male infertility caused by sperm dysfunction is often linked to mutations to mt genes (Kao et al., 1995; Spiropoulos et al., 2002; Baklouti-Gargouri et al., 2014). Moreover, in a study of individuals with mitochondrial disease not primarily related to reproduction, affected men had only 65 percent of the reproductive success of men with no known mitochondrial disease, while affected women had nearly the same reproductive success as women with no known mitochondrial disease (Martikainen et al., 2017). These effects on male fitness were not entirely due to mating success but were also caused by subtle dysfunction of male reproductive systems. Many additional studies have documented an association between mt DNA variants and impairment of male reproductive systems in non-human species including mice (Nakada et al., 2006), fruit flies (Clancy et al., 2011; Yee et al., 2013), seed beetles (Dowling et al., 2007d), and poultry (Froman and Kirby, 2005).

The effect of genomic conflict on patterns of evolution has also been demonstrated experimentally in model species in the laboratory, most notably in seed beetles and fruit flies. In a study of laboratory populations of fruit flies in the genus *Drosophila*, two types of lineages were created: (1) inbred lines in which the only source of males was from within the population, or (2) outbred lines in which the source of males was only from a larger outside population (Patel et al., 2016) (Figure 4.7). The logic of this design is that, in the inbred lines, male fitness might start to affect mitochondrial fitness because the only source of males for sexual reproduction and the eventual propagation of mt genomes is from within the population. Even if males are a genetic dead-end for females, male fitness can affect mitochondrial fitness through effects on the population as a whole. In the outbred lines, however, the fitness of males within the population in any given generation had no effect on overall population fitness or the fitness of females because new males were recruited each generation. After thirty-five generations, there was no change in male fertility in the inbred lines, but in one of the outbred lines males had reduced fertility that could be attributed to a novel mitochondrial mutation on the COXII gene. Thus, the evolution of a male-harming allele was experimentally demonstrated, and it only happened where the effect on population fitness was diluted (Figure 4.6).

Evidence for genomic conflict leading to loss of fitness for males is perhaps most dramatic in plants (Horn et al., 2014; Touzet and Meyer, 2014). For many lineages of flowering plants the ancestral condition appears to be hermaphroditic, with both

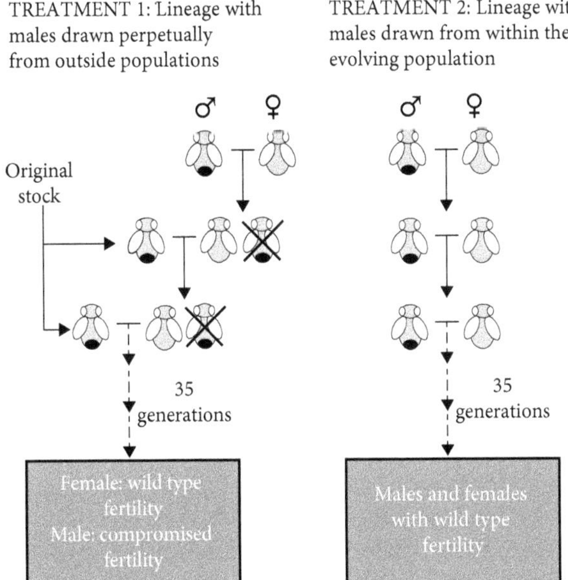

Figure 4.6 An experiment with lab populations of *Drosophila melanogaster* to test the conditions under which male-harming mt genes can evolve. One set of lineages was allowed to produce no males, and females were mated to a large outbred population each generation. Females in the other set of lineages could mate only with males from within the closed population. Male-harming mutations evolved in the population in which male fitness had no effect on female or mitochondrial fitness. Adapted from Patel et al. (2016).

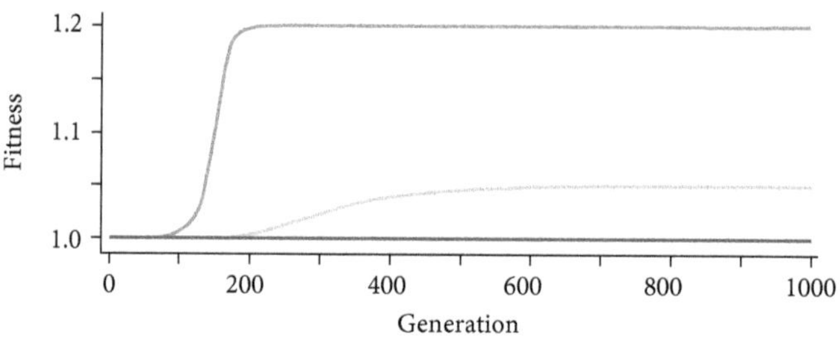

Figure 4.7 The predicted change in gene frequency and fitness under different levels of mt and N conflict and cooperation. The top line depicts genomic cooperation without sexual conflict; the middle line shows genomic cooperation with weak sexual conflict; the bottom line shows the outcome of strong genomic and sexual conflict. According to this simulation, high fitness is achieved only through cooperation. Adapted from Wade and Drown (2016) where details of the simulations can be found.

male and female reproductive systems in each individual. However, gynodioecious mating systems have repeatedly evolved in species within otherwise hermaphroditic clades. In these gynodioecious lineages, there are two types of individuals: (1) those that produce both male and female gametes (hermaphrodites), and (2) those that produce only female gametes. In all cases, these gynodioecious mating systems evolved from hermphoditic systems through the loss of male function in individuals (which become female) via mitochondrial mutations that cause male reproductive systems to be sterile, a condition known as "cytoplasmic male sterility" (Moison et al., 2010). In at least some species with cytoplasmic male sterility, the reproductive success of females—individuals that carry the mitochondrial male-sterilizing gene—is higher than hermaphrodites. Thus, cytoplasmic male sterility is the result of mutations to mt DNA that benefit females but that are highly detrimental to males. More than 140 species of plants are known to exhibit cytoplasmic male sterility (Laser and Lersten, 1972).

While most studies of mother's curse in humans focus on male reproductive performance, there is new evidence that the widely observed reduced lifespan of men compared with women may also be a result of mitonuclear genomic conflict (Wolff and Gemmell, 2013). Moreover, the observation that females live longer than males is not just a human phenomenon. Across many species of animals, the pace of senescence is greater in males than in females (Williams, 1957). And, across a range of animals, premature aging is associated with a deleterious mutation in a mt gene (e.g. Trifunovic et al., 2004). Taken together, these observations suggest that there are sex-specific mitochondrial effects on aging.

The evolution of a mother's curse arising from mt genes affecting longevity was demonstrated experimentally in a study of fruit flies in the genus *Drosophila* (Camus et al., 2012). This study looked at patterns of male and female senescence in a lab population in which variable mt genotypes could be paired to a common nuclear background (Box 4.2). The different mt genotypes had been collected in *D. melanogaster* from around the world, but because all *melanogaster* populations are derived from an ancestral African population within a few thousand years, these were recently diverged populations (Lachaise et al., 1988). Despite the shallow divergence among mt genotypes in the study, the researchers found clear sex-specific effects on longevity that could be attributed to mt genotype. Consistent with the predictions of genomic conflict, the effects on longevity were always negative for males compared with females (Camus et al., 2012). Moreover, further genetic analysis implicated many genes, each with small effects, contributing to loss of male fertility (Wolff and Gemmell, 2013).

Nuclear restorer genes

To this point in my discussion of genomic conflict, I've focused on mt mutations that spread because they benefit females, and hence mitochondria, even if they reduce the fitness of males or the N genome. It is important to point out, however, that the N genome is not doomed to be a passive victim to mitochondrial manipulation. The N genome can counteract detrimental changes imposed by mt genes with its own "restorer" genes. Such nuclear evolution is perhaps best demonstrated in the case of cytoplasmic male sterility in plants that I described above. In response to the spread of mt genes that sterilize the male component of hermaphroditic plants, the N genomes of many species of plants have evolved "restorer" genes capable of counteracting the mt cytoplasmic male sterility genes and restoring male function (Fujii et al., 2011; Caruso et al., 2012). The result is an evolutionary race in which suppression of the effects of mt genes is imposed or lifted depending on the net costs and benefits of the mt gene.

Restorer genes in the N genome are certainly not limited to those directed at cytoplasmic male sterility in plants. Restorer genes seems to be widespread if not ubiquitous in eukaryotes (Johnson, 2010). For instance, in the experiment described above (Figure 4.7) in which the evolution of a male-harming mt DNA mutation involving a missense mutation of COXII appeared in a laboratory lineage of fruit flies, researchers followed up with experiments in which they expressed mt genomes carrying the male-harming allele in diverse nuclear backgrounds from wild-type fly populations from around the world. Through these genetic crosses, they found a naturally occurring N genotype that completely suppressed the lab-generated male-damaging mt DNA mutation (Patel et al., 2016). It seems that this mt mutation had appeared before in at least one lineage of wild-type flies and that it had been suppressed via a nuclear restorer gene. The authors suggested that genetically dominant nuclear suppressors of male-harming mt DNA mutations are common in natural populations of fruit flies. Supporting this view, Dowling et al. (2007a) created lines of seed beetles for which different mt genotypes were expressed against common N genotypes and looked at effects on sperm function. They found that genes that controlled sperm function were inevitably N genes, presumably including restorer genes, and that mt genotype had no effect on sperm performance. Thus, genomic conflict seems to be resolved when N genes override the detrimental effects of mt genes.

Which dictates eukaryotic evolution: Cooperation or conflict?

Is there any way to predict whether mt and N genotypes are more likely to evolve to be in conflict or to be cooperative? Wade and Drown (2016) used what they called a "gene's eye view" to model the evolution of interacting mt and N-mt genes under scenarios of conflict and cooperation. In various iterations, they allowed for X linkage, Z linkage, and different mating systems. They ran one set of simulations in which mt genes acted to promote females by harming males and, indirectly, the N genome. Alternatively, they ran simulations in which mt genes acted in cooperation with N genes to promote overall organism fitness. To summarize a lot of modeling outcomes in simple terms, these simulations showed that cooperation triumphs over conflict (Figure 4.7). Models are approximations of natural processes that are necessarily founded on simplifying assumptions, but the results of Wade and Drown make sense given the world we see around us. Eukaryotes exist as highly functional organisms typically with balanced sex ratios. If genomic conflict ran rampant through eukaryotic populations we would expect to see more than the subtle male-damaging effects that have been revealed in humans and a few model species. This emphasis on cooperation over conflict is certainly not to belittle the importance of genomic conflict to understanding genomic evolution as emphasized by Rice (2013). Genomic conflict is real and it does affect the interactions of mt and N-mt genes and the mt and N genomes. I simply point out that there is strong justification for emphasizing coadaptation, coevolution, and cooperation in studies of mitonuclear ecology.

Within-individual conflict: Mito vs mito

Genomic conflict can manifest not just between mt and N genomes, but also among mitochondria themselves when there are divergent genotypes within the body of a eukaryote (Haig, 2016). The potential for within-individual genomic conflict arises because within each cell in the body of the eukaryote there are hundreds or thousands of mitochondria, and within each mitochondrion there are several copies of the genome. Among these thousands of copies of the mt genome, there are always variants because, as emphasized in early chapters, mt DNA is typically subject to a high mutation rate and to Muller's ratchet. Whether or not any particular mt genotype is passed along as cells divide is largely a product of the replication speed of the mitochondrial variant (Beekman et al., 2014). Thus, replication speed is a key to fitness of mitochondria but faster mitochondrial replication likely comes at a cost to efficiency in cellular respiration, which means that it comes at a cost to the N genome (Haig, 2016). As a consequence, there is a perpetual risk of genomic conflict within each cell within the body of a eukaryote.

Avoidance of intra-individual genomic conflict is proposed to be a primary reason for a germ line in animals, fungi, and some plants, and for the extreme bottleneck in mitochondrial numbers that is characteristic of reproduction of many eukaryotes. Conflict among mitochondria is a topic that will be taken up in Chapter 5, which provides a detailed treatment of sexual reproduction, mating types, and the evolution of germ lines.

Endosymbionts

Conflict arising from third genomes

A major complicating factor to the topics of co-transmission and genomic conflict is the addition of a third genome to a eukaryote. There is remarkably little written about the potential for chloroplast–mitochondria genomic conflict. Typically, both mitochondria and chloroplasts are passed only through the maternal lineage, which means that they are co-transmitted. Complete co-transmission would greatly reduce the rate at which mitochondria-harming chloroplast genes or chloroplast-harming mitochondria could evolve. Avoidance of mito–chloro genomic conflict is presumably *why* both organelles are almost always exclusively maternally transmitted. Co-transmission apparently works to reduce or eliminate chloro–mito genomic conflict.

There are sources for a third genome other than chloroplasts. Some prokaryotes, such as *Rickettsia*, *Wolbachia*, and *Spiroplasma*, exist as symbionts within eukaryotic cells adding an additional genome to the cellular environment. Symbiotic prokaryotes are especially common in insects, but such symbionts have been documented in many invertebrate taxa as well as protozoa and plants (Moran and Baumann, 2000; Wernegreen, 2012). The total absence of obligate endosymbionts from vertebrates (except perhaps a single species of salamander that hosts a symbiotic algae; Burns et al., 2017) is, to my knowledge, unexplained. Prokaryotic symbionts provide important or even critical services for the eukaryotic cell and in some cases they can become obligate symbionts that are required for the survival of the eukaryotic (Moran and Wernegreen, 2000; Moya et al., 2008). Tightly coadapted prokaryotic symbionts therefore can exist in eukaryotes as organelles with their own genomes, and they create new opportunities for genomic conflicts. When prokaryotic endosymbionts are transmitted vertically in a host lineage, the transmission is exclusively through the maternal lineage. Hence endosymbionts are co-transmitted with mitochondria, and the interests of the mt genome and the endosymbiont genome become closely allied.

As is the case for a mt genome, males are a genetic dead-end for maternally inherited endosymbionts. And, as is the case with mitochondria, any allele that is beneficial to the endosymbiont but detrimental to males in the host population can spread. The reliance of endosymbionts on female transmission is very relevant to mitonuclear considerations because any male-harming elements carried by an endosymbiont that is co-transmitted with mitochondria will necessarily affect mitonuclear dynamics. The role of endosymbionts in the introgression of mitochondrial types between populations will be taken up in Chapter 7 (on speciation). Here I'll briefly consider the very large literature on the evolutionary outcomes of genomic conflict involving the genomes of the host and endosymbiont.

The most common outcome of genomic conflict involving obligate endosymbionts like *Wolbachia* is sex ratio distortion, whereby females are promoted and males are eliminated (Figure 4.8). In different invertebrate hosts, *Wolbachia* can distort the sex ratio of offspring in a number of ways. They can cause the feminization of genetic males thereby changing a male lineage, which is of no use to either *Wolbachia* or

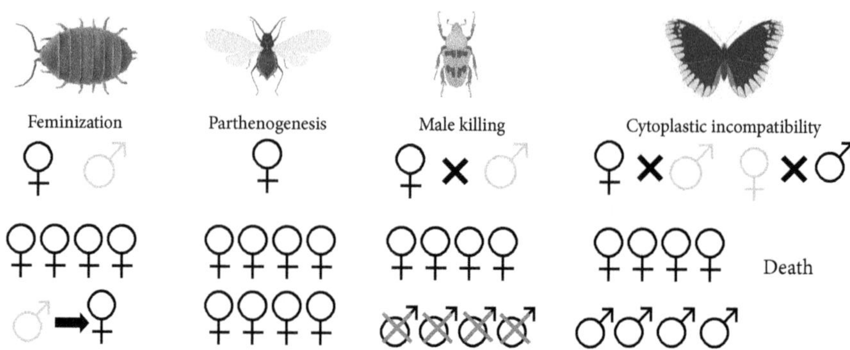

Figure 4.8 Various manifestations of male-harming effects that evolve to benefit maternally transmitted endosymbionts with illustrations of some of the arthropods that show the effect. Infected individuals are coded dark gray; uninfected individuals are coded light gray. In feminization (left), parthenogenesis (middle left), and male killing (middle right), mitochondria also potentially benefit because they are co-transmitted with the endosymbiont. In cytoplasmic incompatibilities (far right), the interests of the endosymbiont are not the same as either the mitochondrial or nuclear genome. Adapted from Werren et al. (2008).

mitochondria, into a mitochondrial and bacterial-transmitting female lineage. Endosymbionts can also sometimes induce parthenogenesis, thereby eliminating males; or they can simply cause the death of all male offspring (Figure 4.8). The most interesting and perhaps the most insidious resolution of genomic conflict is cytoplasmic incompatibility in which infected males are rendered incapable of fertilizing females that lack the same *Wolbachia* type. In the first three examples above, mitochondria can potentially gain a benefit equal to the endoparasite given that both should benefit from promoting the propagation of females at the expense of males. In the case of cytoplasmic incompatibility, the interests of the endosymbiont are pursued to the detriment of both mt and N genomes—individuals carrying the wrong *Wolbachia* type that are eliminated as mates by the endosymbiont might be the best mates from the standpoint of the mt and N genomes. Endoparasites complicate mitonuclear coevolution and coadaptation by making genomic conflict more likely and perhaps by making mitochondrial introgression more likely.

Summary

William Rice, who has spent a career studying the effects of genomic conflict on evolutionary processes, entitled a recent paper "Nothing in biology makes sense except in the light of genomic conflict," a clever play on the famous Theodosius Dobzhansky title "Nothing in biology makes sense except in light of evolution" (Dobzhansky, 1973; Rice, 2013). From the title (and content) of the paper, it is clear that Rice views genomic conflict as a dominant force in evolution, perhaps on a par with natural selection or drift. I agree that the theory that underlies genomic conflict is beyond reproach and that genomic conflict plays out constantly in all eukaryotes (and maybe that is Rice's point). Where I might disagree with Rice and other scientists who view genomic conflict as pervasive in shaping the eukaryotic world is in the ultimate outcome of genomic conflict. What we observe in nature is efficient organisms that are well matched to their environment. Far more prevalent than genes in competition are genes that are highly coadapted and that co-function flawlessly. At least among bilaterian animals, losses of male fertility and subtractions in male lifespan are so slight that they can typically be detected only with careful measurements. This world, with scant harm done to males, appears to be the result of N genes effectively counteracting the negative effects of genomic conflict arising from mt genes and perhaps, also, of natural selection eliminating entire lineages when genomic conflict reduces overall fitness. Even among plants, where cytoplasmic male sterility presents a much more conspicuous outcome of genomic conflict, restorer genes can counteract male-harming mt genes when the costs of male-harming alleles become too great. I will move forward in this book with a focus on coadaptation and cooperation among N and mt genomes, but with an appreciation for the effects of genomic conflict.

5

The evolution of sex and two sexes

When driving toward the rim of a basket, LaBron James is phenotypic perfection. He is huge and powerful, bigger than nearly all other men, but still blazing fast, remarkably coordinated, and even nimble. The magnificence of his phenotype was recognized before he was 15 years old, and the adult LaBron James has not disappointed those who saw future greatness in the boy from Akron, Ohio. As I write this book, he has dominated professional basketball for a decade, but he would dominate most sports that are based on size, strength, and eye–hand coordination. He may be the greatest athlete among the world's 7.5 billion people. He may be the greatest athlete who has ever lived. And yet, there is no guarantee that his children will be standouts in basketball or any other sport. They almost surely will not match the athletic prowess of their father. Albert Einstein's children did not excel in math or physics; Winston Churchill's children were not great statesmen; Marie Curie's children contributed nothing to science. Sexual reproduction scrambles and recombines sets of genes every generation, and in so doing, sex makes phenotypes like LaBron James' unique and ephemeral.

Explaining the evolution of sexual reproduction is among the most significant and long-standing problems in evolutionary biology (Williams, 1975; Maynard Smith, 1978). Sexual reproduction has proven so formidable a challenge for evolutionary biologists that it is commonly spoken of as "the paradox of sex" (Otto and Lenormand, 2002). Aside from breaking up adaptive phenotypes, sex dilutes genetic representation in offspring. If an individual reproduces simply by asexually duplicating its genotype, it gains twice the genetic representation in offspring compared with a sexually reproducing individual. This loss of genetic representation in offspring is termed the "two-fold cost of sex," and there must be substantial benefits to compensate for so great a penalty associated with sexual reproduction.

In a related but fundamentally distinct line of investigation, researchers have long pondered why the existence of two mating types—males and females—is a nearly universal condition for eukaryotes (Fisher, 1930). Theoretically, mating types are not necessary for sex, and yet essentially all eukaryotes have mating types. Two mating types necessitates that half of the individuals in a population are inappropriate as mates. On first consideration, it would seem that, if you must have mating types, the most beneficial strategy would be to have numerous mating types so that most individuals encountered would be a suitable mate. A system with two mating types seems like a losing strategy for sexual reproduction, and yet two mating types is the nearly

Mitonuclear Ecology. Geoffrey E. Hill, Oxford University Press (2019). © Geoffrey E. Hill 2019.
DOI: 10.1093/oso/9780198818250.001.0001

universal state across eukaryotes. As with sexual reproduction, two mating types is an evolutionary outcome that begs for an explanation (Lane, 2005).

New thinking places the interactions of mt and N genomes at the center of explanations for the evolution of both sexual reproduction and two mating types (Lane, 2005). Incorporating a consideration of the fundamental necessity of mitonuclear coadaptation into existing theory has re-invigorated research into both the evolution of sex and the evolution of mating types. In this chapter, I will consider the implications and evidence for these new mitonuclear-based theories of key evolutionary ideas.

The evolution of sex

The necessity of recombination

Sexual reproduction is a specific form of biotic replication, and it is unique to eukaryotes. Prokaryotes certainly replicate their genetic material and engage in various forms of gene exchange, but no prokaryotic lineage engages in reciprocal exchange of genetic material (Xu, 2004; Narra and Ochman, 2006). Sexual reproduction in eukaryotes involves assortment and recombination of chromosomes in a process that entails the fusion of haploid gametes to produce a diploid individual (Goodenough and Heitman, 2014) (Figure 5.1). With very few exceptions (Box 5.1), all eukaryotes engage in sexual reproduction (Vrijenhoek, 1998; Whitton et al., 2008). When judged on an evolutionary time scale, the few eukaryotic lineages that exist as obligate asexuals rarely persist for long (Neiman et al., 2009).

Sexual reproduction with recombination appears to be a primitive trait for eukaryotes. Despite approximately 2 billion years of divergent evolution in lineages such as amoeba and elephants, the proteins required for recombination have been conserved across eukaryotes (Ramesh et al., 2005; Speijer et al., 2015). Conserved protein structure indicates that there has been persistent and consistent selective pressure for recombination across vast evolutionary time and that recombination has been a necessity for eukaryotes throughout their history (Garg and Martin, 2016). The simultaneous evolution of sexual reproduction with recombination and a two-genome cellular structure suggests that something about genomic architecture of eukaryotes made sexual reproduction a necessity (Lane, 2005; Garg and Martin, 2016). It is rather surprising, therefore, that it was only after a hundred years of contemplating the evolution of sex and fifty years after the discovery of the mt genome that mitonuclear coevolution was invoked as an explanation for sexual reproduction.

Theories that have been proposed to explain sexual reproduction do not focus specifically on the exchange of genes between individuals—that can be accomplished with lateral gene transfer as in prokaryotes (Figure 5.1). Rather, the key benefit of sexual reproduction is invariably proposed to be the recombination of sets of N genes within chromosomes (Otto, 2009). Reciprocal recombination of genotypes is only possible through sexual reproduction. The "recombination" in which prokaryotes

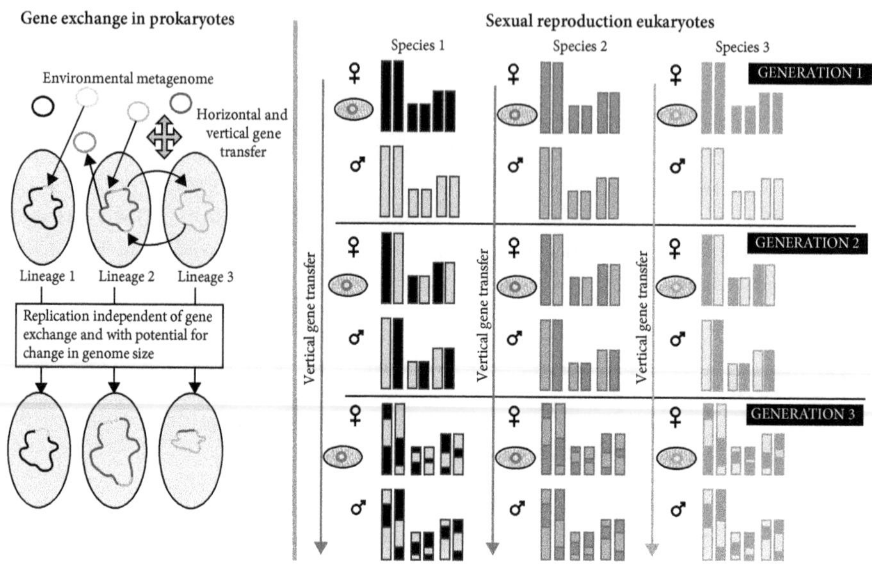

Figure 5.1 The contrast between gene exchange in prokaryotes (left) and sexual reproduction in eukaryotes (right). Gene exchange in prokaryotes is not tied to replication and is not reciprocal. Sexual reproduction in eukaryotes occurs during DNA replication and is reciprocal. Eukaryotes have fixed gene content within a lineage that evolves slowly. Prokaryotes draw genes from a metagenome so gene content can change rapidly.

engage is a one-way lateral transfer of genes and is fundamentally different than the reciprocal recombination of eukaryotes (Ochman et al., 2000; Lang et al., 2012) (Figure 5.1). To enable reciprocal recombination, individuals of at least one sex must exist as diploids for at least part of a life cycle so that haploid gametes of one individual join with the haploid gametes of another individual to produce an offspring with a diploid genotype drawn from two individuals. Reciprocal recombination is the exchange of genes between the two parental genotypes (Figure 5.1). The simple process of recombination creates the opportunity for separating linked genes and thus for isolating specific genes for elimination via negative selection or for proliferation via positive selection (Maynard Smith, 1978).

Two specific benefits of recombination have been the focus of hypotheses for why sexual reproduction is so prominent across eukaryotes. The first and most frequently stated hypothesis for the evolution of sexual reproduction is that there is an urgent and perpetual need for genetic variation on which selection can act (Otto, 2009). Unstable and changing environments is frequently stated as the reason that genetic diversity is essential, and pathogens are often presented as the most significant environmental factor that create a need for diverse gene combinations on which natural selection can act (Hamilton et al., 1990; Howard and Lively, 1994). However,

Box 5.1 Obligate asexuality in eukaryotes

Sexual reproduction is a defining feature of eukaryotes, having emerged in the proto-eukaryote that gave rise to all extant eukaryotes. Paradoxically, however, there are many examples of eukaryotic lineages that have abandoned sex and that exist as obligate asexuals.

Most biologists have heard of parthenogenetic lizards (Fujita and Moritz, 2010) and asexual water fleas (Lynch et al., 2008) and, to be sure, these are well-documented obligate parthenogenetic lineages in animals. But in animals, obligate asexuality almost always occurs amongst the twigs at the tips of evolutionary branches (Schwander, 2016). There is no successful lineage of lizards or any other vertebrate that dates back more than a few million years in the 2 billion-year history of eukaryotes. There are a very few evolutionarily old asexual animal lineages, sometimes referred to as "ancient asexual scandals" because they appear to break the rule of asexuality being short-lived in animals (Schon et al., 2009). Even these generally asexual animals, however, seem to occasionally engage in sex (Schwander, 2016). The bottom line is that obligate asexuality does evolve in animals, but it almost never lasts. In this sense, the rare cases of obligate asexuality actually prove the rule of the fitness advantages of sexual reproduction (Lively and Morran, 2014).

Protozoa seemed to present a much more serious challenge to the idea that sexual reproduction is an essential characteristic of eukaryotes. Although sex is very well documented in many clades of protists, such as ciliates, diatoms, and green algae (Weedall and Hall, 2015), for the majority of groups of protists, sexual reproduction has never been observed and, until recently, most of these protists were presumed to be obligately asexual (Dunthorn and Katz, 2010). If the majority of protists really did lack sexual reproduction, then the statement that essentially all eukaryotes engage in sex would not be defensible. To better assess the possibility of sexual reproduction in protist taxa in which sex had never been observed, Speijer et al. (2015) conducted a genomic analysis of three groups of protists that were long considered to be ancient obligate asexuals: *Jakobida*, *Glaucophyta*, and *Malawaimonadida*. They found that individuals in each of these distantly related groups carry functional genes involved in gamete fusion, which would only be present if the species actively engaged in sexual reproduction. These meiosis genes are widespread among protist lineages in which sexual reproduction has never been observed (Schurko and Logsdon, 2008). The implication of these studies is that sexual reproduction can be very hard to detect in protists through simple observation, but that obligate asexuality may be as rare in protists as it is in animals. Eukaryotes, it seems, really do need sex.

as pointed out by Otto (2009) there are two fundamental problems with the assertion that genetic diversity is beneficial enough to make nearly all eukaryotes sexual. First, recombination is not necessary to generate genetic diversity. Prokaryotes derive tremendous genetic diversity via horizontal gene transfer without sexual reproduction (Figure 5.1). Second, there is a significant cost to genetic recombination that should counteract and potentially outweigh the benefits of genetic diversity. Recombination not only generates new gene combinations that present the opportunity for novel adaptation; recombination also shuffles fit gene combinations making it harder to maintain coadaptive gene complexes. The fact that most genome replication in

Box 5.2 Episodic sex amidst clonal reproduction

There is a pervasive inclination among biologists to characterize multicellular eukaryotes as obligately sexual but to label single-celled eukaryotes as primarily clonal with instances of sex (Speijer et al., 2015). The interesting perspective articulated by Speijer et al. (2015) is that there is less difference in these two patterns of reproduction than is generally presented when their life histories are contrasted. Just as in a protozoan, the vast majority of DNA replication in any multicellular plant, animal, or fungi is clonal propagation, as trillions of cells are replicated. Even in the animal germ line, tens to hundreds of clonal cell divisions can occur before meiosis and sex (Stewart and Larsson, 2014). Thus, it is more accurate to view sex as episodic in all eukaryotes including vertebrate animals, because, by any criterion, clonal reproduction is the dominant means of replicating genomes. Thinking of sex as an episodic but essential mode of reproduction in eukaryotes reinforces the key themes of this chapter. There are huge advantages to rapid clonal replication of cells and this form of reproduction predominates in both the prokaryotic and eukaryotic world. But only in eukaryotes, sexual reproduction with reciprocal recombination is also essential, even if taxa turn to it only sporadically.

eukaryotes is asexual replication (Box 5.2) supports the assumption of significant costs associated with sexual reproduction.

The second set of hypotheses to explain the evolution of sexual reproduction focuses on the accumulation of deleterious genes within a genome (Maynard Smith, 1978; Kondrashov, 1988). As I discussed in previous chapters, without recombination, Hill–Robertson effects arising from absence of genetic recombination can lead to the fixation of slightly deleterious mutations within the genome (Figure 4.1). In addition, once a deleterious allele is fixed within a non-recombining genome, selection has no means to remove it without eliminating the entire genome (Muller, 1964; Felsenstein, 1974). Thus, there is a perpetual accumulation of deleterious alleles via the process known as Muller's ratchet. The result can be mutational erosion and loss of fitness over time (Maynard Smith, 1978). Because sexual reproduction scrambles gene sets, it enables the selective elimination of bad alleles without the elimination of beneficial alleles. The hypothesis that sexual reproduction and recombination are necessary to combat Muller's ratchet is decades old, but new thinking recasts this hypothesis in terms of compensatory coevolution between the N and mt genomes.

The evolution of sex in light of mitochondrial evolution

In the first two chapters of this book, I promoted the hypothesis that the evolution of eukaryotes and the evolution of mitochondria were the same event; in other words, eukaryotes were born of the chimeric fusion of two prokaryotes (Koonin, 2010; Speijer, 2015). I'll now extend this origin theory a step further: the evolution of eukaryotes, the evolution of mitochondria, and the evolution of sexual reproduction occurred simultaneously (Lane, 2015a). Essentially all modern eukaryotes engage in sexual reproduction with recombination and evidence indicates that sexual reproduction is a primitive trait in eukaryotes (Goodenough and Heitman, 2014; Garg and Martin, 2016). It would seem that either sexual reproduction was a key to a successful fusion of the genomes of two prokaryotes, or else the two-genome architecture of the proto-eukaryote made sexual reproduction an indispensable reproductive strategy.

The ideas that I will summarize regarding the evolution of the N genome and sexual reproduction are new and, at present, these new hypotheses are largely speculative. However, these new explanations for the evolution and maintenance of sexual reproduction are consistent with genomic data, and I find them compelling. For simplicity in the following discussions, I'll refer to the cell that gave rise to the nucleus as the "host" and the cell that gave rise to the mitochondrion as the "endosymbiont." As I've stated repeatedly throughout this book, I advocate the hypothesis that these cells were equal partners in the origin of eukaryotes, but "host" and "endosymbiont" labels make the explanations for the evolution of sex much simpler.

The conditions and circumstances that are proposed to have given rise to the evolution of sexual reproduction emerge at the very origin of eukaryotes. When an archaeon and a bacterium fused to form a single organism with two genomes, there would have been immediate problems for the archaeon host partner, whose genome existed in one copy surrounded by multiple copies of the bacterial endosymbiont's genome. Individual endosymbionts inevitably died within host cells without the host cell itself dying, and these deceased endosymbionts would have released DNA into the cytosol. As a consequence, host DNA would have been bombarded by pieces of endosymbiont DNA. These segments of the mt genome would have inserted themselves into the host genome and disrupted coding for essential genes (Lane, 2005, 2015a; Rogozin et al., 2012) (Figure 5.2). Such corruptions of the host genome by insertion of endosymbiont genes created what was essentially a very high mutation rate—an unprecedented rapid and massive alteration of the nucleotide sequence of the host (Timmis et al., 2004). It would also have led to a potentially unstable genome as segments of endosymbiont DNA were inserted into the host genome (Lane, 2011a). It is easy to imagine that bombardment of the host genome by endosymbiont DNA would have put a quick end to the proto-eukaryote before it had much of a chance to get started. One can imagine that bombardment of the host DNA by endosymbiont DNA did put a quick end to a multitude of false starts at eukaryotic evolution. After what likely was an unimaginably large number of failed attempts, a series of highly improbable events fell in sequence and a chimeric cell survived.

There is substantial evidence to support the hypothesis that endosymbiont DNA was inserted into host DNA very early in eukaryote evolution. To begin with, the N genome of all eukaryotes is structured as coding sequences disrupted by non-coding sequences—"genes in pieces" (Koonin, 2009). The coding nucleotide sequences are called exons, and the non-coding sequences that sit between the exons are called introns. Most introns are of mitochondrial origin (Martin and Koonin, 2006), and numerous introns are conserved among diverse eukaryotic lineages, indicating that the introns in universally important and conserved genes (like citrate synthase) are of ancient origin, dating to the origin of eukaryotes (Rogozin et al., 2003). Incorporation of mt DNA into the N genome continues commonly to the present day because the

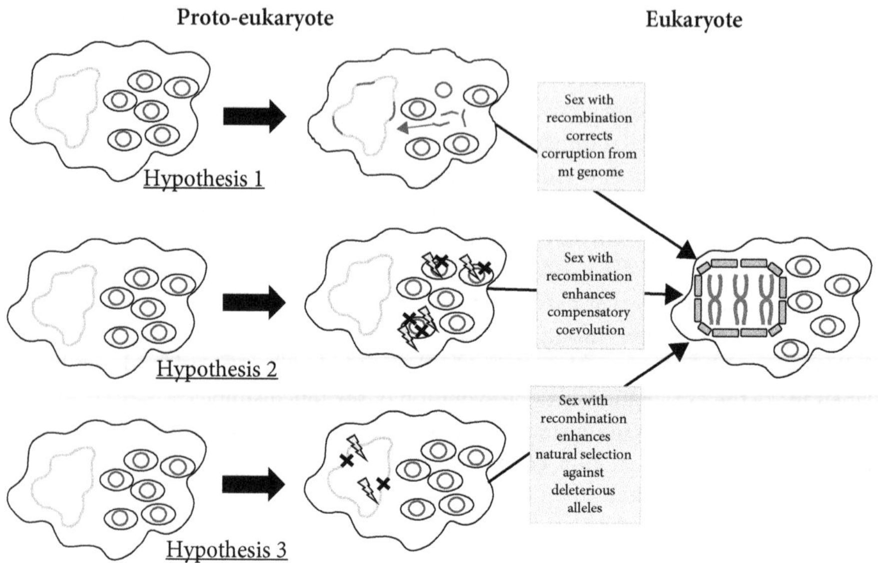

Figure 5.2 Pictorial sketches of three hypotheses for a role of mitonuclear interactions in the evolution of sexual reproduction and recombination. The light grey strands represent the archaeon DNA that gave rise to the nucleus. The circles represent the bacterial genome of the proto-mitochondrion with its own cell membrane (black oval). Lightning bolts represent free radicals, and black Xs represent sites of mutational changes caused by free radicals. Hypothesis 1: Insertion of mt DNA into the N genome selects for sex with recombination. Hypothesis 2: increased mutation rate of the mt genome selects for sex with recombination of N genome to enable compensatory coevolution. Hypothesis 3: Increased mutation rate of N genome selects for sex with recombination to enable selection against deleterious alleles. These models are not mutually exclusive and it is likely that all three proposed effects contributed to the evolution of sexual reproduction in eukaryotes.

problem of endosymbionts deconstructing and releasing DNA into the cytosol has not gone away (Hazkani-Covo et al., 2010). Indeed distinguishing between true mt genes located in the mt genome and copies of mt genes that have inserted into the N genome is an ongoing challenge for modern molecular biologists (Sorenson and Quinn, 1998).

With the random insertion of endosymbiont DNA into the host genome, mutational meltdown would have occurred at an unprecedented rate, jeopardizing the survival of a proto-eukaryotic lineage (Garg and Martin, 2016). There certainly would have been urgent need for mechanisms to edit out inserted DNA as well as to select against deleterious changes to genotype (Lane, 2015a). The need for gene editing is proposed to have given rise to the eukaryotic spliceosome and the nuclear membrane, two adaptations that enabled post-transcriptional editing of the chunks of endosymbiont DNA that had been inserted into functional sequences of the host genome (Lane, 2015a; Garg and Martin, 2016). According to these hypotheses, insertion of endosymbiont DNA into the host DNA in the early evolution of eukaryotes gave rise to introns as well as the potential for alternate splicing—combining exons in different combinations to produce different gene products. Alternate splicing, in turn, created enormous opportunities for evolutionary novelty (Gilbert, 1978). The evolution of the spliceosome, the nuclear membrane, introns, and alternate splicing are each seminal events in the evolution of eukaryotes (Lane, 2015a; Garg and Martin, 2016). Their origins are the subject of current investigation and debate, but these topics are somewhat peripheral to the evolution of sex. Interested readers can easily pick up the threads of these investigations from the literature I've cited. The focus in this chapter is on the evolution of sexual reproduction.

Gene editing is proposed to be a key adaptation that enabled a proto-eukaryotic nucleus to deal with endosymbiont DNA insertion. But in the face of mutational meltdown of the host genome resulting from rampant insertion of endosymbiont DNA, it is reasonable to propose that there was also strong selective pressure for the evolution of recombination, which is the mechanism currently recognized as the best protection against the accumulation of deleterious alleles (Barton and Charlesworth, 1998; Otto and Lenormand, 2002). Even if introns could generally be accommodated through spliceosome editing of DNA transcripts, there would still be important variation among individuals in the position of particular introns (Lane, 2015a). Moreover, with many novel genes being generated through the lateral transfer of genes, there would be a critical need for natural selection to both remove deleterious variants and promote useful novelties. These factors are proposed to have led to the evolution of sexual reproduction involving recombination (Lane, 2015a).

This hypothesis for the origin of sexual reproduction explains why the mechanism of gene exchange that had served prokaryotes for 2 billion years was not adequate for the proto-eukaryote. The massive mutational load that came with a second genome within the proto-eukaryotic cell would have created a heretofore unprecedented destabilizing force on the host genome (Lane, 2015a). The eukaryotic cell required a new form of gene exchange that provided a means to avoid Hill–Robertson effects and Muller's ratchet. Sexual reproduction with recombination restored genome stability, perhaps rescuing the failing lineage that was being overwhelmed by rapid genomic alterations. The incorporation of mt genes into the N genome would have begun immediately following chimeric fusion, so selection for recombination would have arisen during the initial formation of the first eukaryote, explaining why sexual reproduction is a shared and unique feature of eukaryotes. Presumably, because it remains an ongoing issue for all eukaryotes, lateral transfer of mt DNA to the N genome remains a reason why sexual reproduction is maintained in essentially all eukaryotic lineages (Timmis et al., 2004). All other benefits of sexual reproduction—such as the need for genetic diversity to adapt to changing environments—that have been widely discussed as explanations for the evolution of sexual reproduction would only reinforce the need for sexual reproduction in eukaryotes driven by a need to deal with insertions of mt DNA into the N genome.

Another key consideration regarding the evolution of the N and mt genomes of eukaryotes is that sharing genes via a metagenome, which was the mechanism for generating genetic diversity in both of the prokaryotic lineages that gave rise to eukaryotes, was no longer workable. As soon as the core respiratory machinery of mitochondria became dependent on products of both the N and mt genomes, which likely occurred during the early evolution of eukaryotes, coadaptation and compatibility became key factors in individual fitness. N gene sets and mt gene sets had to be transmitted vertically so that, each generation, compatible sets of genes could be matched correctly and the two sets of genes could coevolve. Pulling genes from a common metagenome could not accommodate tight coevolution toward coadaptation and thus it ceased. Recombination of diploid chromosomes became the mechanism for reshuffling of genes without endangering mitonuclear compatibility. Choice for compatible genomes via mate choice prior to sexual reproduction, the topic of Chapter 8, ensures that coevolved and coadapted gene sets are transmitted together across generations. In this way, exchange of genes via sexual reproduction was limited to individuals with a shared coadapted mitonuclear genotype, and thus, this new form of reproduction was also the origin of species boundaries (see Chapter 7).

Avoiding mutational meltdown

Among the many seminal changes that occurred following the chimeric fusion of two prokaryotes to form one eukaryotic organism, aerobic respiration on a heretofore unachievable scale was undoubtedly the most significant (Lane, 2015b). The flow of energy provided by aerobic respiration in many individual mitochondria, which existed as compartmentalized and duplicated power centers within a eukaryotic cell, enabled a massive expansion of the eukaryotic genome (Lane and Martin, 2010). This expansion included an increase in the number of proteins produced by eukaryotes by an order of magnitude or more compared with prokaryotic ancestors, and it trivialized the cost of the transfer of genetic elements from the endosymbiont to the host in terms of the energetic cost of DNA replication (Lane, 2014, 2015a). The clear benefit of an oxygen-fueled respiratory furnace is massive energy production. But, as the manager of any public utility can attest, energy production is a messy business, and the messy part of aerobic respiration via an electron transport system (ETS) in mitochondria is the production of free radicals. These free radical products are unstable and react with whatever molecules they contact, causing a host of cell damage that includes mutating DNA. Indeed, ionizing radiation is mutagenic largely because it creates free radicals through the dislodging of electrons from water molecules within biological systems, thereby creating free radicals that damage DNA as well as other cellular components (Lane, 2002).

The mt genome is positioned immediately adjacent to the ETS, which is the primary source of reactive oxygen species (ROS) production in the cell, so mt DNA exists on the front lines of the redox danger zone. It has long been proposed that exposure to free radicals produced during aerobic respiration is the reason that mt DNA has a much higher mutation rate than N DNA in many eukaryotes (Figure 1.7). However, new evidence indicates that copy error and not exposure to free radicals is the source of most mt DNA mutations (Lagouge and Larsson, 2013; Itsara et al., 2014). I'll take up the discussion of the source of mt DNA mutation in detail in Chapter 6. For now, we need only recognize that mutations of mt DNA lead to fixation of deleterious alleles because, in a haploid and non-recombining genome, Hill–Robertson effects cannot be avoided, and once deleterious mutations are fixed, lack of recombination shields slightly bad genes from natural selection. Accumulation of deleterious alleles in the mt genome is fundamentally bad because mt genes code for core respiratory processes. Hence, it is proposed that N genes perpetually evolve to compensate for deleterious mt genes—these considerations were the topic of Chapter 4. What I did not consider in detail in Chapter 4 is the mechanism that enables the N genome to generate sufficient genetic diversity to plausibly compensate for deleterious genes in the mt genome.

Havird et al. (2015a) proposed that compensatory coevolution by the N genome (Chapter 3) is only effective if the N genome can evolve adaptive changes at a pace that keeps up accumulation of deleterious alleles in the mt genome. Therefore, they proposed that sexual reproduction and recombination evolved in eukaryotes in direct response to the need for substantial and perpetual genetic variation in the N genome to enable compensatory coevolution to keep pace with the high mutation rate of mt genes (Figure 5.2). Sexual reproduction with recombination generates the needed variation in N genotypes. This hypothesis proposes that the benefits of compensation for mutational erosion of the mt genome through the perpetual generation of novel gene sets via sexual reproduction with recombination more than offset the substantial cost of scrambling of successful gene combinations.

Havird et al. (2015a) recognize two critical assumptions of their hypothesis. First, their model assumes that early in eukaryotic evolution and continuing in the lineages of most extant eukaryotes, bi-parental transmission of mt genomes was not an evolutionary option. If eukaryotes could evolve bi-parental transmission of mitochondria then there would be no need for compensatory coevolution by N genes. Bi-parental inheritance of mt genomes creates opportunities for recombination of mt genes and, as I've discussed extensively, recombination stops Muller's ratchet and mutational erosion in the mt genome. The evolution of bi-parental versus uniparental inheritance of mt DNA is the topic of the second half of this chapter, but here I will note that uniparental inheritance of mt DNA is the common pattern of mitochondrial transmission observed in eukaryotes.

Second, the hypothesis proposed by Havird et al. (2015a) requires that a high mt DNA mutation rate was the ancestral condition for eukaryotes. In support of this assumption, these authors point out that a high mt DNA mutation rate is the common state among eukaryotes (Figure 1.7) and that mt DNA is replicated much more than N DNA, providing more opportunity for mutations arising from replication error (Melvin and Ballard, 2017; Szczepanowska and Trifunovic, 2017). Moreover, the DNA of endosymbionts living in eukaryotic cells shows elevated rates of mutation compared with N genomes (Itoh et al., 2002). The rates of substitution are faster in the bacterial lineage that gave rise to mitochondria than in the archaeon lineage that gave rise to the nucleus (Sung et al., 2012). mt DNA is highly derived relative to bacterial DNA suggesting a rapid increase in mutation rate early in eukaryotic evolution (Gray et al., 1989). And finally, as previously noted, mt DNA sits next to the redox furnace with massive exposure to mutagenic free radicals. So, it is certainly plausible that from the early phases of eukaryotic evolution there was a fundamental need for compensatory coevolution by the N genome to bail out the mt genome.

To this point, I've presented two hypotheses for the evolution of sexual reproduction that involve mitochondria: endosymbiont bombardment of the host genome, and the need for N genomic diversity to enable compensatory coevolution (Figure 5.2). Speijer (2016) proposed a third hypothesis for the evolution of sexual reproduction in eukaryotes that involves mitochondria. This hypothesis focuses on the mutagenic effects of the free radicals that came along with aerobic respiration at the origin of eukaryotes in the evolution of sex. In contrast to the Havird et al. (2015a) hypothesis that focused on mutations in the mt genome, however, Speijer (2016) considered the integrity of the N genome. The basic premise is the same as in the hypothesis put forward by Havird et al. (2015a): the evolution of massive respiration via mitochondria created more free radicals and more DNA damage in eukaryotic cells (Lane, 2011b). Where Havird et al. (2015a) focus on mutational meltdown in the mt genome and propose that a need for nuclear compensation drove the evolution of sexual reproduction, Speijer (2016) focuses on maintaining the integrity of the N genome itself as the driving force in the evolution of sexual reproduction with recombination. According to this hypothesis, the high mutation rate of N genes caused by free radical damage resulted in selection for sex and recombination to maintain the integrity of the N genome (Figure 5.2). These two free radical-based hypotheses—the Havird et al. (2015a) hypothesis focused on mt mutations and the Speijer (2016) hypothesis focused on mutations in the N genome—are not mutually exclusive hypotheses. As Speijer (2016) explains, compensatory coevolution to reverse problems in the mt genome would be an added benefit to advantages attained through better editing via sexual reproduction of the N genome. Both of these hypotheses that are founded on the premise that mt DNA mutations arise primarily from the effects of free radicals are challenged by new data showing that, in modern eukaryotes, free radicals contribute little to mt DNA mutation rate, and that mt DNA mutations arise primarily from copy error (Melvin and Ballard, 2017; Szczepanowska and Trifunovic, 2017). However, free radical-induced mutation of both mt and N DNA may have been more of a problem in early eukaryotic evolution before more sophisticated DNA editing processes evolved. Also, the rate of mt DNA replication would have increased exponentially to accommodate the energy needs of the larger and more complex eukaryotic cell, so copy error of mt DNA would have increased mt DNA mutation rate even if free radicals did not induce significant mutation.

At present, all hypotheses for the evolution of sex that invoke mitonuclear coevolution and coadaptation are untested. Going forward, a consideration of mitonuclear dynamics will certainly be a part of any discussion of the evolution of sex.

The evolution of two sexes

The evolution of anisogamy

While the paradox of sex is known to all evolutionary biologists, the paradox of two sexes is a major puzzle in eukaryotic evolution that attracts much less attention (Billiard et al., 2011). For many biologists, the question of why there are two sexes was settled more than 40 years ago by evolutionary theorist Geoffrey Parker and his colleagues when they proposed a game theory model of the likely selective forces at work on a primitive population of sexually reproducing organisms with one gamete type (Parker et al., 1972) (Figure 5.3). In this hypothetical population, all individuals produce gametes that are acceptable as partners for gametes from all other individuals in the population, and there is scramble competition to find a mating partner. It is assumed that there will be variation among individuals in the size of gametes they produce, just as there is variation in expression of any trait in a population. Moreover, it is assumed that there would be a necessary tradeoff between gamete size, the likelihood of gamete survival, and gamete numbers. Large gametes require lots of resources that increase the survival prospects of both the gamete and the zygote that results, but only a few large gametes can be produced because they require substantial investment in resources. Small gametes, on the other hand, are less costly to produce enabling the production of more, but because such small gametes carry few resources, their prospects for survival and the survival prospects of the zygote they produce are low. Thus, some individuals in the hypothetical ancestral population produce fewer but larger gametes and some produce more but smaller gametes.

Initially, there would be mostly gametes of intermediate size, motility, and vitality, but Parker et al. (1972) showed in simulations based on game theory models that this scenario sets the stage for disruptive selection for the two extreme morphologies. The population was predicted to rapidly evolve to be anisogamous—to have two types of individuals that each produce one extreme form of gamete: egg or sperm (Figure 5.3).

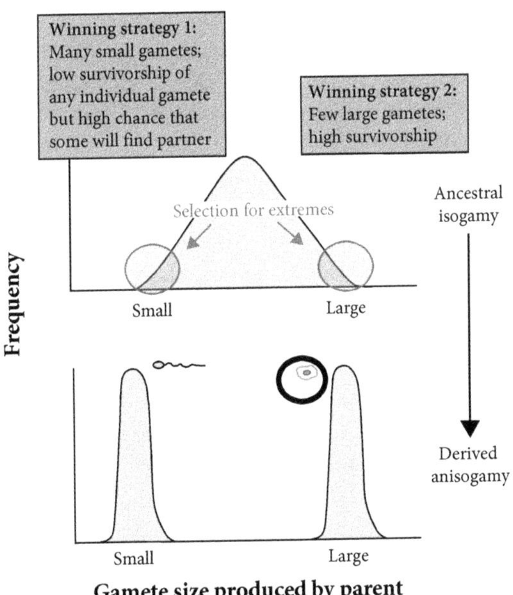

Figure 5.3 A pictorial summary of the game theory model originally proposed by Parker et al. (1972) to explain the evolution of anisogamy.

The reason that extreme morphologies win out is that small gametes gain the huge benefit of being mobile. They are more likely to find a mating partner, but they pay the cost of low survival. Large gametes, in contrast, gain the huge benefit of high survival, but they pay a cost of low motility and low chance of encountering another immobile mating partner. The small gametes ensure that large gametes encounter a sexual partner and the large gametes ensure that sufficient resources are available for the zygote to survive (Parker et al., 1972).

The model created by Parker et al. (1972) is so compelling that it is easy to look past what it does not explain. One glaring omission in the model is the transmission of cytoplasmic elements such as mitochondria. Perhaps it goes without saying that mitochondria are transmitted by the bigger gamete because they have more room to haul around mitochondria, but in real-world anisogamous eukaryotes, sperm also carry mitochondria. Even more fundamentally, the Parker et al. (1972) model does not explain the evolution of mating types. The endpoint of the model is simply two gamete types: small and mobile and big and immobile (Charlesworth and Charlesworth, 2010). In the Parker et al. (1972) model, big gametes can fuse with big gametes and small can fuse with small. As a matter of fact, Parker et al. (1972) considered the likelihood that big gametes would benefit by fusing with other big gametes and eliminate the benefits of small gametes, but an isogamous population of large gametes is subject to invasion by small, mobile gametes and, in these models, anisogamy inevitably evolves.

Thus, at the endpoint of the game theory model put forth by Parker et al. (1972), we are still left with one mating type with two morphological strategies. In the eukaryotic world in which we live, however, there are always mating types, and for nearly all eukaryotes there are exactly two mating types (Billiard et al., 2011). The preponderance of eukaryotes with two mating types presents an evolutionary puzzle potentially as great as the paradox of sex (Hurst and Hamilton, 1992; Lane, 2005).

The paradox of two mating types is easiest to understand if we consider the limitations that are imposed by such a system. When there are two mating types, half of the individuals in the population are unsuitable sexual partners. This is not a trivial consideration. For essentially all organisms, there are real costs involved in searching for a suitable sexual partner (Parker, 1978), and so it is easy to imagine the benefits to be had if a third mating type evolved in a population with two mating types. Assuming that all mating types except one's own mating type are suitable sexual partners, a third mating type would be expected to spread in the population until three mating types existed at equal frequency. With the addition of a third mating type, two out of every three randomly encountered individuals would be a suitable mate, odds that are considerably better than the one-out-of-two success rate when there are two mating types. Addition of a fourth mating type would, in turn, further reduce the cost of mate choice—with three out of four rather than two out of three suitable. Based on this simple assessment, we might expect the evolution of many mating types (Lane, 2005). The other potential winning strategy would be no mating types, whereby every randomly encountered individual would be a suitable mate. Two mating types is the worst system with regard to the cost of mate searching (Box 5.3). It is therefore remarkable that, with few exceptions, eukaryotes have two mating types (Billiard et al., 2011).

Box 5.3 A bird with four sexes

In this chapter, I've reiterated the point that the system of two mating types is costly in terms of the complications of searching for a mate. Following the language of Lane (2005), I dubbed two sexes the worst strategy from the perspective of mate searching. It turns out, however, that two mating types is not actually the worst strategy. That title goes to the system of mating types employed by white-throated sparrows (*Zonotrichia albicollis*), a small bird with a big voice that nests in the cold northern forests of the Canadian Shield. Like all birds, white-throated sparrows with two Z chromosomes are male, and those with one Z and one W chromosome are female. But to this standard two-sex system, white-throated sparrows add two morphs that differ in feather coloration and behavior—a dark morph and a light morph—which are the result of a chromosomal inversion that created a 100 mega-base supergene that acts like a second sex chromosome (Tuttle et al., 2016). Dark-feathered sparrows have one form of the supergene and they mate only with light-feathered sparrows. Light sparrows have an alternate form of the supergene and mate only with dark sparrows (Tuttle, 2003). Strong behavioral aversions to same morph maintain this mating pattern (Houtman and Falls, 1994). Even when rare mistakes are made, the fitness of the offspring of dark/dark or light/light pairs is low (Knapton and Falls, 1983). So, for any given sparrow, only one-fourth of other birds in the population are suitable mates. This four-mating-type system of the white-throated sparrow appears to be unique in the bird world and across all eukaryotes, and how it resists invasion by a normal, two-sex system is poorly understood. It should truly be the worst of all sex determination strategies.

One unexplored aspect of this white-throated sparrow system is mitochondrial divergence between the morphs. The system of two mating types of white-throated sparrow creates two subpopulations of females. If any N-mt genes that co-function with mt genes are part of the supergene that determines morph, then one would expect substantial divergence in mt genotype between the two color morphs. This should be a straightforward prediction to test (Figure Box 5.3).

Figure Box 5.3 Light morph (left) and dark morph (right) white-throated sparrows. Photos by Geoffrey E. Hill.

Anisogamy, mating types, and mitochondrial inheritance

The connection between uniparental inheritance of mitochondria and two mating types is fundamental (Figure 5.4). If there are no mating types such that any individual in a population is a suitable sexual partner, then there is no mechanism to ensure that, during sexual reproduction, one individual will transfer a mt genome and one will not. Uniparental inheritance of mitochondria becomes a possibility only if there are two mating types and transmission of mitochondria is restricted to just one mating type. This argument also comes back to the evolution of isogamy/anisogamy: uniparental inheritance of mitochondria is most commonly achieved in eukaryotes when one mating type (female/egg) receives N DNA from the other mating type (male/sperm) and the DNA-receiving gamete does not permit the transfer of mitochondria. It is not a coincidence that nearly without exception across eukaryotes, the DNA-receiving sex is also the sex that exclusively transmits mt DNA to the next generation. Anisogamy is not essential for uniparental inheritance of mitochondria—there are diverse mechanisms for ensuring uniparental inheritance of mt DNA (Sato and Sato, 2013)—but anisogamy provides a ready mechanism for transmission of mt DNA by only one parent. Moreover, it is certainly possible to have two mating types and to still allow for bi-parental transmission of mitochondria—all that is required is weak gatekeeping by the egg—and indeed such bi-parental transmission in systems with two mating types is not rare in eukaryotes (Birky, 2001). Most eukaryotes, however, have two mating types and uniparental inheritance of mitochondria that is enforced by the sex with larger gametes.

Genomic conflict within an individual

For decades, explanations for the evolution of two mating types in eukaryotes focused on genomic conflict among cytoplasmic elements, including particularly mitochondria (Cosmides and Tooby, 1981; Hoekstra, 1987; Hurst and Hamilton, 1992; Hutson and Law, 1993). These hypotheses are all founded on the importance of avoiding heteroplasmy, which is having multiple mt DNA types within a single individual. Heteroplasmy is proposed to be a problem because, with multiple mt genotypes replicating within a single eukaryotic individual, there can be selection for traits that promote a mt genotype at the expense of the N genome and the eukaryotic organism as a whole. While there is strong theoretical support for the problems that can be caused by heteroplasmy, only a few empirical studies have directly confirmed a loss of organism fitness due to competition among mitochondria (e.g. Sharpley et al., 2012).

The basic argument for the potential dangers of heteroplasmy can be explained most easily with a simple verbal model. Begin with an ancestral population of a simple eukaryote with no mating types and bi-parental transmission of mitochondria, chloroplasts, and other cytoplasmic elements. During sexual reproduction, the nuclear material from the two parents would combine to form one N genotype in the offspring, but mt genomes would be transmitted as multiple copies with no fusion or recombination. Thus, divergent mt genotypes would exist within a single organism.

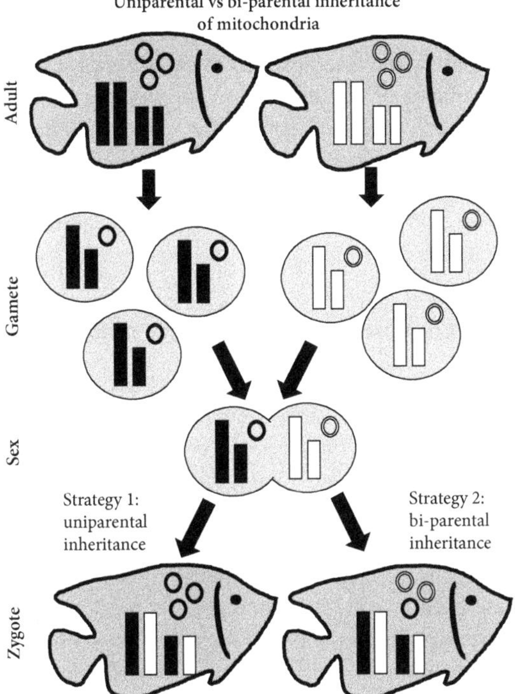

Figure 5.4 Simple illustration of bi-parental versus uniparental inheritance of mt genomes. Bars represent N chromosomes and circles represent mt DNA. Bi-parental inheritance of mt leads to heteroplasmy.

In such a situation, genomic conflict is theoretically inevitable (Cosmides and Tooby, 1981; Hurst and Hamilton, 1992). Any variant mt genotype that promoted its own replication at the expense of the replication of any other mt genotype would rapidly go to fixation. Such self-serving strategies would be good for that particular mitochondrial lineage, but they would likely be bad for the N genome and for the organism as a whole. This concept can be recast in mathematical models, and the conclusions are the same: bi-parental inheritance of mitochondria inevitably leads to conflict among mitochondria within an individual (Hurst and Hamilton, 1992; Christie et al., 2015).

How does a lineage solve the problem of conflict among mt genomes within an individual? The most basic answer is to allow only one parent to transmit mitochondria; in other words, to have uniparental inheritance of mitochondria (Figure 5.4) (Hurst and Hamilton, 1992). The mitochondria within an individual will inevitably be more genetically similar than mitochondria between individuals. Limiting transmission of mitochondria to one parent goes a long way toward eliminating conflict among mitochondria within an individual. The most familiar manifestation of uniparental transmission is to allow only females to transmit mitochondria (Hutson and

Law, 1993). There can still be genetic variation among mitochondria even within the germ line of a single individual (a topic that will be taken up in Chapter 6), but restricting between-generation transmission of mitochondria to a single mating type vastly reduces the problem of conflict within the population of mitochondria in a zygote. Thus, a widely accepted explanation for why a system of two mating types predominates in eukaryotes is that this provides a mechanism to enable uniparental inheritance of mitochondria and to avoid genomic conflict among mitochondria (Cosmides and Tooby, 1981; Hoekstra, 1987; Hurst and Hamilton, 1992; Hutson and Law, 1993).

Selection against heteroplasmy and selection for mitonuclear coadaptation

Because the mt genome is subject to Muller's ratchet and mutational erosion, theory predicts that mt DNA will be prone to the accumulation of deleterious changes to coding sequences. Paradoxically, however, the coding sequences of mt genes tend to be more conserved than comparable coding sequences of N genes (Popadin et al., 2013) (see Chapter 3). Two factors are proposed to explain how a genome predicted to be eroded by deleterious mutations is observed to undergo little functional change over time. The first is a mitochondrial genetic bottleneck via a sequestered germ line each generation. I'll come back to mitochondrial genetic bottlenecks when I discuss the evolution of senescence in Chapter 6. Here I focus on the second factor that is hypothesized to counteract mutational erosion of mt DNA: the evolution of uniparental inheritance of mt genomes.

As I reviewed in Chapter 3, a prominent theory for how the ETS remains functional in the face of mutational erosion of mt DNA is compensatory coevolution: N genes are proposed to evolve functional changes that compensate for deleterious mutations in mt genes. Even the most ardent supporter of the compensatory coevolution hypothesis, however, acknowledges that compensatory coevolution with a N genome can at best be an emergency bailout of the mt genome once deleterious alleles become fixed. The first line of defense against mutation erosion must be strong and perpetual natural selection against deleterious mitochondrial mutation each generation. A serious limitation for natural selection acting on mt genotypes, however, is that unlike a N genome that always exists as one genotype, a single individual can carry multiple and divergent mt genotypes (Chinnery et al., 2000).

The optimal situation for efficient purging of deleterious mitochondrial alleles is to have little variation in the mt genotypes within an individual and substantial variation in mt genotypes between individuals. In this way, selection against poorly performing individuals will effectively remove mitochondrial alleles that bestow low fitness. Conversely, if there is heteroplasmy and thus multiple mt genotypes within each individual, then variation in mt genotypes within an individual can be as great as variation in mt genotypes among individuals and natural selection on any one genotype is less effective (Figure 5.5). As a consequence, when there is heteroplasmy, the purging of deleterious alleles becomes less effective and Muller's ratchet ensues

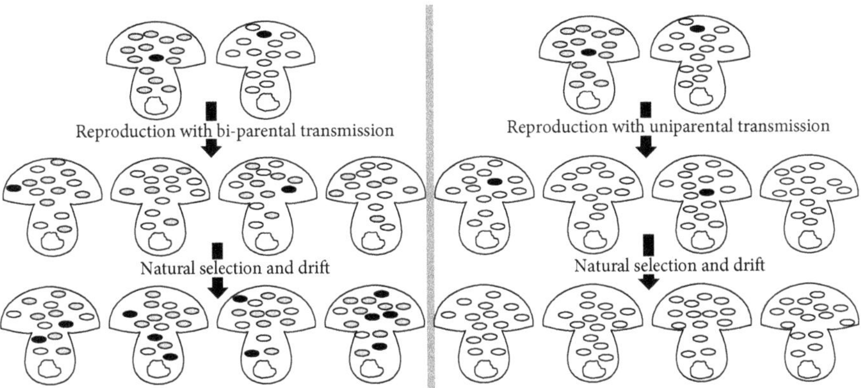

Figure 5.5 A simplified illustration of how heteroplasmy can shield deleterious mutations from selection. Depicted are eukaryotic organisms with no mating types. Ovals represent mitochondria with different shades showing different genotypes. White and gray shaded mitochondria produce equal fitness but through different phenotypes and black carries a deleterious allele that slightly lowers fitness. With uniparental inheritance, all offspring have a common background mitochondrial genotype and the only variation is whether or not the deleterious allele is present. Selection efficiently eliminates individuals with the deleterious allele. With bi-parental inheritance, offspring have two functional but different parental mitochondrial genotypes plus the deleterious allele. Selection is less efficient at removing the deleterious allele in the diverse population of mitochondria and the deleterious mutation increases in frequency.

(Bergstrom and Pritchard, 1998; Roze et al., 2005; Christie and Beekman, 2016). Christie and colleagues (Christie et al., 2015; Christie and Beekman, 2016) used simulation models to study the importance of uniparental versus bi-parental inheritance for natural selection on mt genotypes. They found that even when the mutation rate was relatively low and with or without mating types, the benefits of homoplasy led to the evolution of uniparental inheritance of mitochondria.

The simulations by Christie and colleagues focused on the purging of deleterious mt genes. But there is potentially more to selection on mt genotypes than simply eliminating bad and promoting good variants. For mitochondrial variants, good versus bad must always be considered in light of the N genes to which they are paired. Each generation, new N genotypes are produced through sexual reproduction, and these new N genotypes are matched to mt genotypes that are transmitted to offspring either from the mother, from the father, or from both. It is essential that combinations of mt and N genes with poor compatibility are revealed to selection so they can be eliminated. Without perpetual selection for mitochondria that are properly coadapted to N genes, such that they enable full respiratory function, then there will be a loss of mitonuclear coadaptation and a decline in the respiratory efficiency and fitness of a lineage (Lane, 2011c). It is in consideration of the need to purge deleterious mitochondrial alleles and to maintain mitonuclear compatibility that Lane (2005) and

Hadjivasiliou et al. (2012) proposed a new hypothesis for the evolution of two mating types and uniparental transmission of mitochondria. According to this new idea, uniparental inheritance of mitochondria via two mating types evolved specifically for the maintenance of mitonuclear coadaptation that can be eroded when there is heteroplasmy.

As with hypotheses focused exclusively on mutational erosion of mitochondria, this coadaptation hypothesis is centered on the effectiveness of selection when there is too much variation in mt genotypes within individuals. When mitonuclear function arises from diverse mt genotypes within an individual, selection does a poor job of either promoting good genotypes or eliminating bad genotypes. This is because natural selection is necessarily limited to assessment of average mitochondrial function, and bad mt genotypes can be carried along by good mt genotypes when there is heteroplasmy (Figure 5.5). As a result, when there is heteroplasmy the coevolution of co-functioning mt and N-mt genes falters and coadaptation declines (Lane, 2005). In contrast, when a single mt genotype is matched to a single N genotype within cells, selection is much more effective at eliminating the poorly functioning, poorly coadapted combinations and at promoting highly functioning, fully coadapted combinations. These authors proposed that the necessity for tight mitonuclear coadaptation might be an even more important selective force in the evolution of mating types and uniparental inheritance of mitochondria than genomic conflict among mitochondria with different genotypes (Hadjivasiliou et al., 2012).

Is this advantage of better mitonuclear coadaptation enough to drive the evolution of two mating types and uniparental inheritance of mitochondria? This question is trickier than it seems on first consideration. Mitochondrial mutation rate is a big factor. Mutation rate determines how rapidly variant mt genotypes will be generated and hence how valuable it is, each generation, to test the compatibility of the N genotype and single mt genotypes. Hadjivasiliou et al. (2012, 2013) used simulation models to study the conditions under which two mating types and uniparental inheritance evolve. They found that, under diverse conditions, uniparental inheritance of mitochondria enhances mitonuclear coadaptation and leads to higher population fitness over time. However, the model really only worked after uniparental inheritance of mitochondria was already established. Moreover, when the model started with a population with bi-parental inheritance of mitochondria, uniparental inheritance would only spread through a minority of the population no matter how high they made the mutation rate. The costs of mate searching were simply too high. The results of these models could explain why the degree of heteroplasmy is so variable in many groups of eukaryotes (Hadjivasiliou et al., 2013).

Thus, it seems that avoidance of Muller's ratchet is the primary benefit of uniparental inheritance of mitochondria (Christie et al., 2015; Christie and Beekman, 2016), but that promotion of mitonuclear coadaptation can play an important role (Hadjivasiliou et al., 2012, 2013). This is an area of active research and more empirical studies to validate the modeling outcomes are needed.

Conflict versus coadaptation

A theme that runs through the chapters of this book is the dichotomy from which evolutionary biologists view the interactions of mt and N genes. Many evolutionary biologists focus on the conflicts that can arise when independently replicating genomes exist within a single individual. Genes that are beneficial to one genome might proliferate even if they detrimental to other genomes and to the organism as a whole (Burt and Trivers, 2008). As discussed above, resolving genomic conflict is the leading explanation for the evolution of two mating types and uniparental transmission of mitochondria. But even as there is a potential for genomic conflict, there remains the fundamental necessity for cooperation among mt genomes and between the mt and N genomes. The products of mt genes and N-mt genes must be able to function together with high efficiency; otherwise, cellular respiration is compromised and there is loss of fitness. Models suggest that the benefits for maintaining mitonuclear coadaptation can indeed be a major force in the evolution of two sexes. But conflict and cooperation are not alternative explanations; they are complementary explanations and both are perpetually in play during eukaryotic evolution. In the end, the necessity for genomic cooperation should perpetually reel in a tendency toward genomic conflict (Connallon et al., 2018).

Summary

For many decades, evolutionary biologists pondered the evolution of sexual reproduction from the perspective of extant eukaryotes with highly integrated systems. But sexual reproduction seems to have evolved in the very early stages of eukaryotic evolution not only before genomic systems were fully integrated but when the state of the organism was probably closer to genomic chaos. The fusion of two formerly independent prokaryotic organisms to form a proto-eukaryote with two genomes necessitated rapid and dramatic restructuring of both genomes. This process of genomic restructuring appears to have coincided with the evolution of sexual reproduction with recombination, and it seems likely that these events were causally linked. The bombardment of the N genome with genes and gene fragments from the mt genome drove selection for mechanisms that enabled cutting and splicing DNA sequences as well as for recombining sets of genes to better enable selection against deleterious changes to the genome. According to this hypothesis, the origin of a eukaryotic cell with mitochondria necessitated the evolution of sexual reproduction with recombination, a nuclear membrane, and the spliceosome. The prokaryotic genome that was evolving into the mt genome also faced challenges even as it became streamlined. Because the symbiont genome existed as multiple copies, competition among symbiont genomes within an individual posed a grave risk to the emerging eukaryote. In addition, because the mt genome remained a single-copy genome that was asexually transmitted, Muller's ratchet leading to mutational meltdown posed an ever-present risk to the eukaryote. And, as a backdrop to all of these interactions, there was a perpetual need to maintain coadaptation between the N and mt genomes. Two mating types promoted the transmission of a single mt genotype to offspring enabling stronger selection for mitonuclear coadaptation. According to these hypotheses, both conflict and cooperation played key roles in the evolution of two mating types.

6

Life eternal in the face of senescence

Why do we get old and die? Since humans became sapient, we have been fixated on death. In the age of modern medicine, the literal search for the Fountain of Youth has given way to a quest for a biochemical catholicon to aging, pursued by both mainstream scientists and profiteers working at the shadowy edge of academia. The result is a seemingly endless parade of miracle panaceas for aging, all vaguely linked to aging research but each no more effective than an elixir purchased from the back of a tent at a nineteenth-century medicine show. Inexorably, death comes to us all. And despite our better understanding of the physiological basis for death, human aging continues to follow approximately the same timetable that it did in Ancient Egypt or Victorian England (Box 6.1).

A related and equally challenging question is: how can a germ line be immortal? Consider a 20-month-old mouse that suffers from osteoporosis, cardiovascular disease, and diabetes, 4 weeks from death due to failing organs. This aged mouse gives birth to its last litter, and miraculously, the pups are born as disease-free, vital, and pure as the mother was when she was born. The newborn mouse pups have no hint of the maladies of age that are debilitating and killing their mother, but they are destined to follow her path, to decline with age, and to die. Eukaryotic organisms move through time with disposable somas protecting immortal germ lines (Figure 6.1).

The evolution of aging has been a focus of research by evolutionary biologists for decades. The maintenance of a germ line that is shielded from the effects of aging is a more recent topic of discussion among evolutionary biologists, with even more recent realization that the processes that shield the germ line from aging are also the processes that hold mutational erosion of mt genomes somewhat in check (Liu and Demple, 2010). New theories founded on the need for mitonuclear coadaptation and respiratory function are reshaping the way biologists think about aging in individuals, the evolution of germ lines, and the critical need for intense natural selection on the germ line. Many of these topics are discussed most commonly in the biomedical literature, but evolutionary ecologists should pay attention as well. The screening of mt genomes at the germ line may have played a pivotal role in enabling the evolution of complex life.

Mitonuclear Ecology. Geoffrey E. Hill, Oxford University Press (2019).
DOI: 10.1093/oso/9780198818250.001.0001

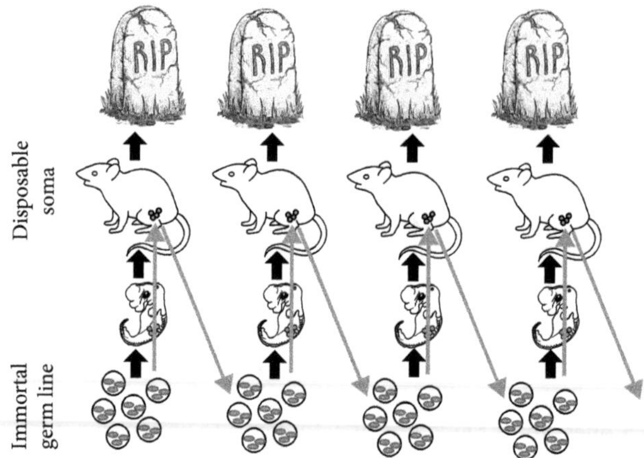

Figure 6.1 In bilaterian animals, the soma is a disposable vehicle for perpetuation of the germ line, which is potentially immortal. Changes to mt or N genotypes within somatic cells affect organism function, but not the evolution of a lineage. Evolution occurs via changes to the nucleotide sequence of germ line cells.

Box 6.1 The search for the Fountain of Youth gene

The key to eternal youth was just discovered in the DNA of a nematode. No, hold on. A set of genes in mice might not just keep you young, these forever-young genes might actually reverse aging. Wait a minute. The real key to eternal youth was just discovered in the N DNA of an Amish community in Indiana.

Even a clumsy Google search turns up endless claims for aging breakthroughs from clinical researchers with confident statements to the effect that cheap therapies for much prolonged life are only a few years away. It is hard not to be drawn in and become hopeful until one notices that many of these proclamations of an imminent end to aging were made 8, 10, even 15 years ago. I haven't notice any Fountain of Youth pills at my nearby CVS Pharmacy.

There is a fundamental disconnect between aging research over the past 50 years that is focused on decline in mitochondrial function as the basis for senescence and the search for a Fountain of Youth gene in clinical medicine. I think that the tendency for many biomedical researchers to ignore the literature on mitochondrial function and aging and play up the latest single-gene cure is not fundamentally different than the tendency of politicians to ignore the vast scientific literature on climate change and pursue an energy policy that allows voters to continue to drive big cars. The mitochondrial theory of aging and the greenhouse gas theory of climate change are both "inconvenient truths" for anyone dreaming of quick and cheap fixes to these complex problems.

Why is an aging theory based on antagonistic pleiotropy and mitochondrial decline bad news for aging therapy? The mitochondrial theory of aging proposes that aging is not an error in the genetic code that can simply be patched over, like a point mutation that causes a digestive disorder. Rather, aging is an evolved response that optimizes a balance between early onset benefits and late-onset problems that sets the time of existence of our disposable

Box 6.1 *Continued*

somas. After millions of years of evolutionary tweaking, a multicellular individual like a human being is defined by its optimal trajectory of senescence. Because aging is a fundamental property of humanity, the only way that humans could stop aging would be essentially to stop being human. No worries though—I don't think we will be challenged with the ethical dilemma of an eternal youth pill any time soon.

mt DNA mutation

What underlies mutations in the mt genome?

All leading theories to explain both aging and the evolution of a germ line focus on decline in mitochondrial function associated with an accumulation of mutations in the mt genome. Thus, before we go further in a discussion of aging or germ line evolution, it is critical to consider the factors that underlie mutational changes to mt genotypes. Age-dependent accumulation of mutations in mt DNA can be explained by two mechanisms: unrepaired damage and replication error. The conventional view is that mt DNA mutation is primarily a product of unrepaired damage caused by exposure to free radicals that are produced during OXPHOS (Harman, 1956; Barja, 1998; Lane, 2011c). Because mt DNA is located near the source of free radicals in mitochondria (Figure 6.2), there has always been an intuitive appeal to the argument that the high mutation rate of mt DNA is a function of oxidative damage. There is growing

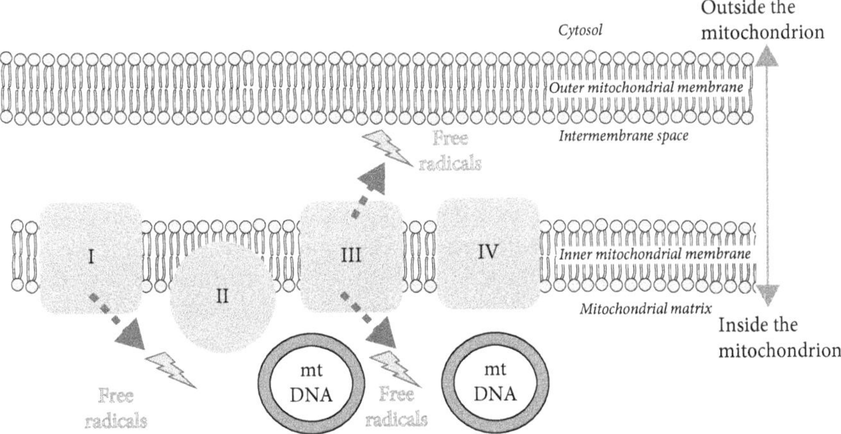

Figure 6.2 Simple sketch of a few key components of a mitochondrion. Mitochondrial DNA, depicted as a circular chromosome, is located inside mitochondria in the mitochondrial matrix. This location makes it potentially vulnerable to damage from free radicals produced during OXPHOS. Numbered shapes represent ETS complexes.

empirical evidence, however, that free radical damage may be only a minor player in the accumulation of mt DNA mutations (Lagouge and Larsson, 2013; Itsara et al., 2014).

If mutation is not driven by free radical damage, then what is the basis for the high rate of mt DNA mutation observed in many eukaryotic lineages including animals? There is growing consensus that replication error (also called copy error) is the source of most mt DNA mutation (Larsson, 2010; Melvin and Ballard, 2017; Szczepanowska and Trifunovic, 2017). Replication error is exactly what it sounds like: as mt DNA is copied to make more mt DNA, mistakes are made. The accumulation of errors in a DNA sequence as it is repeatedly copied is analogous to the accumulation of errors in the texts of books when they were copied by scribes before the invention of the printing press. The original book was not always available as a template; rather, a copy was used as the template for the next copy, and then the newest copy as the template for the next. Human scribes invariably made mistakes as they copied letters, words, and sections of the text, and any errors in a copy were retained in all subsequent copies that used the slightly erroneous copy as a template. Just as with hand-copied texts, the rate of copy errors in the replication of mt DNA is partly a function of the quality of the machinery (or scribe) that reads the original, partly a function of how carefully the new copy is edited for errors, and partly a function of the rate at which new copies are made.

Itsara et al. (2014) published an analysis of mt DNA mutation in fruit flies that provided compelling evidence that free radical damage was not a major source of mt DNA mutation. First, they found that in both young and old animals, there were few G:C to T:A transversions in mt DNA; the most common mutation type was a transition from G:C pairs to A:T pairs. The reason that this observation is significant is that the damage caused by free radicals to a DNA strand inevitably causes a conversion of a G:C base pair to a T:A base pair (De Bont and van Larebeke, 2004). Such changes could be thought of as signature of free radical-induced mutation, and a strong signature of mutations caused by free radicals simply wasn't present in the fruit fly mt DNA. In addition, Itsara et al. (2014) found that the G:C to T:A transversions that were present did not increase with age. A rapid increase in mutations from oxidative damage as individuals get older is a key prediction of the hypothesis that mt DNA mutations are primarily caused by free radicals, which I will come back to in a discussion of the free radical theory of aging. And finally, Itsara et al. (2014) observed a "strand bias" whereby the strand of the mt DNA that was copied the most showed the highest rate of mutation. Again, this observation is not consistent with the hypothesis that most mt DNA mutation is caused by free radical damage because both strands of a DNA would be subject to the same effects of free radicals. Overall, Itsara et al. (2014) made a compelling case that DNA damage from free radicals was not the major cause of mutations in mt DNA, at least in fruit flies. Their analysis strongly implicated copy error as the source of mt DNA mutation.

A critical but rarely noted facet of replication error is that correct or incorrect copying of DNA results from processes that depend on mitonuclear coadaptation (Ricchetti, 2018). First, the replication of the mt genome involves the functional interaction of the N-encoded protein mtDNA polymerase γ and the promotor region of the mt DNA (Szczepanowska and Trifunovic, 2017). Error in the replication of mt DNA might emerge, in part, due to mitonuclear incompatibilities (Ellison and Burton, 2010; Cline, 2012). In addition, the main repair mechanism in mammalian mitochondria is base excision repair, which relies on DNA repair enzymes that are encoded in the nucleus (Boesch et al., 2011). It is proposed that the efficiency of mt DNA repair is a key factor in the accumulation of mutations in mt DNA (Boesch et al., 2011), and again, this process involves the functional interaction of a mt gene sequence and a N-encoded protein. And finally, all else being equal, the higher the respiratory rate of the organism, the more copies of mt DNA that are needed and the higher the rate of copy errors. This last argument becomes central to new ideas about the evolution of germ lines.

The evolution of germ lines

Strong selection on germ cells before proliferation and atresia

A reoccurring theme in science fiction novels is the asexual cloning of people—the transformation of a somatic cell, like an epithelial cell from a cheek swab, into a zygote and then a person. In the era of Dolly the Sheep and mitochondrial replacement therapy, the idea of recasting a somatic cell as a zygote actually seems quite plausible, especially if a stem cell rather than an epithelial cell is used (Magnúsdóttir and Surani, 2014). But regardless of what type of somatic cell is chosen, the big problem with procreation via somatic cells lies in bypassing the germ line. A growing understanding of the role of germ lines in controlling mitochondrial evolution provides new insights into why bypassing the germ line in reproduction is likely to be a dicey proposition for any eukaryote that moves around under control of a nervous system.

A germ line is a lineage of cells that remains eternally undifferentiated even as their descendent cells specialize to form elephants, mosquitos, and velvet worms. The germ line in my body is the current end of an unbroken lineage of germ cells that have existed, with substantial genetic change but with no differentiation, for 500 million years. Germ cells are sex cells in the sense that they are the cells that undergo meiosis and engage in sexual reproduction. All differentiated cells—somatic cells—are strictly asexual in bilaterian animals. Germ cells are immortal while all somatic cells are subject to decline with age.

In contrast to the ancient origins of sexual reproduction and recombination of N genes, germ lines are obviously not an ancestral trait in eukaryotes. Eukaryotes were exclusively single-celled organisms through their early history, and it was only after the rise of multicellular eukaryotes that one group of cells could be designated as a germ line. Even then, most multicellular eukaryotes did not evolve a germ line. Among modern eukaryotes, only bilaterian animals and ctenophores typically have sequestered germ lines that play no role in somatic development and that transmit both the N and mt genomes between generations (Extavour, 2007). Plants and "lower" animals lack a clear division of somatic and germ line cells; in these eukaryotes, cells that become gametes also participate in somatic development (Seipel et al., 2004). Given the potential advantages of using any cell in the body as a reproductive cell, a fundamental question is why did germ cells evolve? Following the recurring theme in this book, new theories and observations suggest that the answer may lie in the need for mitonuclear coadaptation and the ever-present threat of mutational erosion of the mt genome.

Through the great majority of time, female germ line cells are held in a state of metabolic quiescence with no OXPHOS activity (Figure 6.3). This quiescent state begins with the formation of primary oocytes in the embryonic body of the future mother. It continues through the life of the female. When a primary oocyte is fertilized to form a zygote, it continues through many cell divisions well into embryonic development in the next generation as the numerous copies of mitochondria that are transmitted to the zygote via the primary oocyte are dispersed among multiplying cells (Stewart and Larsson, 2014).

Abruptly, about 2 weeks following conception, mitochondria in both somatic cell lines and in the germ line of the embryo have their metabolic engines powered to life. In somatic lines, mitochondria will be metabolically active for the life of the individual. In the germ line of females, the metabolic awakening is brief but intense as a new generation of primary oocytes is created. Then, as abruptly as they were brought to life, mitochondria in the germ line are powered down again, and remain quiescent through the remainder of the life of the female (Stewart and Larsson, 2014). Through this process, mitochondria in the germ line of animals ride through evolutionary time with brief bursts of metabolic activity and long periods of functional quiescence. In humans with a generation time of 20 years, the germ line cells are metabolically activity only about one-tenth of 1 percent of the life of an individual (Figure 6.3).

The mitochondria in the male germ line, in contrast, are not excluded from aerobic activity; the male germ line cells engage in OXPHOS through the life of the individual. Thus, female but not male germ line cells are shielded from mt DNA mutations

Figure 6.3 The relative proportion of the life of a human female in which the mitochondria in an individual's germ line cells are metabolically active. Humans have overlapping generations so the mitochondria in the germ cells of the offspring of focal woman would also be active at an early embryonic stage.

via cessation of metabolic activity and replication of mt DNA. It is not surprising, therefore, that the mitochondria of egg and sperm are phenotypically distinct (De Paula et al., 2013). In mammals, female gametes generated from germ line cells have mitochondria that are small and structurally simple, while the mitochondria of sperm from the same species are as large and structurally complex as the sperm of any somatic cell. Holding primary oocytes in a state of respiratory quiescence between generations seems to be a strategy to reduce accumulation of mutations in the germ line. However, shutting down OXPHOS during these periods is not sufficient to hold mt DNA mutations in check. It is essential that germ line cells also be subjected to strong purifying selection each generation.

To understand the evolutionary forces that shape the rate at which mutations accumulate in the mt DNA in a germ line, it is important to consider the points that I emphasized in Chapter 4: theoretical models suggest that because mt genomes are transmitted asexually across generations, they should be subject to accumulation of deleterious alleles (Lynch et al., 2006; Neiman and Taylor, 2009; Greiner et al., 2015). This is the process of Muller's ratchet (see Chapter 4; Figure 4.2). Paradoxically, in many animal taxa, the rate of non-synonymous changes to the mt genotype is lower than the rate of non-synonymous change in the N genotype (Popadin et al., 2013; Zhang and Broughton, 2013; Cooper et al., 2015), and the rate of accumulation of deleterious mutations appears lower than predicted. The explanation for this low rate of functional changes to mt genes is that Muller's ratchet is countered by strong purifying selection. With no recombination, however, purifying selection has to be at the level of the entire mt genome (Radzvilavicius et al., 2017); there is no mechanism for individually parsing good mt genes from bad mt genes because all mt genes act as a single linkage group (Chapter 4). In light of these considerations, Radzvilavicius et al. (2016) proposed that germ lines evolved specifically as a means to subject mt genomes to intense purifying selection and to thereby limit the accumulation of mutations in mt DNA. This hypothesis builds from growing evidence that a key component of the maintenance of mitonuclear compatibility and mitochondrial function is strong selection on mt DNA in the germ line (Krakauer and Mira, 1999; Fan et al., 2008; Stewart et al., 2008; Dowling, 2014; Radzvilavicius et al., 2016).

In mammals, where most of the research on germ lines has been conducted, primary oocytes each contain hundreds of thousands of mitochondria (Wai et al., 2008) (Figure 6.4). Indeed, the mammalian primary oocyte has the greatest number of mitochondria recorded in any eukaryotic cell (Wai et al., 2010). Having so many copies of mitochondria transmitted to an offspring is necessary because mitochondrial biogenesis is suspended during early embryonic development (Mishra and Chan, 2014), and the huge number of mitochondria in the oocyte ensures that there are enough mitochondria to distribute among rapidly proliferating cell lines as a zygote begins to develop into an adult (Ebert et al., 1988; Shoubridge and Wai, 2007). (Why it is critical to distribute identical copies of mitochondria to cell lines in early development is a topic I'll come back to in a few paragraphs.) Studies in frogs (el Meziane et al., 1989), mice (Ebert et al., 1988), and pigs (El Shourbagy et al., 2006) have produced evidence that replication of mt DNA does not occur until embryogenesis has proceeded to a stage with dozens or hundreds of cells. The partitioning of mitochondria carried

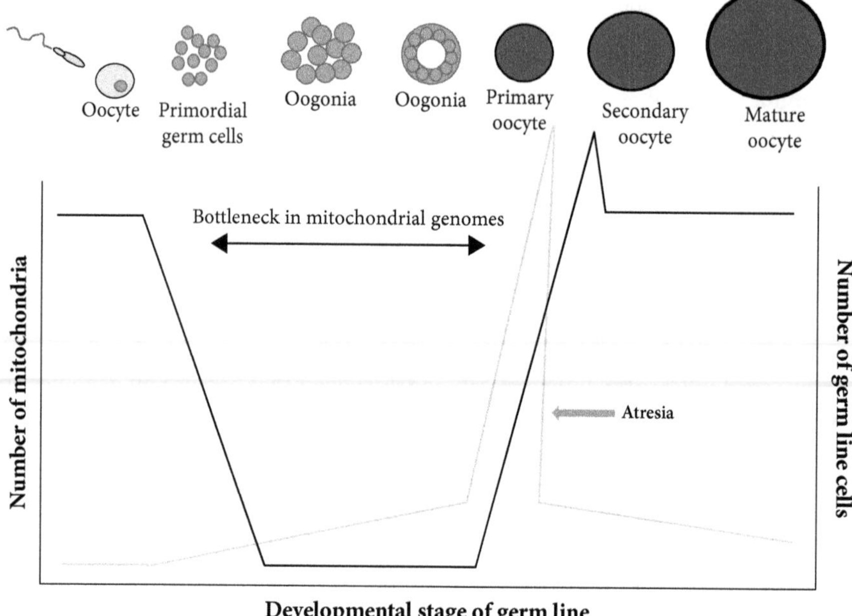

Figure 6.4 Changes in the number of germ line cells (the line with the sharp peak) and the number of mt genomes (the line that starts and ends high) during embryonic development of a mammal. A bottleneck in the number of mitochondria per germ line cell occurs as mitochondria from the zygote are dispersed without mitochondrial replication. Mitochondrial copy number is then massively increased when mitochondrial biosynthesis is re-initiated. Atresia begins with the culling of primary oocytes and continues through the life of the individual.

by the primary oocyte among the many cells produced during early embryonic development leaves only a few mitochondria per cell. By the end of this process there may be as few as ten mitochondria in each of the primordial germ cells of a human (Shoubridge and Wai, 2007). The exact number of mitochondria per cell at this stage is still debated, but in mammals it is somewhere between 10 and 100 mitochondria, meaning that only around 0.01 percent of the mitochondria in the primary oocyte actually contribute to the mt DNA gene pool in offspring of the next generation (Shoubridge and Wai, 2007). Clearly there is potential for intense natural selection as mt genotypes are squeezed through this numerical bottleneck, but not all authors agree that it is at this the point in the germ line life cycle where selection on mt DNA is most likely to occur (Wai et al., 2008).

At the end of this process of partitioning maternal mitochondria, a bilaterian embryo is a mass of cells each with a set of N chromosomes contributed by a male and female parent as well as a mt genome contributed by the mother. The 200,000 or so mitochondria that exist at the start of development depend on mitonuclear compatibility *in the mother*, not in the offspring (because the mitochondria to this point were all synthesized in the mother's body), but as soon as mitochondrial biogenesis begins, new mitochondria are produced with the products of the *embryo's* mt and N-mt genes. Discussions of selection on mt genotype is typically focused on processes that occur in the germ line following the bottleneck in which the number of mitochondria per cell drops from about 200,000 to about 10 in a primordial germ cell. But a critical arena of selection also exists during very early development of the soma, which relies on performance of the new mitonuclear gene combinations that operate in the development of each new cell line (Lane, 2005). With few mitochondria per cell, the growth and development of the new individual is critically dependent on highly functional mitochondria—emerging from compatible and functional mt and N-mt gene combinations—to meet the metabolic needs of the rapidly dividing cells as well as to generate new mitochondria to populate the many new cells being created (Nakada et al., 2006). Poor performance by the mitochondria at this stage will lead to developmental problems and such poorly performing embryos are likely to be terminated (Van Blerkom, 2011).

Developmental problems at early embryonic stages are well documented in humans. Embryonic arrest during pre-implantation development is a major source of infertility in women and, in many cases, the cause of such embryo failure is insufficient mt DNA copy number (McFarland et al., 2007). As described in detail in Chapter 2, replication of mt DNA is dependent on N-encoded proteins co-functioning with mt DNA. In a study of pigs, Cagnone et al. (2016) experimentally demonstrated that problems with mt DNA copy number could be reversed by replacing poorly functioning mitochondria with fully functional mitochondria. Thus, the entire developing embryo, germ line and all, can be culled by selection if mitochondrial function is poor, and such culling may be a major selective sieve for highly functional combinations of mt and N-mt genes.

Strong selection on germ cells after proliferation and atresia

There remains a critical need for selection on the mt genotype in the germ line after the individual carrying the germ line has proceeded with an acceptable developmental trajectory. Indeed, providing the opportunity for such selection on mt genotype is proposed to be the *raison d'être* for the germ line (Radzvilavicius et al., 2016) and why the germ line enables each generation of bilaterian eukaryotes to start life fresh and unhindered by the somatic mutational load that beset mature individuals in the previous generation. This selection on the germ line involves massive replication of mt genomes that remain in the germ line following the bottleneck in early embryonic development and then massive culling of the huge population of mt genomes that is produced (Figure 6.4).

Why is such massive selection needed if mitochondria within an individual are identical clones of each other? If mothers and offspring share mt genotype, where is the variation on which selection can act? For readers focused on mt genome evolution in the face of perpetual mutational erosion, this might seem like a rhetorical question. But consider that until very recently, the most widely accepted explanation for the evolution of a germ line was the avoidance of conflict among cells within the body (Michod, 1996; Michod and Roze, 2001). Elimination of mt DNA mutations in the germ line was typically not even mentioned as an alternative hypothesis (Radzvilavicius et al., 2016). It turns out that small variations in the mt nucleotide sequence are ubiquitous, even among the cells within an individual (Payne et al., 2013). Inevitably, the primary oocytes that exist in early development draw their mt DNA from a heteroplasmic pool in the maternal germ line (Stewart and Larsson, 2014). Consequently, there are often mt DNA variants among the different primary oocytes, and these mitochondrial variants have nowhere to hide—they are revealed to selection through the process of proliferation (Cree et al., 2008).

In the above paragraph, I emphasized the strong selection to which the mt genome is subjected during proliferation, but mitochondrial function during the proliferation of germ line cells is a product not only of the mt genotype; it arises from the combination of mt and N-mt genotype. Selection at this stage is for *mitonuclear* function and hence mitonuclear compatibility and coadaptation. The test to which each potential germ cell is subjected is presumably the capacity to replicate (Lane, 2005), although, startlingly, this remains entirely unknown. From the founding population of 100 or so primordial germ cells, divisions produce millions of oocytes each with about 10,000 mt DNA molecules per cell; by the end of this proliferation of oocytes, approximately 10^{14} copies of the mt genome exist. Then the culling begins. In mice, the number of oocytes is cut from about 7 million to 2 million, in the first important stage in the process of atresia (Stewart et al., 2008). Atresia is sometimes narrowly defined as the degradation of follicles that do not ovulate, but I follow the broader definition that it is the culling of primary oocytes from the height of abundance following proliferation to the final culling at ovulation (Figure 6.4).

Only a tiny fraction of the oocytes that persist following this first stage of atresia will eventually ovulate, and this culling of germ line cells provides the ideal arena in which oocytes can be screened based on their metabolic integrity (Cree et al., 2008; Wai et al., 2008). Lane (2005) and others have hypothesized that only those oocytes with the greatest capacity for efficient respiratory function, requiring compatible mitonuclear genotypes and mt genotypes purged of deleterious alleles, survive the process of proliferation and then atresia (Dumollard et al., 2007; Dowling, 2014; Stewart and Larsson, 2014). Strong selection on the integrity of mt genotype in the germ line is central to eukaryotic evolution (Morrow et al., 2015). Intense selection on germ line cells is a key mechanism preventing inter-generational accumulation of the deleterious mt DNA mutations predicted by theory (Stewart et al., 2008; Cooper et al., 2015). It is also likely the explanation for why life springs from zygotes perpetually fresh and new. The germ line of bilaterian animals carries the template for mt genotype and this template is checked and double checked for accuracy each generation.

Why plants and most other eukaryotes don't have a germ line

Natural selection is not very efficient when it is forced to assess linked sets of genes. Selection on linked genes creates Hill–Robertson effects and is a recipe for accumulation of slightly deleterious alleles. Natural selection works even less efficiently when it is forced to assess units carrying multiple genomes, such as cells that carry multiple mt genomes. With multiple genomes within the cells under selection, the high function of good genotypes can obscure the poor function of poor genotypes leading to the accumulation of poor genotypes (Larsson, 2010). In the cell lines within a multicellular organism, each cell carries many mt genomes, so selection to maintain the integrity of mt genomes in these lines is ineffective. As a consequence, mt DNA in somatic cell lines inevitably accumulate mutations, and these mutations cause a loss of optimal function. You can see the truth of this statement in the slow gait and aged face of any senior member of society. With the decay of function in somatic cell lines, multicellular individuals are inherently mortal.

Because of the inescapable accumulation of deleterious mutations in somatic cell lines, the lineages of multicellular organisms absolutely must, periodically, return to a single cell with one N genome and a small number of mt genomes. The return to a single cell with one or a very few copies of genomes is the only mechanism by which natural selection can sort good genotypes from bad. It may appear that a few bilaterian animals like the naked mole rat (*Heterocephalus glaber*) have immortal somas (Box 6.2), but I'm quite confident that they do not. The disposable soma is more than a catchphrase. It is an inescapable necessity for multicellular life.

Box 6.2 Naked mole rat: The living sausage that never grows old

There may be no stranger animal on Earth than the naked mole rat (*Heterocephalus glaber*), a burrow-dwelling mammal that looks remarkably like a sausage with legs. To begin with, the naked mole rat is the only known vertebrate that is eusocial, like an ant or a termite; only one individual in a group of mole rats reproduces and the rest of the group work to assist (Bennett and Faulkes, 2000). But for sheer oddity, the social behavior of naked mole rat takes a backseat to the claim that mole rats are immortal (Ruby et al., 2018). That claim was made recently by gerontologist Rochelle Buffenstein and colleagues based on 30 years of observations of laboratory colonies of naked mole rats. For all other animals studied in detail, the risk of death increases steadily with age, a characteristic of life called Gompertz law. All bilaterian animals eventually show signs of senescence, as they well should, given the arguments that I make in this chapter. All animals grow old and die, that is, with the possible exception of the naked mole rat. Based on the 30 years of observation, mole rats actually show slightly decreasing risk of death as they grow older, and they are reported to never die of old age (Ruby et al., 2018). This claim of immortality for mole rats got a lot of press coverage and was widely stated as a well-documented fact, but it would be an understatement to say that I'm skeptical of the claim that naked mole rats are immortal. They may live to be hundreds of years old, making it hard for a human to discern old age in these animals, but naked mole rats have somatic cell lines that accumulate mutations and you can rest assured that their somas are as disposable as those of any other bilaterian animal.

Naked mole rats are also significant because their very existence is a falsification of the free radical theory of aging. Because they are so long-lived and show such minimal indications of senescence across decades of life, if the free radical theory of aging is correct, then

(*Continued*)

Box 6.2 *Continued*

Figure Box 6.2 Naked mole rats show a very slow rate of aging but paradoxically they show high levels of oxidative damage of tissues. These features of mole rats are hard to reconcile with the free radical theory of aging. Photo by Roman Klementschitz.

naked mole rats would have to have very, very low levels of oxidative damage to cells relative to short-lived rodents. Paradoxically, however, naked mole rats produce large quantities of free radicals and they show high levels of oxidative damage to cells (Andziak and Buffenstein, 2006; Andziak et al., 2006). The key to their long lives in the face of high free radical production is that free radicals are not responsible for most mt DNA mutations (the free radical theory of aging is wrong) and mole rats have very efficient mechanisms for repair of mt DNA copy error. Even with good repair mechanisms, however, Muller's ratchet will click away and bad genes will hide among good genes and the mortality of naked mole rats is certain to catch up to them.

Multicellular somas must periodically be abandoned for single-cell reproductive units, but this theory does not necessitate a sequestered germ line. Why can't individual somatic cells simply be subjected to strong selection? In plants and other multicellular eukaryotes except bilaterian animals and ctenophores, somatic cells are used as reproductive cells, and natural selection on these reproductive cells is adequate to counterbalance deleterious mutations. The pertinent question, therefore, is why did a germ line evolve in bilaterian animals? Radzvilavicius et al. (2016) proposed that two unique features of bilaterian animals drove the need for a germ line: a high mt DNA mutation rate resulting from a high rate of mt DNA replication error and a need for many mitochondria in the germ cells of bilaterian animals.

Let's start with the mutation rate of mt DNA in different eukaryotes. The rate at which mt DNA mutates is actually difficult to measure, but there is general agreement that the mt genomes of plants and "simple" animals like sponges mutate at a much lower rate than the mt DNA of bilaterian animals. Radzvilavicius et al. (2016) proposed that the high mutation rates, and the germ lines, of bilaterian animals evolved during the Cambrian explosion half a billion years ago—an explosion of biodiversity in which all animal phyla burst into existence. The Cambrian explosion led to, among other big evolutionary changes, a transition from sessile animals to active predators and fleeing prey. An active and mobile body plan necessitated many specialized cell types and the need for substantial energy production to support the movement and the synthesis of the complex organism. Radzvilavicius et al. (2016) proposed that this fundamental redesign of animals created the two key distinctions between eukaryotes with a germ line and those without: the degree of cellular specialization and rate of mitochondrial replication error. With a need for more energy to produce motor neural systems and muscular activity, there would have been a significant increase in the replication of mt genomes (Clayton, 1982; Lane, 2005; Kovalchuk, 2016). More replication opens the door for more replication errors. And a higher mt DNA mutation rate increased the chances of heteroplasmy in the mitochondria carried by germ cells.

Why does complexity create a greater chance for heteroplasmy? The amount of variation in mt genotypes within a population will be an outcome of mutation–selection balance (Charlesworth, 2008). Mutations perpetually bring variant genotypes into the population and because almost all such changes are negative, selection perpetually takes them out. Complexity increases the rate of replication errors and hence the rate of mutation in mt genomes. By this line of reasoning, there would have been a corresponding need for stronger selection on germ cells. In addition, the gametes of complex eukaryotes had to carry more copies of mt DNA. Radzvilavicius et al. (2016) proposed that as they evolved more complex phenotypes, animals were faced with the dual problem of the need for many mitochondria per germ cell and increased risk of heteroplasmy among the mitochondria resulting from a higher replication error (Figure 6.5).

Another challenge for highly complex eukaryotes with many specialized cell types is that there would have been a critical need to transmit many mitochondria in gametes. If only one or a few mitochondria are transmitted in germ cells, then the mt DNA that gives rise to the mt DNA in each of the specialized cell lines in the complex animal is necessarily not an original maternal mt DNA but a copy of a copy that can be dozens of replications away from the maternal type. If there is a high mt mutation rate, then the cells that give rise to each of the specialized cell lines might be slightly different, carrying unique mutations, and that is trouble for organism integrity. With a high mt DNA mutation rate, it is critically important that all of the cell lines in the body descend from a common progenitor mt DNA. Uniformity in founder mitochondrial type is only possible if there is no mitochondrial replication—and hence no

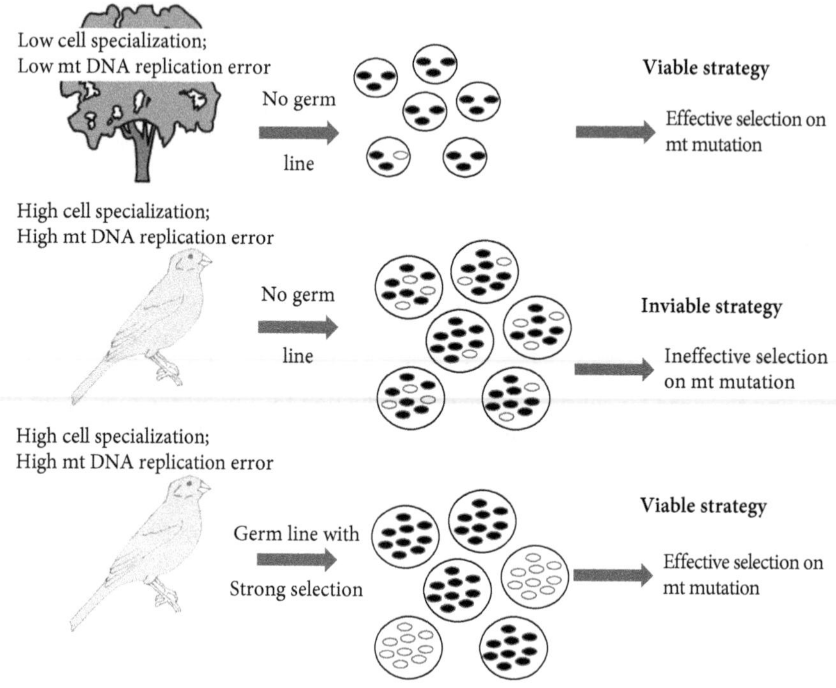

Figure 6.5 An illustration of the hypothesis for why plants (symbolized by a tree) do not need a germ line and why bilaterian animals (symbolized by a bird) do. Bilaterian animals have a high rate of DNA replication errors introducing more deleterious mt DNA mutations each generation, and the extreme cell specialization in bilaterian animals necessitates many mitochondria per gamete. This combination necessitated a sequestered germ line that could be subjected to intense purifying selection.

opportunity for replication error—until major cell lines are established. Thus, the gamete transmitting mitochondria needs to carry a sufficient number of mitochondria to establish all of the cell lines of the new individual.

How is it hypothesized that cellular complexity and a higher rate of mt DNA mutation drive the evolution of a germ line? It all goes back to the basic principles of the effectiveness of selection on asexual lineages with no recombination that I laid out in Chapter 4. If there are multiple mitochondria in each individual in a population of cells under selection (whether it is a population of germ cells within a multicellular organism or individual single-celled eukaryotes), then Muller's ratchet and Hill–Robertson effects will come into play because selection cannot effectively assess any specific mt genotype. Selection will necessarily be on the collective performance of all of the mt genotypes within each cell, and bad genotypes will hide among good genotypes. Selection only becomes efficient enough to counteract Hill–Robertson effects and oppose Muller's ratchet if each cell in the population carries only one mitochondrial type. The answer to the dilemma was to create a unique line of cells that could be subject to intense selection before hundreds of thousands of copies of the mt genome are put in a gamete.

Why are plants and "simple" animals not subject to the same problems? How do they succeed without a germ line? First, plants and lower animals do not share the high respiratory needs of bilaterian animals. They have fewer mitochondria, produce fewer ATP, and do not copy mt DNA as rapidly as animals. Replication error, and the mt DNA mutation rate, is thus lower. Moreover, they lack the degree of cell specialization of bilaterian animals. That means that gametes do not need to carry as many copies of the mt DNA, because they are establishing fewer specialized cell lines. With less heteroplasmy and fewer mitochondria per gametic cell, natural selection can be efficient in eliminating mt DNA mutations in germ cells derived from somatic cell lines (Figure 6.5).

Radzvilavicius et al. (2016) modeled production of gametes from a germ line versus from somatic cell lines, and they found that in plants and other eukaryotes with lower mt genome replication errors, the low rates of replication error reduced the rate of mutation of mt DNA and reduced the risk of heteroplasmy in gametes generated from somatic cells. Moreover, with fewer specialized cell lines, gametes needed to transmit fewer mitochondria, further reducing the risk of heteroplasmy. With low rates of heteroplasmy, selection could be effective on germ cells derived from somatic cell lines. As replication errors increased with organismal complexity, however, selection for mitochondrial competence within somatic lines was no longer sufficient to limit the accumulation of deleterious mt alleles and hence to maintain mitochondrial function. The solution to these problems that accompanied greater complexity and cell line specialization was a designated germ line where intense natural selection could be imposed on mt genomes and mitonuclear combinations.

Selection on the male germ line

In bilaterian animals, sperm rarely contribute mitochondria to offspring. Cells in the male germ line of many animals undergo many more replications than cells in the maternal germ line and hence mutation resulting from replication error is expected in both the mt and N genomes in the male germ line (Wilson Sayres and Makova, 2011). On the other hand, sperm contribute half of the N genes for the offspring, and thus selection on the male germ line can also play an important role in maintaining mitonuclear coadaptation. James Crow famously mused "that the greatest mutational health hazard in the human population at present is fertile old males" (Crow, 1997). There is at least circumstantial evidence for intense screening of sperm for genes that retain the capacity to produce energy, which will include N-mt genes that are free of mutations and compatible with mt genes.

At least half of the cases of human infertility are caused by male dysfunction and the most common type of male infertility is poor sperm motility (Thangaraj et al., 2003). A huge biomedical literature has linked sperm dysfunction to mutations in mt DNA (Rajender et al., 2010). And interestingly, sperm dysfunction can occur due to mitochondrial effects that are male specific (so-called mother's curse effects;

Vaught and Dowling, 2018). Sperm are already evolutionary dead-ends for mt DNA, so selection against deleterious mt DNA in sperm is likely of little consequence to accumulations of mt DNA mutations in a population. However, sperm dysfunction is also linked to dysfunction of the electron transport system (ETS), which is a product of both mt genes and the N-mt genes, the latter being passed to offspring (Ruiz-Pesini et al., 1998; Moraes and Meyers, 2018). So long as the male that produces the sperm carries the same mt genotype as the female that contributes the mt genome (see Chapter 7), then proper mitonuclear compatibility in sperm could translate to proper mitonuclear compatibility in the offspring. The focus in the biomedical literature on sperm genotypes has been on the dichotomy between individuals who are fertile or infertile. But among the 200 million sperm in a human ejaculate, it is not hard to imagine that there might be N-mt genotypes that bestow better or worse respiratory function. The respiratory challenge to which sperm are subjected—swimming speed, ability to cope with the chemical environment of the egg, and capacity to penetrate membranes to complete fertilization (Simmons, 2001; Snook, 2005; Pizzari, 2009) could be an important but overlooked source of selection for mitonuclear compatibility.

Selection across developmental stages

In introducing germ line selection, I noted that, at the inception of mitochondrial biogenesis in early embryonic development, poorly functioning individuals along with their potentially defective mt genes would be culled by natural selection. Such selection for mitochondrial function and hence for compatible and functional mt and N-mt genotypes is not, however, restricted to embryonic development. It continues throughout the life of the organism.

In mammals, spontaneous abortion can occur at any developmental stage up to stillbirth. For instance in humans, 30–50 percent of all conceptions and 15–20 percent of clinical pregnancies end in spontaneous abortions (Gupta et al., 2007). In birds, about 15 percent of eggs that are laid fail to hatch (Koenig, 1982). At present, mutations in mt genomes have not been implicated as major factors in these developmental failures, but deleterious mutations in mt DNA have yet to be the focus in studies of spontaneous abortions or unhatched eggs. After birth or hatching, mitochondrial diseases can trigger dysfunction and death or loss of reproductive potential.

The importance of considering all life stages when attempting to assess the fitness effect of different combinations of mt and N-mt genes has been clearly demonstrated in a lab study of mice. Latorre-Pellicer et al. (2016) created mice with the N genotype of one lab strain and the mt genes of another. They then followed the mice throughout their lives "using transcriptomic, proteomic, metabolomic, biochemical, physiological and phenotyping studies" to document the effects of different combinations of mt and N-mt genes. They found no differences in activity, growth, or fertility among the treatment groups, but they found clear effects of mt DNA genotype on production of free radicals, protein stability, obesity, and aging (Latorre-Pellicer et al., 2016). The individual effects combined to have a significant effect on longevity. All of these mitonuclear fitness effects would have been missed in most conventional studies of small mammals, which would have measured only growth and fertility in young adults.

Even among fully functional adult individuals, only a portion of individuals in many species will procure a mate, and mate choice is plausibly linked to mitochondrial function (see Chapter 8). Among those individuals that succeed in attracting a mate, only a fraction will produce viable offspring. There is the potential for selection to purge mt genomes carrying deleterious alleles at each of these life stages and the cumulative effect is a very tight selective sieve through which mt genotypes must pass.

Evolution of senescence

Antagonistic pleiotropy

I can clearly remember the first time when, as an undergraduate, I heard a professor suggest that senescence can evolve. Actually, the message took a few hours to penetrate because I didn't know what the word "senescence" meant when I heard the lecture. After the lecture when I looked up the term, I was at first certain that I must have written it down incorrectly. Senescence is deterioration with age. How could old age evolve? Didn't we just wear out like a refrigerator or an automobile? But then I thought about some obvious challenges to the wear-out-like-an-old-refrigerator hypothesis that I had always simply accepted as part of the way the world works: my friend's dog died of heart failure at age 14 with gray hair, cataracts, poor bowel control, and just about all of the symptoms of old age in people. I was older than the dog at the time and I hadn't even reached puberty. Different animals clearly grew old at different rates, but still, it was a shocking revelation to me that the pace of aging could evolve.

The theme of the lecture that introduced me to the topic of senescence was antagonistic pleiotropy. This is an idea proposed by the great evolutionary theorist George Williams (who I met unexpectedly at the toaster in Bob Montgomerie's lab in Kingston Ontario when I was a postdoc). The concept of antagonistic pleiotropy builds from the simple argument published a decade earlier by Medawar (1946) that risk of death makes the strength of natural selection against deleterious traits stronger on young versus older individuals within a population (Figure 6.6). For instance, consider a mouse in a field that is hunted by cats, hawks, and owls such that its life is at risk of death every hour of every day. Even if the mouse remained healthy and robust, the chances of it living to be 5 years old might be astronomically low because it will inevitably be caught and eaten. Thus, a mutation that had a negative effect on mouse survival at age 3 months would be subject to much stronger natural selection than a mutation that caused the same effect when the mouse was 5 years of age (Figure 6.6). Medawar pointed out that if there is such a decline in the strength of natural selection with age, then mutations that are neutral early in life but deleterious late in life could accumulate in a population. Natural selection is blinded to the late-onset mutations because so few individuals live long enough for them to have a meaningful effect

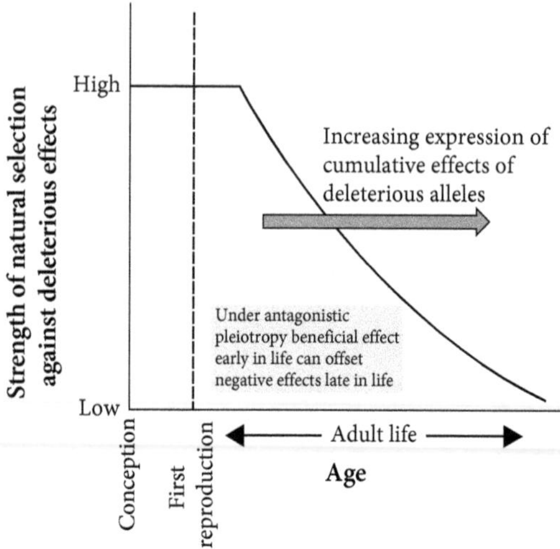

Figure 6.6 The relationship between the age of an individual and the strength of natural selection to oppose deleterious genetic effects. Alleles that decrease function and hence contribute to senescence are weakly opposed by selection following peak reproduction. Such alleles will spread rapidly if they are beneficial early in life.

(Figure 6.6). Medawar's hypothesis relied on drift to increase the frequency of alleles that were shielded from natural selection (Medawar, 1946).

Williams' (1957) antagonistic pleiotropy hypothesis extended the basic ideas of Medawar so that the process could be driven by natural selection. Pleiotropy describes multiple effects arising from a single allele. Antagonistic pleiotropy results when a single allele causes antagonistic effects—both beneficial and harmful. Williams hypothesized that a genetic change that both enhanced immediate reproductive output (the good part of the pleiotropic interactions) and caused a negative effect later in life (the bad part of the pleiotropy) could spread by natural selection because the weak negative selection on the harmful effects of the allele in the old individual would be offset by strong positive selection for the beneficial effects early in life (Figure 6.6). Williams proposed that, through this process, deleterious genes that cause harmful effects at ages beyond the average reproductive life would pile up. From a human perspective, this means that traits that diminish function, such as cardiovascular disease or breast cancer, are far more likely to manifest well after the peak reproductive years of life. By the fifth or sixth decade of human life, an age class into which few individuals would have ventured through most of the evolutionary history of *Homo sapiens*, expression of deleterious alleles manifest in all individuals and function begins to conspicuously decline.

Antagonistic pleiotropy remains the foundation of all modern theories of aging. As I will discuss, current theories of aging focus primarily on mechanisms that cause declines in cellular function that underlie senescence, and some scientists studying aging might suggest that these mechanistic theories of aging supplant William's theories of antagonistic pleiotropy. In fact, hypotheses that describe the biochemical basis for aging provide the mechanisms on which natural selection can act to achieve the optimal balance between improved performance early in life versus deleterious effects late in life (Galtier et al., 2009a). Antagonistic pleiotropy remains the cornerstone theory for the evolution of senescence (Austad and Hoffman, 2018).

The mitochondrial theory of aging

The mitochondrial theory of aging proposes that senescence is a consequence of a decline in the function of mitochondria (Cui et al., 2012; Long et al., 2014). It is an umbrella hypothesis that encompasses different proposed mechanisms that might underlie a decline in mitochondria, and it is often invoked because there is broad consensus among aging researchers that mitochondrial dysfunction is the root cause of aging (Bratic and Larsson, 2013; Sun et al., 2016; Theurey and Pizzo, 2018). The primary debate in this robust field of study concerns the specific mechanisms that cause mitochondrial function to decline in the lifetime of an individual. Thus, much of the research on the evolution of senescence is focused on the evolution of mt DNA mutation rate and hence the maintenance of mitonuclear coadaptation across generations.

Through the latter half of the twentieth century, the free radical theory of aging was the most widely stated mechanistic hypothesis for the decline of mitochondria (I'll describe the free radical theory of aging in detail in the next section). The dominance of the free radical hypothesis in the aging literature has caused the mitochondrial theory of aging and the free radical theory of aging to frequently be conflated. The key point is that free radical damage is not the only mechanism for increasing mitochondrial dysfunction through the lifetime of an individual. As I outlined in the opening section of this chapter, there is growing evidence that mt DNA replication error plays an even larger role in mitochondrial decline than free radical damage to mt DNA. So, as I present overviews of the free radical theory of aging and the replication error theory of aging, keep in mind that these two hypotheses provide different mechanisms for the mitochondrial theory of aging.

Another issue that tends to get confused in discussions of the mitochondrial theory of aging is where and when antagonistic pleiotropy applies. I'll reiterate that antagonistic pleiotropy remains the foundational theory that underlies all of the mechanistic hypotheses that are currently being debated. Galtier et al. (2009a) make this point in a discussion of the evolutionary forces that might shape the mechanism underlying mitochondrial dysfunction. They point out that selection for longer lifespan would lead to selection for lower somatic mutation rate. They propose the name "longevity hypothesis" for selection on mt DNA mutation rate and hence on the rate of mitochondrial decline, but the hypothesis that rates of DNA editing or DNA resistance to mutation evolve according to the risk of death of the individual is simply a restatement of antagonistic pleiotropy with more specific suggestions for the targets of selection.

The free radical theory of aging

The free radical theory of aging is the oldest theory of senescence that is founded on the idea that aging is a consequence of molecular damage leading to cellular and organism dysfunction. Interestingly, Denham Harman proposed the free radical theory of aging in 1956, 1 year before Williams (1957) published his ideas about antagonistic pleiotropy in the context of senescence (Harman, 1956), although I doubt that either of these scientists was influenced by the work of the other. Even in a restatement of his hypothesis in 1994, Harman (1994) focused exclusively on biochemistry and does not mention the ideas of Williams or other evolutionary biologists who considered the age-specific fitness consequences of cellular decline. Williams also never cited Harman in any of his reconsiderations of the evolution of senescence (Williams, 1966). The gulf between reductionistic cell biology and biochemistry and evolutionary ecology was vast in the twentieth century.

Because free radicals are primarily produced by the ETS in mitochondria, the free radical theory of aging was the first hypothesis to invoke mitochondrial dysfunction in the aging process. This hypothesis begins with the basic observation that mt DNA is located in the mitochondrial matrix, in the same cellular neighborhood where the electron transport chain carries out OXPHOS (Figure 6.2). Even when it is running with maximum efficiency, OXPHOS produces free radicals, which react with and can damage molecular components of the cell including DNA. Thus, mt DNA is exposed to the mutagenic effects of the free radicals produced by OXPHOS. Mutations to mt DNA potentially can cause functional problems with subunits of the respiratory chain complexes, which will in turn cause functional problems with ETS complexes. Dysfunction in the machinery of aerobic respiration then leads to increased production of free radicals, which cause more mt DNA damage, more dysfunction, and so forth (Bandy and Davison, 1990) (Figure 6.7). The crux of the free radical theory of aging is that a vicious cycle involving mt DNA damage and declining mitochondrial function is the fundamental mechanism underlying aging and death (Harman, 1994) (Figure 6.7). The free radical theory of aging is very intuitive, and it remains the most widely stated hypothesis for the basis for decline and mitochondrial function and senescence.

The free radical theory of aging played a key role in focusing aging theory on mitochondria (Stuart et al., 2014), but it is unlikely to be a correct hypothesis for why eukaryotes decline with age (Galtier et al., 2009a; Lagouge and Larsson, 2013). A litany of empirical observations that are not supportive of the free radical theory of aging have been published in recent years. Stuart et al. (2014) recognized three basic predictions that emerge from the free radical theory of aging with regard to capacity of long-lived or short-lived species to deal with oxidative stress. First, long-lived species should produce fewer free radicals than short-lived species. Second, the capacity to neutralize free radicals should be greater in long-lived compared with short-lived organisms. And third, mechanisms for repairing and preventing free radical damage should be superior in long-lived versus short-lived species. Stuart et al. (2014) concluded that none of these predictions is consistently supported by data.

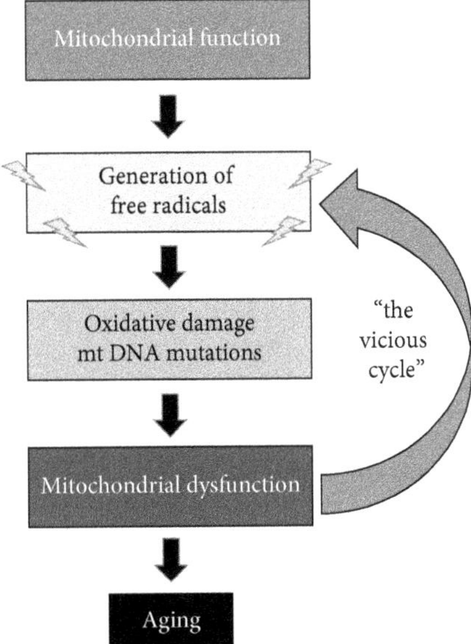

Figure 6.7 The proposed steps involved in the free radical theory of aging. This hypothesis proposes a positive feedback loop that leads to what is commonly referred to as "the vicious cycle" of declining mitochondrial function and increasing release of free radicals.

Observations of the naked mole rat revealed particularly egregious deviation from expectation (Box 6.2).

One of the most telling experiments to test the free radical theory of aging involved the creation of a mutator line of lab mice that carry a point mutation in the proofreading domain of DNA polymerase-γ, a key component of the mt DNA editing system. This mutation causes faster accumulation of mt DNA mutations (Trifunovic et al., 2004b). Such mutations should greatly accelerate the vicious cycle of more free radical damage and more mt DNA mutations and lead to very rapid aging. However, researchers failed to observe the predicted rise in free radicals or an exponential increase in the accumulation of mutational changes to mt DNA (Lagouge and Larsson, 2013). The basic concept of the free radical theory of aging is a positive feedback loop (Figure 6.7), in which each mutational step leads to greater respiratory chain dysfunction and the release of more free radicals, which has a positive effect on the rate of mutation and further dysfunction. Physiological systems that evolved to protect homeostasis are invariably constructed around negative feedback loops for the very reason that positive feedback gets out of control very fast. If the free radical theory of aging were correct, the mt DNA editor deficiency in the mutator mouse should have ignited a rapid system collapse as dysfunction from mt DNA mutations resulted in release of more free radicals, which in turn caused more mutations that caused increasingly greater release of free radicals. However, these mutator mice accumulated mt DNA mutations at a pace consistent with a constant background mutation rate and they showed no increase in free radical production (Kujoth et al., 2005). As inconsistent and contradictory data piled up, many authors delivered eulogies for the free radical theory of aging (e.g. Bonawitz and Shadel, 2007; Gems and Doonan, 2009; Pérez et al., 2009; Lapointe and Hekimi, 2010; Van Raamsdonk and Hekimi, 2012).

As I outlined at the beginning of the chapter, the most direct evidence that the free radical theory of aging is incorrect comes from analysis of the types of mt DNA mutations that are observed in mutator mouse as well as fruit flies and other animals. These observed nucleotide changes were not consistent with the sorts of mt DNA mutations predicted if they were the result of reactive oxygen species (ROS) damage. Instead, the types of mutations that are observed bear the signature of replication error.

The replication error theory of aging

The second major hypothesis under the umbrella of the mitochondrial theory of aging is the idea that the decline in mitochondrial function that underlies aging is a consequence of errors incurred in the process of replicating mt DNA (Larsson, 2010; Lagouge and Larsson, 2013). I will call this the replication error theory of aging to emphasize that it is distinct from the free radical theory of aging. The free radical theory of aging and the replication error theory of aging are similar in that they both focus on mt DNA mutation as the fundamental cause of senescence, but unlike the free radical theory of aging, the replication error theory of aging does not rely on a positive feedback loop. Changes to the mt DNA are proposed to accumulate at a constant rate that is set by the rate of replication error over a lifetime (Galtier et al., 2009a; Szczepanowska and Trifunovic, 2017) (Figure 6.8).

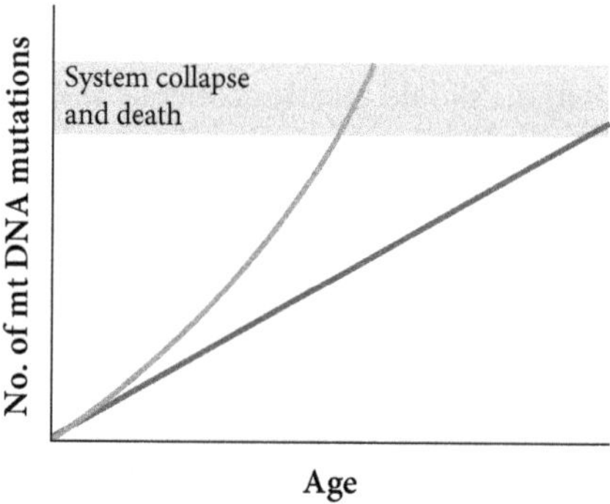

Figure 6.8 The predicted effects of genetic defects in a mt DNA editor protein in a mutator line of mice. Under the free radical theory of aging, an exponential increase in the accumulation of mt DNA mutations is predicted (top line). Under the replication error theory of aging a linear increase in accumulation of mt DNA is predicted (bottom line). The latter was observed (see text).

Strong support for the replication error theory of aging came from the same experiments with mutator lab mice described above (Thompson, 2006). Mice carrying an allele that increased the rate of replication error in mt DNA did not show a predicted accumulation of mt DNA mutations caused by free radicals or an exponential increase in free radical damage of mt DNA, but they did show an accelerated rate of aging (Trifunovic et al., 2004b). The rate of aging was accelerated but it remained constant throughout a lifetime (Figure 6.8). Moreover, the rate of accumulation of point mutations in the mt genome was approximately linear and not exponentially increasing (Trifunovic et al., 2004a; Kujoth et al., 2005).

If aging is a function of accumulation of mutations to the mt genome in somatic cell lines that erode mitochondrial function over time, then why isn't aging constant among all eukaryotes? The answer must be that lineages of eukaryotes differ in the rate at which changes to the mt DNA sequence occur or in the efficiency with which errors are corrected (Melvin and Ballard, 2017). Is there empirical support for a relationship between mt DNA mutation rate and longevity? For birds and mammals there is a strong association between per year rate of mt DNA mutation and rate of senescence, supporting the replication error theory of aging (Nabholz et al., 2008, 2009, 2016) (Figure 6.9). However, this pattern of longer life being associated with a lower rate of mt DNA mutation appears not to hold when tested more broadly across animals, including vertebrates and invertebrates (Allio et al., 2017). The latter observations have yet to be integrated into the aging literature. More tests of replication error theory of aging are clearly needed.

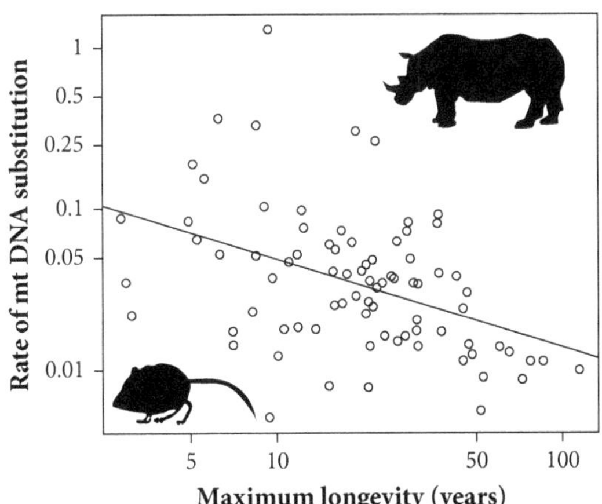

Figure 6.9 The per-year rate of mt DNA nucleotide substitution versus maximum longevity for mammals. Each point is a species of mammal. Mammals with a greater maximum lifespan tend to have a lower rate of mt DNA mutation, consistent with the mt DNA replication error theory of aging. Data from Nabholz et al. (2008).

Apoptotic threshold

In previous sections of this chapter, I built a case that a germ line enables strong selection on the cells that transmit mitochondria between generations both to eliminate mt DNA mutations and to ensure proper mitonuclear coadaptation. But how functional does a germ line cell have to be to avoid culling during oogenesis? According to a theoretical argument proffered by Lane, (2011b), the strength of selection on germ line cells—or, stated conversely, the degree of dysfunction permitted in germ line cells—is a function of the respiratory needs of the animal.

Animals that require high output of ATP—the vertebrates capable of powered flight are always put at the top of this list—have the least tolerance for imperfections in the respiratory chain (Figure 6.10). Terrestrial mammals, on the other hand, are expected to have greater tolerance for less-than-perfect function and to assent to weaker selection on germ line cells during oogenesis.

The core premise for Lane's argument is that there is a direct connection between proton leak, the conditions that signal cell death (apoptotic threshold), and longevity. Cell death is triggered by release of cytochrome *c* from mitochondria, which in turn is a response to increased proton leakage and reduced ATP production, both signals of poor mitochondrial function. But the level at which apoptosis is signaled is not

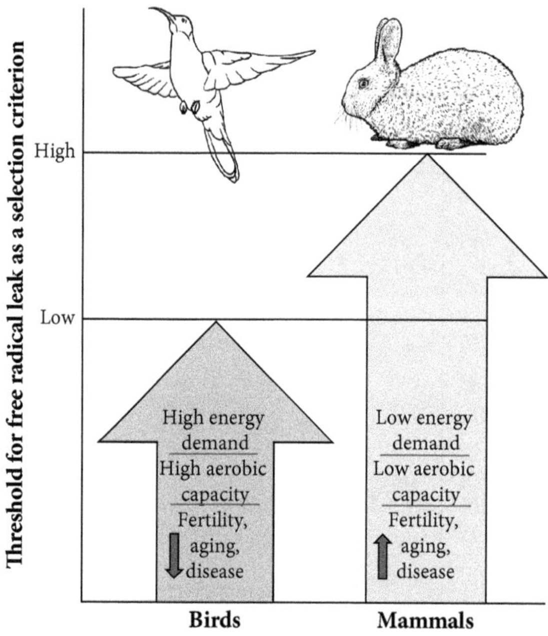

Figure 6.10 A hypothesized tradeoff between efficiency of OXPHOS, measured as free radical leakage, and the energy demands of the taxon. Because of their higher energy demands, birds are proposed to tolerate less leakage than mammals. Redrawn and modified based on Lane (2011b).

fixed; it can evolve in response to the life history of the animal. Lane argued that animals with less demand for high energy output from the mitochondria but with greater potential to benefit from the production of many offspring would trade mitochondrial quality for more gametes. On the other hand, animals that depended on highly functional aerobic respiration, such as birds that fly across oceans, would hold to strict quality control of mitochondria, even if that meant production of fewer offspring (Figure 6.10). These arguments essentially provide a mitochondrial and biochemical mechanism for the old r selection and K selection arguments from freshman ecology. Animals with high mortality rates that produce numerous offspring are called r selected. Lane is basically arguing that r-selected animals have a low apoptotic threshold. Animals with low mortality rates that produce few offspring are called K selected. K-selected animals are proposed to have a high apoptotic threshold.

These arguments dovetail nicely with the growing focus on mt DNA replication error and aging. The strength of selection on functional errors each generation will determine the rate of accumulation of replication errors in mitochondrial lines. Thus, the apoptotic threshold proposed by Lane is mechanistic restatement statement of William's antagonistic pleiotropy. Animals tradeoff a more rapid accumulation of replication errors via a higher apoptotic threshold for greater fecundity early in life. Various observations support a variable apoptotic threshold in different groups of animals with different expected lifespans, but the hypothesis has yet to be rigorously tested.

Summary

The union of an archaeon and a bacterium in the evolution of eukaryotes brought with it the potential for the production of essentially unlimited energy and the evolution of multicellular complexity (Lane and Martin, 2010), but multicellular complexity came at a cost of mortality (Jones et al., 2014). Somatic cell lines within individuals accumulated mutations, particularly mt DNA mutations, making it necessary to forever return to a single-cell starting point with each generation. Only when genes were limited to a few copies in a single cell could purifying selection keep pace with mutational erosion. As evolution pushed the boundaries of complexity to greater cell specialization and the evolution of a nervous system and movement, the need for mitochondrial energy became vast, leading to greater replication of mt genomes and a greater mt mutation rate. The extremely complex bilaterian body plan also called for many highly specialized cell types. With mutation increasing the genetic variability among germ cells derived from somatic cells and the need for transmission of many mitochondria between generations, selection on somatically derived gametes became inadequate to counter mutational erosion. The evolutionary answer to this problem was the sequestration of a germ line, which could be shielded from mt DNA mutation and subjected to intense purifying selection each generation. Thus, bilaterian animals exist as mortal individuals carrying immortal germ lines that are sequestered away to be used as an (almost) static reference in the creation of the next disposable soma.

We must grow old and die to cast off the mutations accumulated in a lifetime, but why must we die at 75 years of age? Why not 105 like a Galapagos tortoise or 205 like a bowhead whale? Even if we must be mortal, what sets the pace of life? Old theory proposes that benefits early in life trade off against deleterious effects late in life and that the risk of death in a population of bilaterian animals sets the rate of senescence. Before people constructed boats and devised commercial whaling, a bowhead whale had very low risk of death and individuals could expect to live for centuries. As Paleolithic hunter-gatherers, when our patterns of aging evolved, we usually made it through about three or four decades. Our pace of aging is set to a rhythm for about four decades of high function. The notion that a Fountain of Youth gene is just waiting to be discovered ignores the robust theory of aging that has been developed. We are inescapably a population of disposable somas.

7

Mitonuclear speciation

The two outcomes of evolution by natural selection are adaptation and speciation, and I contend that speciation is the more consequential. It was not happenstance that Darwin named his great treatise on natural selection *On the Origin of Species* and not *On the Origin of Adaptation*. Speciation is the engine that drives biodiversity. It enables diversification and adaptation by dividing lineages so that each population can better adapt to local environments. Gene flow homogenizes while speciation sets populations free to diverge and rise to local adaptive peaks.

Adaptation is trivial to understand; speciation is not. We can literally watch organisms evolve adaptations to changing environments. In their classic studies on the Galapagos Islands, Peter and Rosemary Grant and their students demonstrated how wet and dry cycles changed the availability of seeds of different sizes, which in turn led to natural selection for different beak sizes in populations of seed-eating Darwin's finches (Grant and Grant, 2008). Natural selection shaped populations to be better adapted to their environment (Grant, 2006). The basic elements of this process of adaptation evolving through natural selection can be easily taught and quickly grasped by students.

In Darwin's finches, as in all other life, understanding speciation has proven far more formidable (Mckay and Zink, 2015; Hill and Zink, 2018). The nature of species has been debated continuously since Darwin first proposed that all life arose from a common ancestor. Against the backdrop of this century of discussion of the fundamental nature of species, evolutionary biology is suddenly awash with new genomic information. We have ever more detailed data on the genomic structure of populations, including comparative data for sister taxa and across hybrid zones. Rather than clarifying species concepts and species boundaries, however, new genomic data present complex and confusing patterns that challenge current theories of speciation (Toews et al., 2016b). In particular, expanding genomic data sets often reveal discordance between the divergence and variation in mt and N genomes as well as between genes on sex chromosomes versus autosomes. It was discovered that populations long recognized as species regularly exchange genes, in contradiction to the leading hypothesis for defining a species. Moreover, in what began as a literature detached from speciation theory, a method for diagnosing species based on mt DNA sequences, known as DNA barcoding, is proving to be a key line of investigation for understanding the fundamental nature of species. And finally, there are new ideas that place the coadaptation of mt and N-mt genes at the center of the speciation process.

Mitonuclear Ecology. Geoffrey E. Hill, Oxford University Press (2019). © Geoffrey E. Hill 2019.
DOI: 10.1093/oso/9780198818250.001.0001

Traditional species concepts

Biologists do not observe the process of speciation; they observe the outcome (Nei and Nozawa, 2011). With the exception of studies involving speciation via the doubling of chromosomes (polyploidy), claims that the process of speciation can be observed directly in macro-organisms within the time scale of a researcher's career rest on questionable species concepts and do not add clarity to discussions of the nature of animal species (Hill and Zink, 2018; Lamichhaney et al., 2018). In practice, evolutionary biologists must deduce the processes that give rise to biodiversity from patterns in biogeography, phylogeny, morphology, and genetic structure of living organisms (Nei and Nozawa, 2011). Understanding speciation has proven to be one of the great intellectual challenges in all of the sciences.

Some of the earliest articulations of the process of speciation and how to define a species remain dogma in modern evolutionary biology. In the middle of the twentieth century, Harvard ornithologist and evolutionary biologist Ernst Mayr proposed the biological species concept, whereby speciation occurs when there is disruption of gene flow leading to accumulation of genetic differences between populations (Mayr, 1942, 1963) (Figure 7.1). This hypothesis defines a species as an interbreeding population that is reproductively isolated from all other populations (Mayr, 2000; Coyne and Orr, 2004). Under the biological species concept, the key element to speciation is reproductive isolation. Mayr (1942) proposed that, for populations to be recognized as distinct species, there should be essentially no gene flow between populations (although there can be instances of individuals hybridizing). By Mayr's definition, observation of more than rare hybridization between individuals from putative species is evidence against their status as legitimate species (Zink and McKitrick, 1995). This hypothesis was proposed before biologists had much information about the genetic structure of populations and before mt DNA was even discovered, so it invokes nothing about mitonuclear interactions or mt genotypes. Seventy-five years after it was proposed, however, the biological species concept remains the most widely stated hypothesis for speciation (Toews et al., 2016a; Bobay and Ochman, 2017).

A host of alternative species concepts have been proposed (De Queiroz, 2007; Zachos, 2016), but the hypothesis that presents the major conceptual alternative to the biological species concept is the phylogenetic species concept (Eldredge and Cracraft, 1980; Nelson and Platnick, 1981). This species concept focuses not on hybridization or the degree of reproductive isolation but rather on the inferred evolutionary history of the putative species (Wiens, 2004). If individuals in a population can be recognized (i.e. if they are diagnosable) and if they are inferred to have had an evolutionary history that is unique from the evolutionary histories of all other populations, then they are recognized as a species (Cracraft, 1983; Donoghue, 1985). According to this concept of species, current hybridization may affect the future uniqueness of a lineage and it might change future diagnosability, but current hybridization should not affect the species status of lineages if they can be demonstrated to have had a unique evolutionary history (Zink and McKitrick, 1995; Templeton, 2001).

I propose that these and related twentieth-century species concepts are archaic constructs that are the outcome of attempts to describe the pattern of biodiversity

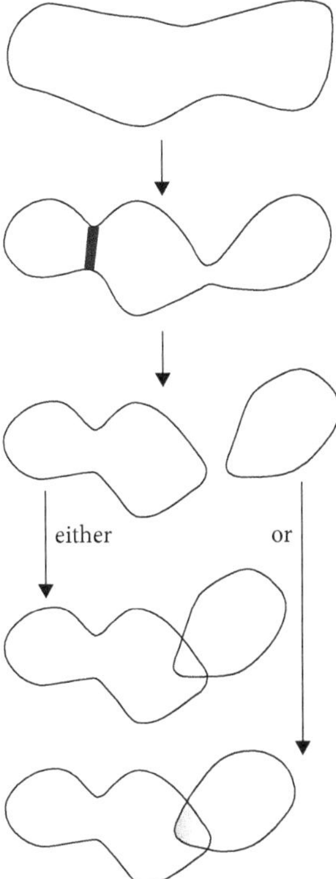

Figure 7.1 Allopatric speciation as depicted by Mayr (1942) in his book *Systematics and the Origin of Species*. In the two alternative outcomes of disruption of gene flow that are illustrated, either divergence leads to daughter populations that remain reproductively isolated following secondary contact (outcome at the end of the short arrow) or daughter populations fuse back into a single population following secondary contact (outcome at the end of the longer arrow). Mayr's concept of allopatric speciation driven by divergence in nuclear genes remains the most widely embraced model for speciation.

solely from the observation of the morphologies, interactions, geographic distributions, and ecologies of extant species, without adequate consideration of the underlying molecular mechanisms of speciation. This is not a criticism of the hard work and intellectual effort that went into studying speciation in the century after Darwin; it is a statement that these species concepts were developed when too little information was available for the process of speciation to be properly understood. The fact that more books on speciation have been written by philosophers than biologists (Morrison, 2018) is testament to the lack of basic grounding in cell and molecular biology of current species concepts. Without basic information on the genomic structure of populations, species debates became little more than esoteric intellectual exercises (Wheeler and Meier, 2000). The biological species concept and the phylogenetic species concept are to speciation theory what Gregor Mendel's descriptions of inheritance based on garden experiments with peas are to genetic theory. Just as with Mendel's description of heredity, the twentieth-century concepts of species are generally accurate descriptions of biological entities based on examinations of external macroscale outcomes of complex molecular interactions. And, as was true in the study of genetics, understanding of the origin of species requires consideration of the molecular mechanisms that give rise to macroscale patterns. More specifically, speciation theory needs consideration of the coevolution and coadaptation of mt and N genes.

Speciation fundamentals: Dobzhansky–Muller incompatibility

Incompatibilities in coadapted sets of mitonuclear genes are one manifestation of a more general phenomenon known as Dobzhansky–Muller incompatibility (Burton and Barreto, 2012). This fundamental evolutionary concept was first proposed by William Bateson (1909) but Theodosius Dobzhansky and Hermann Muller (Dobzhansky, 1936; Muller, 1942) independently rediscovered the idea and made it the centerpiece in discussions of speciation. (See Orr, 1996 for the history of the Dobzhansky–Muller incompatibility model.) The Dobzhansky–Muller model for hybrid incompatibility presented a potential solution to one of the great challenges to Darwin's theory: how does hybrid sterility evolve?

The basic problem that the Dobzhansky–Muller model was proposed to have solved is how to evolve from one fixed genotype to an alternative fixed genotype that is incompatible with the first. Consider an endpoint to the process of speciation with hybrid sterility: one population is fixed for allele A and another population is fixed for an alternative allele a. Both populations (AA and aa) are fertile, but hybrid offspring with the allele combination Aa are infertile. The conundrum that is presented by this simple speciation model is that the endpoint seems impossible to achieve through evolution by natural selection. Whether the ancestor was AA or aa, the first step in the evolution of an incompatible sister population would be a heterozygote, Aa, which would be infertile. The Dobzhansky–Muller incompatibility model presented a theoretical solution to this paradox simply by having infertility result from epistatic interactions *between* loci rather that interactions of alleles *within* a single locus.

The Dobzhansky–Muller incompatibility model begins with an ancestral population that, as a simple example, is often described as being homozygous at two loci with a genotype: AA BB. Following disruption of gene flow, the parent population is divided into two daughter populations with independent evolutionary trajectories. One daughter population evolves so as to fix a new allele at the first locus, becoming aa BB. A and a are compatible so there is no theoretical problem with a transition through Aa. The other population evolves so as to fix a new allele at the second locus, becoming AA bb. B and b are compatible so, again, there is no theoretical problem transitioning through an intermediate heterozygous genotype, Bb. However, throughout the evolutionary history of the populations, allele a has never been matched to allele b. In other words, a and b alleles have never been tested to see how they function together, and hybrid infertility can result if the allele a is not compatible with the allele b (Figure 7.2).

The simple two-locus Dobzhansky–Muller incompatibility model revolutionized the study of speciation by presenting a coherent theory for how species boundaries could evolve through Mendelian mechanisms. It remains the cornerstone of modern speciation theory. Paradoxically, however, even though isolation of gene pools via negative epistatic interaction of *nuclear* genes remains the foundation of modern speciation theory, decades of focused search have turned up very few such nuclear incompatibility genes (Cutter, 2012; Wolf and Ellegren, 2017). As Dion-Côté and Barbash (2017, p. 17) stated in a recent review, "One surprising finding [of modern speciation research] is that there seems to be only rare instances where diverging

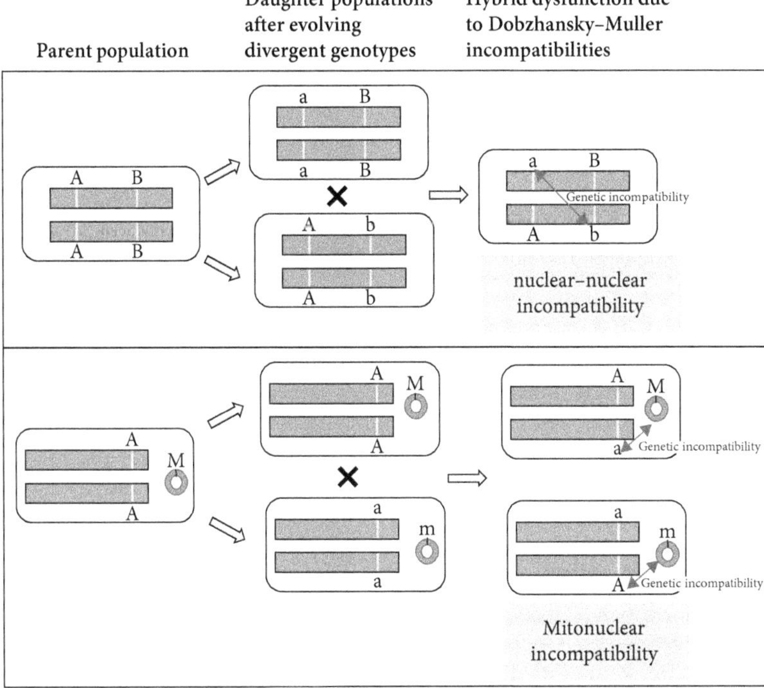

Figure 7.2 A simplistic illustration of Dobzhansky–Muller incompatibilities arising from nuclear–nuclear genomic interactions (top panel) versus mitonuclear interactions (bottom panel). Bars represent diploid chromosomes with loci shown as gray crossbars. Circles represent mitochondrial genomes with loci indicated as black crossbars marks. Different alleles are indicated by different letters.

Box 7.1 Magic traits

In the long and largely fruitless pursuit of N genes that serve as hybrid incompatibility genes, the holy grail has long been the discovery of "magic traits" (Gavrilets, 2004). [*The opening sentence to this box seems so farfetched it is hard to believe that I haven't started a parody of the actual science of speciation, but the term "magic trait" has appeared in over 400 papers in the evolutionary literature. I understand that the term "magic" is being used in a manner meant to be fun and interesting and it is certainly not meant literally, but just the notion that a N-encoded trait that isolates populations seems like magic should tell us something about this line of investigation.*] A magic trait is simultaneously under divergent natural selection for both fitness effects and reproductive isolation (Servedio et al., 2011; Thibert-Plante and Gavrilets, 2013). Such a trait would be invaluable to the effort to make speciation a phenomenon driven by N genes because a nuclear "magic" locus would simultaneously stop gene flow between populations and enable the two new populations to ecologically co-exist. There are numerous putative examples of magic traits, but none has been definitely linked to speciation.

The impetus that led evolutionary biologists to the search for magic traits within a speciation framework that is founded on incompatibilities among N genes is not unlike the current challenges faced by evolutionary biologists contemplating how a speciation framework founded on mitonuclear incompatibility could lead to linkage between incompatibility genes and species-recognition genes (see Figure 8.6). For organisms to avoid the high cost of post-zygotic selection in culling incompatible mt and N-mt gene combinations following random mating, they require some means to assess the compatibility of mt and N-mt genes in mates. Thus, just as it is important to link incompatibility loci and fitness loci in nuclear-based models, it is important to link N-mt (incompatibility) genes to the genes that create a signal of species identity in mitonuclear models. I'll return to this topic of linkage of signals of species identity and N-mt genes in Chapter 8.

populations carry differentially fixed alleles of genes involved in post-zygotic reproductive isolation." Even "magic traits" have not solved the dilemma (Box 7.1).

The mt genome was not discovered until decades after Dobzhansky and Muller proposed their models to explain hybrid incompatibility based on the interactions of autosomal nuclear alleles, but even in the early twentieth century, cytonuclear interactions were implicated as central to the formation of species (e.g. Johannsen, 1923; Chittenden, 1927). What early-twentieth-century plant and animal breeders repeatedly observed was that the interactions of nuclear elements seemed far less important to hybrid infertility and dysfunction than the effects of the cytoplasm (Sapp, 1987). From a modern perspective of the importance of the mitonuclear coadaptation to organism function, it now seems premature for Dobzhansky (1937a) and other Mendelian geneticists to have summarily dismissed a role for cytoplasmic elements in hybrid dysfunction (Orr, 1996) (Box 7.2). This early rejection of a role for cytoplasm in speciation led to the decades-long and largely unsuccessful search for N-encoded speciation genes.

Levin (2003) first presented the hypothesis that coadaptation between the N genome and both mt and chloroplast genomes made it highly likely that epistatic

Box 7.2 The premature execution of mitonuclear speciation

In the decades following the turn of the twentieth century when basic concepts in genetics were being deduced, many leading scientists, particularly in Europe, held that Mendelian genetics could not explain speciation (Sapp, 1987; Orr, 1996). While Mendelian processes had seemingly unlimited success when applied to morphological traits within species, these same Mendelian processes seemed to offer no explanations for the nature of species. For instance, through exhaustive experiments, animal and plant breeders had verified essentially all of the laboratory findings of Mendelian geneticists; the glaring except was the origin of species boundaries. Attempts to breed for hybrid inviability by selection on Mendelian traits consistently led to failure (Sapp, 1987). Thus, the problem faced by geneticists attempting to apply Mendelian genetics to speciation is that the interactions of N genes did not appear to explain hybrid infertility (Johannsen, 1923; Chittenden, 1927). Rather, differences that were the basis for hybrid sterility and hence species boundaries appeared to reside in the cytoplasm rather than in the N chromosomes (historical overview provided by Sapp, 1987). The scientists who looked to the cytoplasm for the basis of speciation are now described as cytocentrists (Orr, 1996). This era of biology was, of course, long before the discovery of a mt genome, so by invoking a cytoplasmic effect, it seemed that the cytocentrists were turning to non-genetic effects and indeed invoking some mysterious property of the cytoplasm that was not amenable to scientific research. This thinking was unacceptable to committed Mendelian geneticists like Theodosius Dobzhansky and Thomas Hunt Morgan: "Except for the rare cases of plastid inheritance all known characters can be sufficiently accounted for by the presence of genes in the chromosomes. In a word, the cytoplasm may be ignored genetically" (Morgan, 1926, p. 491).

With a growing focus on cytoplasmic effects, it seemed that, despite the objections of Morgan, Dobzhansky, and other very influential beanbag geneticists*, early-twentieth-century biologists were on the cusp of correctly deducing that interactions of the cytoplasm and nucleus play a central role in speciation. And then, a game-changing (and in retrospect perhaps unfortunate) event occurred: a line of flies was discovered in the *Drosophila pseudoobscura* populations in Morgan's fly lab at Columbia University that produced viable hybrid females, but inviable males, when crossed to the parent line (Provine, 1981 provides an engaging account of the key people and events surrounding this discovery). This line of flies was later recognized as a new species of *Drosophila*, *D. penimilis*, although ironically Dobzhansky's mentor, Alfred Sturtevant, never recognized it as a species because it was not morphologically distinct. Like the wing flap of a fly setting off a chain of events leading to a devastating hurricane, the accidental discovery of male-specific N-encoded incompatibility factor fundamentally changed the next century of speciation research. The discovery of a line of flies in which inviability could be mapped as a trait in fly crosses using the same beanbag genetic approaches as the study of any other phenotypic traits was a revelation. Dobzhansky proceeded to map the inviability genes that were responsible for infertility in *D. pseudoobscura* × *D. penimilis* crosses, and he wrote a seminal book establishing hybrid incompatibility as simply another Mendelian trait (Dobzhansky, 1937b). It seemed to be of no special concern to Dobzhansky or his followers that the nuclear inviability genes in *D. penimilis* were a rare type of barrier to gene flow when considered across all eukaryotes and that this effect had been documented in only one lab-derived population of flies. The desire to make speciation a Mendelian phenomenon and the brilliance of Dobzhansky's

(Continued)

Box 7.2 *Continued*

book swept aside alternative views. As Orr put it, Dobzhansky "was quickly able to put several popular theories of speciation to the sword. (Indeed, some of these hypotheses were so thoroughly dispatched that their very existence is now forgotten.)." Cytonuclear speciation is one of those ideas that Dobzhansky "put to the sword." This key concept in eukaryotic speciation is only now being rediscovered.

* "Beanbag genetics" is founded on the premise of one gene causing one phenotypic effect such that genes could be studied as if they were colored beans being shuffled in a bag (Mayr, 1963. p. 263).

interaction of cytonuclear gene sets would play a key role in speciation. He supported his hypothesis with extensive examples from plants, which was later expanded by Greiner et al. (2011). Levin's essay was entirely plant-focused, but Gershoni et al. (2009) made a much more animal-focused argument that Dobzhansky–Muller incompatibilities involving not just epistatic interactions of N loci but also coadapted mt and N-mt genes might play an important role in speciation. They emphasized that selection for adaptive divergence in mt genotype could be driven by selection for climate adaptation and other environmental accommodations. This idea that mitonuclear interactions might play a disproportionate role in generating Dobzhansky–Muller incompatibilities in the process of speciation relative to nuclear–nuclear interactions was expanded by Burton and Barreto (2012) (Figure 7.2). Instead of focusing on divergence in mt genotype driven by adaptive changes, Burton and Barreto (2012) focused on non-adaptive divergence in coadapted mt and N-mt genes that results from mutation of mt genes leading to fixation of slightly deleterious alleles via Hill–Robertson effects and compensatory coevolution with N-mt genes (topics that I covered in detail in Chapters 3 and 4). Burton and Barreto (2012) emphasized that epistatic interactions of mt and N-mt genes appear to play a role in hybrid dysfunction across essentially the entire diversity of eukaryotes. A recent detailed study of mt and N genomes from divergent populations of copepods (Barreto et al., 2018) provided substantial additional evidence that compensatory coevolution of mt and N-mt genes underlies genetic divergence and genetic incompatibilities between populations.

The basic concept of Dobzhansky–Muller incompatibilities arising from mitonuclear interactions is fundamentally the same as incompatibilities arising from interactions among different loci in the N genome, with a few important differences (Figure 7.2). The mt genome is typically proposed to exist as a single-copy genome not as a diploid or other polyploid, such that there is only one allele at each mitochondrial locus. Moreover, because mt genomes are typically transmitted exclusively by females, sex-specific effects of hybrid crosses play a more central role in Dobzhansky–Muller incompatibilities involving mitonuclear interactions compared with nuclear–nuclear interactions (Hill and Johnson, 2013; Hill, 2017). And, perhaps most importantly, the necessity of mitonuclear coadaptation provides a precise and universal explanation for the evolution of genetic incompatibilities between isolated populations.

The mitonuclear compatibility species concept

Mitonuclear coevolution when gene flow is disrupted

Under any model, disruption of gene flow is the foundation of speciation (Nei and Nozawa, 2011; Ravinet et al., 2017). Since Dobzhansky's seminal experiments on hybrid inviability in *Drosophila pseudoobscura* and *D. persimilis* (Dobzhansky, 1936) (Box 7.2), consideration of functional changes in genes that could affect the outcome of hybridization in diverging lineages has focused primarily on N genes (Coyne and Orr, 2004). For decades after the discovery of the mt genome, variation in mt genotype was generally dismissed as neutral variation (Ballard and Kreitman, 1995; Galtier et al., 2009b), and neutral variation will not affect gene flow or play a role in speciation. In the last two decades of the twentieth century and into the twenty-first century, a handful of population biologists such as David Rand, James Ballard, and Ronald Burton published the results of experiments showing functional variation in mt genotypes in populations of animals, and slowly the dogma that genetic variation in mt genotypes must be neutral began to be overturned. It is now becoming more generally accepted that there is perpetual functional change in mitochondria genes over evolutionary time (Ballard and Whitlock, 2004; Wolff et al., 2016a), and divergence in genes with functional consequences is exactly what is needed to disrupt gene flow.

In Chapter 2, I outlined in detail the evidence that in comparisons among closely related species, mt and N-mt genes have sufficient functional genetic changes to cause a reduction of respiratory function when the mt genes from one species are forced to co-function with N mt genes from another species. Coevolution by N-mt genes to compensate for the accumulation of deleterious mutations in the mt genome is the primary hypothesis to explain the maintenance of mitonuclear coadaptation over evolutionary time, and I presented the hypothesis of compensatory coevolution in detail in Chapter 3. Putting these ideas together, a hypothesis for the process of genetic divergence of populations takes shape (Gershoni et al., 2009; Burton and Barreto, 2012) (Figure 7.3). Through Muller's ratchet, mutations perpetually accumulate in the mt genome. When these changes are functional, they are nearly always deleterious, but they persist nevertheless because Hill–Robertson effects makes selection on such slightly deleterious genes ineffective (Fan et al., 2008). In response to mutational erosion of the mt genome, N-mt genes that code for products that interact with mt gene products perpetually evolve so as to compensate for deleterious mt genes (Chapter 3).

The course of this change and counter-change is stochastic and unpredictable because it is driven by random mutations in both the mt and N-mt genomes (Gershoni et al., 2009; Burton and Barreto, 2012). Thus, the specific changes to genotype that occur in one population have essentially no chance of being repeated in another population. As a consequence, when gene flow between subpopulations is disrupted, the subpopulations will diverge rapidly in the set of coevolving mt and N-mt genes that gives rise to electron transport system (ETS) function. This divergence in coadapted mitonuclear gene sets will inevitably occur faster than divergence in any sets of

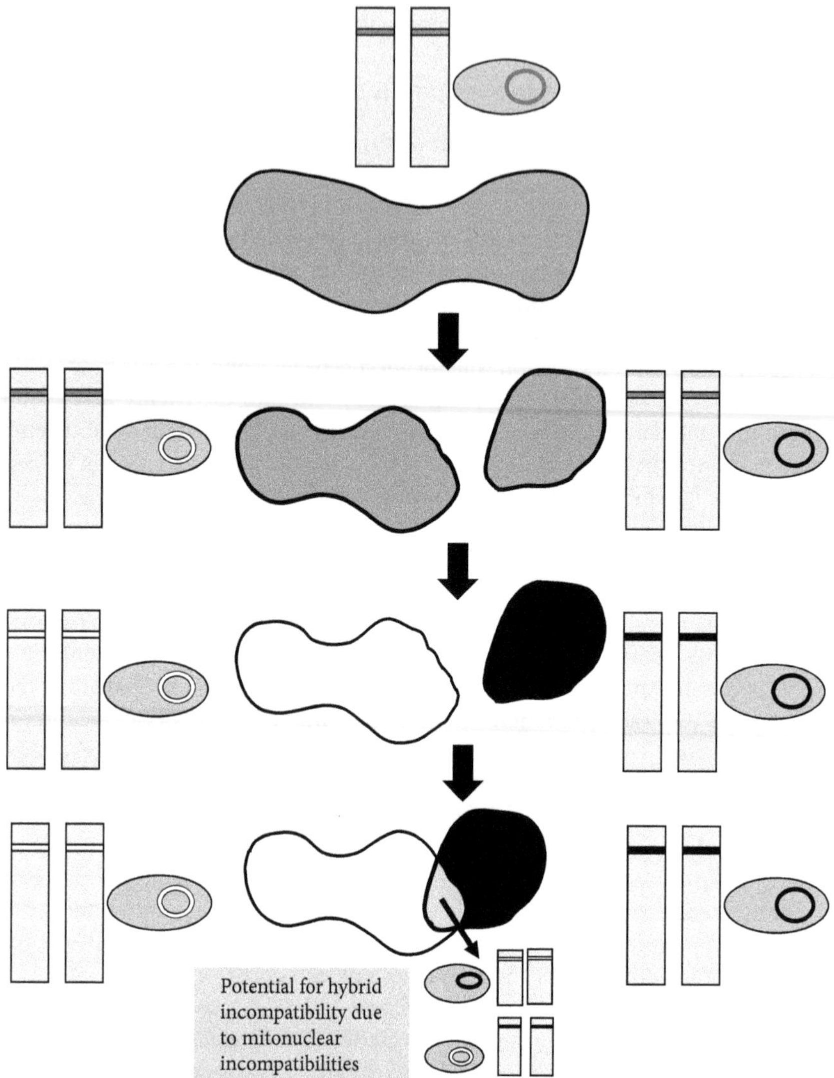

Figure 7.3 A pictorial model of allopatric speciation driven by Dobzhansky–Muller incompatibilities between coadapted mt and N-mt genes. The irregular shapes depict the ranges of hypothetical eukaryote populations (as in Figure 7.1). N genotypes are depicted as diploid chromosomes, with N-mt genes that are coadapted with mt genes shown as the crossbars within. Mitochondrial genotype is depicted as an oval. Shared color indicates shared coadapted mitonuclear genotype. Across generations, divergence in mt genotypes is proposed to occur first, followed by changes to N-mt genes to maintain mitonuclear coadaptation. These divergent pairs of mt and N-mt genes in the two daughter populations may be incompatible if mixed when populations come into secondary contact, creating barriers to gene flow and maintaining the integrity of the new populations.

interacting N genes because of the higher rate of mitochondrial mutation (Gershoni et al., 2009; Burton and Barreto, 2012) and much faster time to coalescence of mt genes compared with N genes (Zink and Barrowclough, 2008). Thus, before there is any meaningful change in most of the N genome, there will be changes in sets of coadapted mitonuclear genes.

This hypothetical process of allopatric divergence in sets of coadapted mt and N-mt genes as a driver of speciation leads to a novel concept of how to define a species (Hill, 2016, 2017). Following on the pioneering work of Levin (2003), Lane (2009), Gershoni et al. (2009), Chou and Leu (2010), Bar Yaacov et al. (2012), and Burton and Barreto (2012), I proposed the mitonuclear compatibility species concept, whereby *a species is a population that is genetically isolated from other populations by incompatibilities in uniquely coadapted mt and N-mt genes* (Hill, 2016, 2017, 2018). The mitonuclear compatibility species concept potentially presents a concept of species by which species boundaries can be objectively demarcated in any clade of eukaryotes (Box 7.3).

Box 7.3 A unifying species concept

Gene flow across the population boundaries of eukaryotes can be physically disrupted when breeding individuals no longer come in contact in time or space (Mayr, 1963; Coyne and Orr, 2004). It can also be disrupted by mating preference (Coyne, 1974). And finally, evolutionary biologists have recognized three primary genetic mechanisms that can disrupt gene flow: (1) disruption of recombination caused by misalignment of chromosomes during meiosis, (2) chromosomal rearrangements, and (3) genetic incompatibilities between divergent alleles (Maheshwari and Barbash, 2011). While recognizing a role for all of these processes by which gene flow between populations can be disrupted, I propose that the evolution of uniquely coadapted mt and N-mt gene sets should be the basis for demarcating species boundaries (Figure Box 7.3) (Hill, 2017). I contend that such an emphasis on mitonuclear interactions is justified because the need for mitonuclear coadaptation is universal among eukaryotes. With the adoption of the mitonuclear compatibility species concept, species boundaries become objective and defensible. The mitonuclear compatibility species concept arises from first principles based on the fundamental genomic architecture of all eukaryotes. Since Linnaeus, biologists have proceeded with certainty that across all eukaryotic life individuals cluster into discrete species. Only a characteristic common to all eukaryotes could generate a universal pattern in eukaryotes, and I contend that that this common characteristic of eukaryotes is the necessity of mitonuclear coadaptation (Lane, 2005).

Other genetic mechanisms for disrupting gene flow are specific to only subsets of eukaryotes and, importantly, any alternative means of disrupting gene flow will eventually lead to divergence in coadapted mt and N-mt genotypes and a speciation endpoint by this definition (Figure Box 7.3). The reverse is not true: the halting of gene flow by mitonuclear processes will not inevitably lead to chromosomal rearrangements or the disruption of meiosis. The necessity of mitonuclear coadaptation is the single ubiquitous and unifying component of speciation across all eukaryotes (and explicitly excluding prokaryotes; Box 7.4) and, in my opinion, it should be the universal criterion for defining species.

I strongly advocate that any hypothesis for speciation and definition of species should emerge from fundamental understanding of the structure of populations in nature, and

(Continued)

Box 7.3 *Continued*

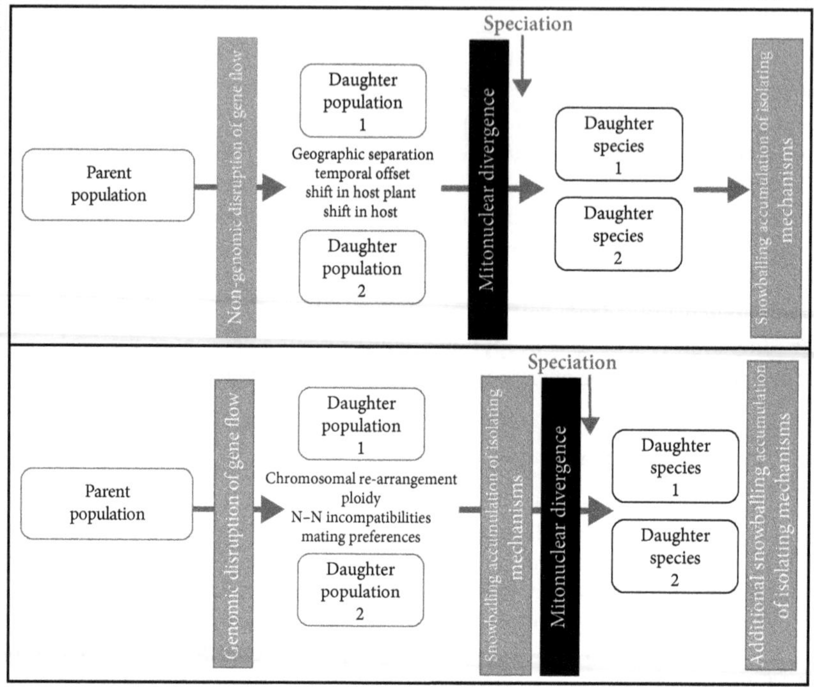

Figure Box 7.3 A summary of the sequence of evolutionary events that might lead to speciation. The top panel focuses on non-genomic time-and-space mechanisms for disruption of gene flow between populations. The bottom panel focuses on genomic mechanisms for disruption of gene flow between populations. The black bar depicts the position at which coadapted sets of mt and N genes in daughter populations diverge to the point of generating hybrid incompatibilities. Mitonuclear interactions can either be the primary mechanism for isolating gene pools or divergence in coadapted mitonuclear genotypes can follow disruption of gene flow caused by other genetic mechanisms such as chromosomal re-arrangement or N-N incompatibilities. Either way, I advocate that the process of speciation is not complete until populations diverge in coadapted sets of mt and N-mt genes.

not from political expediency or consensus. Nevertheless, the operational advantages of a unified species concept are substantial. For the first time, biologists would have an objective, universal definition of a species that applies to all eukaryotes; determining species boundaries would no longer be an esoteric intellectual exercise. In this way, declarations of species boundaries become much easier to defend to non-biologists. Checklist committees, and all of the politics that come with them, will no longer vote species into and out of existence. For a time, subjectivity will persist as researchers debate degrees of mitonuclear coadaptation, but such uncertainty regarding the functional outcomes of different gene combinations should be short-lived. Functional modeling of the molecular systems is advancing so rapidly that I can plausible anticipate that in the near future we will have a pipeline going from N and mt genotype to assessment of coadaptation to objective species decisions.

Box 7.4 "Species" is exclusively a eukaryotic concept

Scientists seek generalizations. The greatest scientific discoveries are those that are most broadly applicable across organisms, such as DNA as the genetic code or OXPHOS as the basis for aerobic respiration. Within evolutionary biology, there is a strong tendency to want to drop the wall of discontinuity between prokaryotes and eukaryotes and find universal themes that apply to all life. Hence, there have been attempts to apply the biological species concept, first conceptualized in birds and most generally applied to animals (Donoghue, 1985), to prokaryotes (Bobay and Ochman, 2017). Given that I define species in terms of interactions of mt and N genes, it should not be surprising that I argue that eukaryotic species concepts cannot be applied to prokaryotes. But I take this distinction between speciation in eukaryotes and prokaryotes a step farther. I think that the term "species" should never be applied to prokaryotes. The mechanisms for generating biodiversity in prokaryotes are so fundamental different than the mechanisms that generate biodiversity in eukaryotes (see Figure 5.1) that it becomes misleading to use the term "species" both for evolutionary lineages in eukaryotes and for collections of genes in prokaryotes. I won't presume to tell microbiologists what they should call their population groupings, but the term "species" is taken by eukaryotic biologists.

Unlike previous species concepts, it is based on first principles of biochemistry and genomic interactions that apply to all eukaryotic lineages.

mt DNA barcodes as evidence for mitonuclear speciation

As the cost of sequencing DNA dropped precipitously around the turn of the twenty-first century, biologists began to consider the prospect of using nucleotide sequence as a tool in taxonomy (Wilson, 1995). This effort was first restricted primarily to prokaryotes and single-celled eukaryotes because they provided the least morphological information for taxonomic study (Pace, 1997; Hamel et al., 2001). In 2003, however, Canadian biologists Paul Hebert and colleagues proposed a broader application of taxonomy based on DNA sequences, with animals, plants, and fungi the primary targets (Hebert et al., 2003a). They proposed that an excellent gene sequence to use for such a DNA barcoding effort was a portion of the gene sequence for mt-encoded subunit 1 of Complex IV of the ETS (COX1). They chose this particular mt DNA sequence as the "DNA barcode of life" because (1) its protein-coding regions were extremely conserved across many life domains and thus primer sets for amplifying this DNA barcode sequence could be derived for taxa for which no genomic data had previously existed, and (2) in preliminary studies COX1 generated informative sequence data relative to other potential barcode sequences (Hubert and Hanner, 2015), although subsequent assessments have shown that the nucleotide sequence of other mt genes, such as cytochrome *b*, can work equally well at delimiting species boundaries (Hajibabaei et al., 2007; Tobe et al., 2010; Nicolas et al., 2012; Lv et al., 2014).

It is important to emphasize that the initiative to assign a unique mt DNA barcode to all eukaryotes was focused on automating and accelerating taxonomy (Hebert et al.,

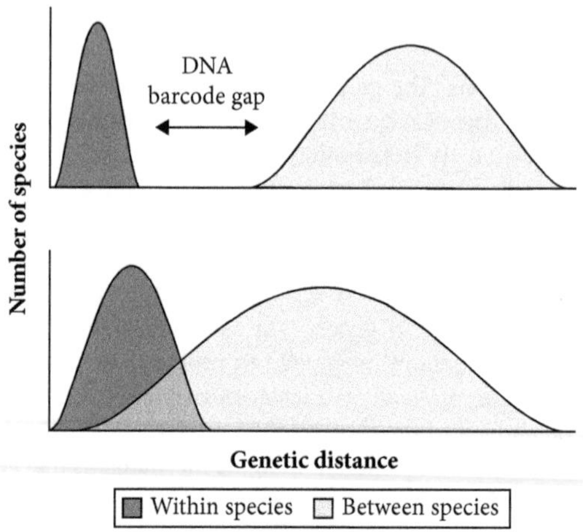

Figure 7.4 The pattern of within- versus between-species divergence in genotype if DNA barcoding is effective (top panel) or ineffective (bottom panel) at distinguishing species. A DNA barcode gap will exist if the degree of divergence within species is consistently lower than the degree of divergence between species.

2003a; Hubert and Hanner, 2015). This effort was completely detached from theory regarding speciation or the nature of species beyond using published taxonomies to assess the efficacy of the DNA barcoding effort. Mitochondrial DNA rather than N DNA was the focus, not because of any specific interest in a potential role of mt DNA in the process of speciation or because speciation theory indicated that mt DNA should be best aligned with species boundaries, but simply because mt DNA worked better than N DNA when numerous potential N and mt barcode sequences were tested (Figure 7.4). In the context of compiling better inventories of biodiversity, the efficacy of barcoding methods was assessed by how well a DNA barcode-generated taxonomy matched conventional taxonomy.

A funny thing happened on the way to creating the Consortium for the Barcode of Life (Schindel and Miller, 2005). In contradiction to neutral theory and despite the fact that it was argued that DNA barcoding could not work based on basic population biology theory (Hickerson et al., 2006), mt DNA barcoding did work. For bilaterian animals, it worked consistently well (Bucklin et al., 2011). With some exceptions, animal species showed little variation in mitochondrial nucleotide sequence within a species, but substantial variation between a species (Figure 7.4). This between-species divergence in mt gene sequence is typically referred to as "the DNA barcode gap" and in taxa such as birds, the barcode gap almost invariably fell where biologists had traditionally recognized species boundaries (Hebert and Gregory, 2005; Kerr et al., 2007; Zink and Barrowclough, 2008). The success of DNA barcoding in delimiting the boundaries of species was anathema to the biologists who based their interpretation of the evolution of the mt genome on the premise that nucleotide sequence evolution is neutral or who resisted replacing traditional morphological taxonomy with genetic analysis (Packer et al., 2009). The capacity for mt nucleotide sequences to consistently and accurately predict species boundaries suggested that mitochondria were more than passive bystanders during speciation (Lane, 2009). It implicated mitochondria as a central player in the process of speciation (Hill, 2016, 2017).

There are, of course, explanations other than mitochondria playing a direct role in the process of speciation for why transitions in mt genotype match species boundaries. In allopatric speciation, when gene flow is stopped, both mt and N genotypes are predicted to diverge between the isolated populations (Toews and Brelsford, 2012; Hung and Zink, 2014) (Figure 7.3). Because mitochondria have a high mutation rate in many eukaryotic taxa including most animals and the effective population size of mt genomes is thought to be less than N genomes, mt genes will evolve faster than N genes (Brown et al., 1979; Li, 1997). This rapid evolution of the mt genome, combined with more rapid coalescence of novel mt genotypes (Zink and Barrowclough, 2008), should make mt genotypes good predictors of species boundaries even if mitonuclear coadaptation plays no direct role in the speciation process (Zink and Barrowclough, 2008; Hung et al., 2016).

This alternate explanation for the coincidence between species boundaries and transitions in mt genotype is a hypothesis for neutral mitochondrial evolution, and it makes clear and testable predictions (Hudson and Coyne, 2002). A simple prediction is that divergence in mt genotypes should be proportional to time since divergence (Orr and Turelli, 2001). According to neutral theory, mutations accrue in a clock-like manner, so degree of divergence in mt genotypes between any two populations should be directly proportional to time since divergence from a common ancestor (Kimura, 1983; Nei et al., 2010). Interestingly, however, time since divergence is a poor predictor of degree of divergence of mt genotype in avian lineages (Kerr et al., 2007; Kerr, 2011). Another key prediction of neutral theory is that genetic diversity should be greater in large populations compared with smaller populations. More variation in mt DNA should make the effort to identify species via mt DNA more effective for small populations because barcoding relies on there being little within-population variation compared with the degree of between-population variation (Hebert et al., 2003b; Stoeckle and Thaler, 2014). For N gene evolution, this assumption holds: species with greater population sizes have more diversity in N genotypes. In contrast, across diverse eukaryotic lineages, this central prediction of neutral theory is not supported for mt genes (Bazin et al., 2006). In Class Aves, where the relationship between population size and diversity of mt genotypes was studied in the greatest detail, variation in the mt DNA barcode sequence was similar and uniformly low despite five orders of magnitude difference in the population sizes of different bird species, in direct contradiction of neutral theory (Stoeckle and Thaler, 2014). The conclusion from these studies is that evolution of mt genotypes is not neutral (Ballard and Whitlock, 2004; Dowling et al., 2008). A barcode gap emerging from the need to maintain coadapted sets of mt and N-mt genes predicts time- and population size-independent evolution of mt genomes and is consistent with these empirical observations.

Mitonuclear speciation driven by mitochondrial-based adaptation

The speciation process that I described above occurs in allopatry, and it invokes no adaptive divergence in populations; accumulation of deleterious alleles in the mt genome is counteracted by changes in N-mt genes with the entire process directed at maintaining established OXPHOS function. As will be discussed in detail in Chapter 9, however, different mitonuclear gene combinations can give rise to different mt phenotypes that vary in how well they perform in different environments. The evolution of new adaptive mitonuclear gene combinations presents the possibility of speciation via adaptive divergence in mitonuclear genotypes (Gershoni et al., 2009) and the potential for speciation without barriers to gene flow other than the interaction of mt and N-mt genes (Hill, 2016).

Theoretically, natural selection could drive the evolution of mitonuclear gene complexes in populations to match local environments leading to the divergence of populations and to speciation. I sketched out how such a process could work in a simple verbal model (Hill, 2016). I used as an example a species with a range that extends into different thermal environments with two adaptive peaks in mitochondrial function associated with warm and cool environments. In this hypothetical population, the ancestral population carried a warm-adapted mt genotype, such that individuals in the cool northern part of the species range were poorly adapted to their thermal environment and had relatively low fitness compared with individuals in warmer regions (Figure 7.5A). Such a population might persist indefinitely in such a state with a portion of the population carrying a mt genotype well matched to the local environment and with declining fit to environment away from that core area. If, by chance, a mutation produced a novel mitochondrial allele that enabled better function in cooler environments, this novel mt genotype would spread via natural selection in the cooler part of the species range. The result of such sorting of mt genotypes by environments could be clinal variation, with high frequencies of the warm-adapted mt genotype in the south, high frequencies of the cool-adapted mt genotype in the north, and a gradual transition from one to the other between (Figure 7.5B; see also Figures 9.9 and 9.10 for real-world examples). Because there would exist no barriers to gene flow, local selection would be constantly eroded by exchange of genes between subpopulations, and no subpopulation would reach a fitness peak.

To this point, what I have outlined is not farfetched: clinal variation in mt genotypes linked to local thermal adaptation has been documented in fruit flies, gastropods, birds, and humans (Mishmar et al., 2003; Balloux et al., 2009; Quintela et al., 2014; Camus et al., 2017; Sunnucks et al., 2017). I'll provide specific examples of adaptive divergence in mt genotype in Chapter 9; here, I simply need to establish the plausibility of such variation. What I proposed next, however, remains completely hypothetical, but I think it holds potential to explain some patterns of biodiversity.

If there is a mutation in a N-mt gene that complements the cool-adapted northern mt genotype, then selection on the sets of N-mt genes can potentially lead to barriers to gene flow that enable speciation without geographical barriers to gene flow. The emergence of two genetically isolated populations could result if the fitness of

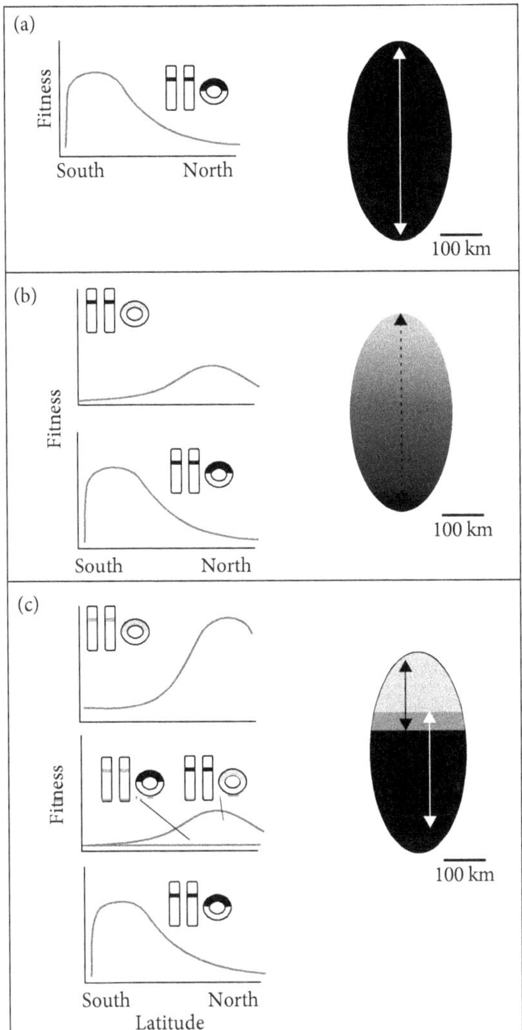

Figure 7.5 A graphical model for speciation via adaptive divergence in mt and N-mt genotypes. Graphs depict hypothetical fitness curves across a latitudinal gradient for different mt/N-mt gene combinations. The shapes on the right depict the ranges (oriented north to the top) of the hypothetical populations through evolutionary time. The focal N-mt genotype is represented by shaded crossbars within paired chromosomes, while the focal mt genotype is represented by shaded sections of the circular mt genome. (A) The ancestral mt genotype (black shaded) has high fitness in the south and low fitness in the north. The ancestral N-mt gene (black shaded) enables the function of the ancestral mt. (B) A mutation produces a novel mt genotype (pale shaded) that bestows high fitness in the north and low fitness in the south. Its function is supported by the ancestral N-mt gene (black shaded) but it fails to achieve fitness in the north as high as the ancestral mt genotype in the south. Natural selection creates a cline in mt genotypes but gene flow erodes local adaptation. (C) A mutation leads to a novel N-mt genotype (pale shaded) that better supports the function of the cold-adapted mt gene (pale shaded) and raises the fitness of individuals in cold climates. The new N-mt gene has low compatibility with the ancestral mt genotype, creating a barrier to gene flow via hybrid dysfunction and initiating speciation. Adapted from Hill (2017).

cool-adapted mt genes matched to cool N-mt genes creates high fitness individuals in the north, and warm-adapted mt genes matched to warm N-mt genes create high fitness in the south. In this model, warm mt genes matched to cool N-mt genes or cool mt genes matched to warm N-mt genes are at a fitness disadvantage compared with mixed genotype "hybrids" (Figure 7.5C).

The result of this hypothetical scenario is two populations, one with a mitonuclear gene combination that enhances fitness in warm environments and a second with a mitonuclear gene combination that enhances fitness in cooler environments. Gene flow between the populations is disrupted by natural selection removing individuals with mixed genotypes because they have low fitness in all environments. This model is hypothetical, but closely related taxa with divergent mt genotypes that show an abrupt geographic transition from one species to another could be explained with such adaptive divergence in mitonuclear genotypes.

Mitonuclear interactions and gene flow

Allele dominance and introgression of mt and N genes

In conventional descriptions of Dobzhansky–Muller incompatibilities the negative epistatic effects of mixing incompatible alleles manifest in the F1 generation because, in the theoretical models as well as in studies with some species of *Drosophila*, inviability or infertility is fully effective in the heterozygous F1 hybrid offspring (AaBb; Dobzhansky, 1937b; Muller, 1942); Figure 7.2). In contrast to the assumptions of the Dobzhansky–Muller model, the common assumption in population genetics models is co-dominance of novel alleles (Teshima and Przeworski, 2006). For many loci in many taxa, however, relative to a novel allele, the wild-type allele is typically dominant to a novel allele (Kondrashov and Koonin, 2004).

When genes are mixed via hybridization, such that the wild-type allele of one population is matched to the wild-type allele of another population, the outcome of the allelic interactions becomes much less predictable. Such unpredictability seems especially true for combinations of N-mt alleles for genes that must co-function with mt genes. In most eukaryotes, females contribute to offspring both the entire mt genome and one complete set of N genes. Thus, at each locus that codes for a N-mt gene that co-functions with mt genes, there will be at least one N allele that is compatible with maternal mt alleles (Figure 4.4). Consequently, regardless of the degree of incompatibility of paternal N-mt alleles with maternal mt genes, if there is dominance of the maternally transmitted functional allele, there can be full system function in the F1 generation (Burton et al., 2006). With complete dominance of the most functional allele in a heterozygous pair, the negative consequences of hybridization arising from Dobzhansky–Muller incompatibilities can be delayed for a generation (Burton and Barreto, 2012).

This generational delay in the negative effects of mating between individuals with incompatible mt and N-mt genes was most clearly documented in studies of marine

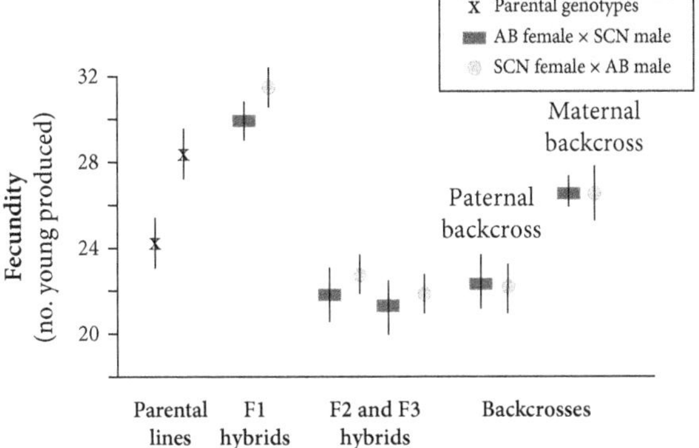

Figure 7.6 The fitness (measured as fecundity) of offspring produced by crossing copepods from two populations (AB and SCN) that have divergent and somewhat incompatible mt and N-mt genotypes. No negative fitness effects are evident in the F1 hybrid offspring. The slight fitness gain of F1 hybrids is due to heterosis involving N–N gene combinations. By the F2 and F3 hybrid generations, however, negative fitness effects are apparent, and these negative effects can be reversed via maternal but not paternal backcrossing. Adapted from Ellison and Burton (2008).

copepods, *Tigriopus californicus*, which have no sex chromosomes (Alexander et al., 2015). Ronald Burton and colleagues observed that when populations with highly divergent and poorly coadapted mt genotypes were crossed, there was an increase in fitness, measured as fecundity, in the F1 generation (Ellison and Burton, 2006; Burton et al., 2013) (Figure 7.6). They attributed this bump in fecundity of hybrid offspring to a complete lack of negative effects of the incompatible paternal N-mt alleles when paired with compatible maternal N-mt alleles in the F1 generation as well as to the benefits of increased heterozygosity across N loci (Burton et al., 2013). The full complement of compatible N-mt alleles from the mother was enough to overcome any negative influence from incompatible N-mt alleles from the father. The story changed dramatically, however, when Burton and colleagues subsequently crossed copepod hybrids to produce F2 and then F3 hybrid generations (Figure 7.6). Due to recombination of N genes, in the F2 and F3 generations there were two paternally inherited alleles at some loci, forcing maternal mt genes to co-function with poorly coadapted and incompatible paternal N-mt genes. The result was what Burton et al. (2006) called "the sorry state of F2 hybrids." The hypothesis that these negative effects of hybridization were the result, specifically, of the interactions of mt and N-mt gene products was dramatically demonstrated using maternal and paternal backcrosses. When F3 hybrid copepods with compromised cellular respiration resulting from incompatibilities in mt and N-mt genes were backcrossed to the maternal line, full fitness was restored in the resulting offspring (Figure 7.6). This fitness rescue via maternal backcross is a consequence of the female once again providing a full complement of N-mt alleles that are coadapted with the maternal line mt genes. The paternal backcross, in stark contrast, did not yield any improvement in fitness because it only added more N-mt alleles that were incompatible with the mt DNA genotype carried by the hybrids (Figure 7.6).

One way to consider the effects of hybridization when the two parent populations carry divergent and non-coadapted mt and N-mt alleles is that the degree of fitness loss of F1 hybrids will depend on the degree of penetrance of the effects of the incompatible allele paired with the coadapted allele (López-Fernández and Bolnick, 2007). In *Tigriopus* copepods, coadapted N-mt alleles appear to have full penetrance and thus provide full fitness effects even if paired with alleles that would yield poor fitness if they were paired with the mt genotype in a homozygous state. In other taxa, however, alleles that are not coadapted may have a variable degree of penetrance such that they share expression with the coadapted allele with which they are paired. With co-expression of alleles, there will be fitness loss in the F1 generation if one of the alleles is poorly coadapted and incompatible with mt genes.

A negative effect of non-coadapted paternal alleles paired with coadapted maternal alleles in F1 hybrids was documented in sunfish in the genus *Lepomis*, which is in a clade of fish that lack sex chromosomes. In hybrid crosses involving various species of *Lepomis* sunfish, F1 hybrids showed a significant loss of fitness (Bolnick and Near, 2005). In crosses between bluegill (*L. macrochirus*) and pumpkinseed (*L. gibbosus*), loss of fitness in F1 fish was linked specifically to mitonuclear incompatibilities that caused increased levels of free radicals and reduced activity of Complex I (Davies et al., 2012; Du et al., 2017). Thus, incomplete penetrance of non-coadapted alleles can affect fitness in the F1 generation in some species. Interestingly, F2 bluegill × pumpkinseed progeny are almost never observed (Du et al., 2017) suggesting that there is severe hybrid dysfunction in F2 and later generations, as predicted if there are mitonuclear incompatibilities. A generational delay in the cost of mating with individuals that do not have coadapted mitonuclear genotypes has big implications for the fluidity of species boundaries, so the degree of penetrance of non-coadapted N-mt genes when paired with coadapted N-mt genes is a very important topic deserving focused investigation.

If there is no fitness loss in F1 offspring, as in *Tigriopus* copepods, there are no immediate costs to either the male or the female to mating outside of species boundaries. Moreover, if the hybrid offspring mate with individuals from their mother's population (i.e. if there is maternal backcrossing), then there is never a cost to hybrid pairing. Negative fitness consequences delayed until at least the F2 hybrid generation should encourage the flow of some N genes (but not N-mt genes) between species (Figure 7.7). Non-coadapted mt and N-mt genes in F2 hybrids will lead to respiratory dysfunction and loss of fitness, and such individuals will be removed by natural selection. These processes are predicted to enable introgression of N genes that do not co-function with mt genes but to ensure that mt and N-mt genes from foreign populations do not rise in frequency (Figure 7.7).

If there are negative fitness consequences from the introgression of coadapted mt and N-mt genes into foreign populations but no negative fitness consequences, or even positive fitness effects, of the introgression N genes into a foreign population, then we would expect discordance in the flow of different classes of genes. Specifically,

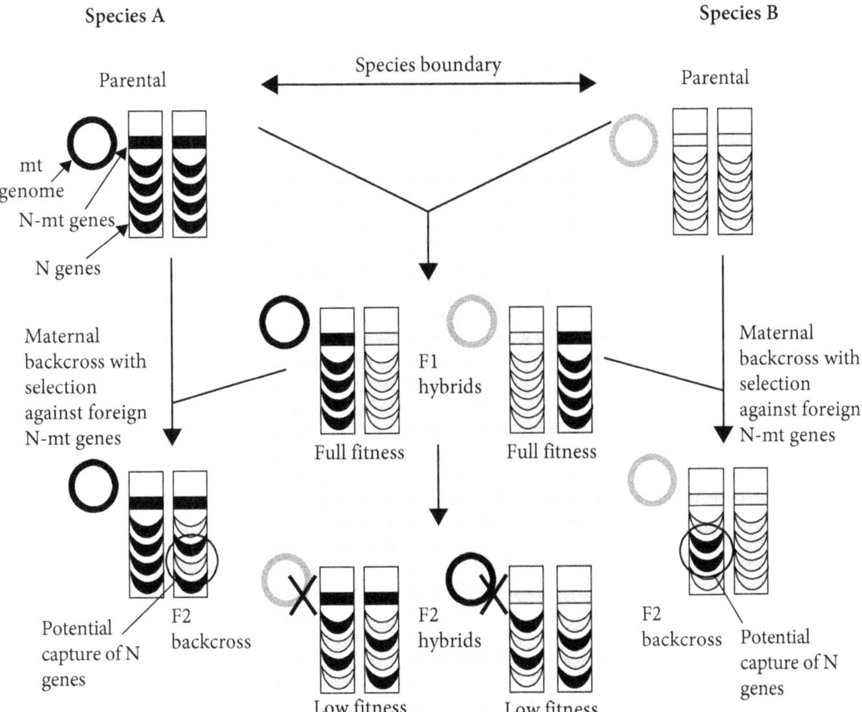

Figure 7.7 A pictorial sketch of pathways for and barriers to gene flow across species boundaries. In this example, there are incompatibilities between the products of the dark-shaded mt genotype (circle) of species A and light-shaded N-mt genotypes (bars within diploid chromosomes) of species B and vice versa, but no negative interactions between any nuclear genes in either population that do not co-function with mt genes (crescents within diploid chromosomes). Due to complete dominance of maternal N-mt alleles, there is no loss of fitness in F1 hybrids or in F2 individuals from maternal backcross, but substantial loss of fitness in F2 hybrids (see text). Under these conditions no introgression of mt or N-mt genes across species boundaries is predicted, but introgression of N genes that do not interact with mt genes is predicted.

mt genomes and N-mt alleles should resist movement across species boundaries (unless they can move together, see below) because each will have low fitness if expressed in a foreign mitonuclear genomic environment (many examples are presented in Chapter 2; Figure 7.7). On the other hand, N alleles that do not co-function with mt genes should frequently flow across species boundaries because novel N alleles often create a fitness benefit either by increasing heterozygosity or due to specific functional benefits of the novel alleles (Figure 7.6). Discordance of gene flow of mt genes and N genes has been well documented, and the patterns that are observed typically match the predictions stated in the above argument: N genes introgress across species boundaries while mt genes do not (hence the effectiveness of a DNA barcode based on mt nucleotide sequence but the failure of N gene sequence to effectively define many species) (Toews and Brelsford, 2012; Hill, 2017; Sloan et al., 2017). This pattern is reversed in the relatively few species that experience complete replacement of their native mt genome by a foreign mt genome, and I'll discuss such mitochondrial introgression in detail in the next section.

The pattern of lack of introgression of mt genes but common introgression of N genes is particularly striking and well documented in birds (Rheindt and Edwards, 2011; Hill, 2017) (Table 7.1; Figure 7.8). An interesting example was brought to light through genomic analysis of the North American wood warblers, the blue-winged (*Vermivora cyanoptera*) and golden-winged warblers (*V. chrysoptera*). These birds are very different in feather coloration, song, and the climatic zone to which they are adapted (Gill, 2004; Confer and Knapp, 2016). Golden-winged warblers and blue-winged warblers hybridize frequently in the relatively small region where their ranges overlap, and the hybrid offspring that result from such pairings can appear as healthy adults that are so well known to birdwatchers that they have their own name—Brewster's warblers (Gill, 2004). The mt genotypes of the two species are about 3 percent divergent in nucleotide sequence, with an abrupt transition in mt genotype at the species boundaries (Gill, 1997). This abrupt transition in mt genotype at the species boundary indicates that there has apparently been no introgression of mt genomes from one species to the other despite hybridization. The pattern of exchange of N genes, however, is strikingly different. Whole-genome analysis of the N DNA indicates that nuclear loci of the two bird populations are only 0.03 percent divergent on average—they share more than 99 percent of nuclear alleles. The authors concluded that there has been extensive N gene flow between the populations for millennia (Toews et al., 2016b). Interestingly, of the six regions of the N genome that show fixed differences between the warbler species, two are on the Z

Table 7.1 A summary of patterns of introgression of mt, Z-linked, and autosomal genes between sister taxa of birds.

Species Compared	Mt Genes	Z-linked N Genes	Autosomal N Genes	Fits Pattern Figure 6.6	Reference
Pied flycatcher, collared flycatcher	M	M	S	Yes	(Saetre et al., 2003; Hung and Zink, 2014)
Golden-winged warbler, blue-winged warbler	N	M	S	Yes	(Gill, 1997; Toews et al., 2016b)
Indigo bunting, lazuli bunting	M	M	S	Yes	(Carling and Brumfield, 2008)
Common nightingale, thrush nightingale	N	M	S	Yes	(Wink and Sauer-Gürth, 2002; Storchová et al.,, 2010)
Mourning warbler, MacGillivray's warbler	N	N	S	Yes	(Irwin et al., 2009)

Table 7.1 *Continued*

Species Compared	Mt Genes	Z-linked N Genes	Autosomal N Genes	Fits Pattern Figure 6.6	Reference
Nelson sharp-tailed sparrow, salt marsh sparrow	M	M	S	Yes	(Walsh et al., 2016)
Mexican duck, black duck	M	M	M	Yes[§]	(Lavretsky et al., 2015)
Greater spotted eagle, lesser spotted eagle	N	M	S	Yes	(Backström and Väli, 2011)
Western willet, eastern willet	N	M	S	Yes	(Oswald et al., 2016)
Gray teal, chestnut, teal	S	M	S	No	(Dhami et al., 2016)
Eastern yellow robin (hot/dry), eastern yellow robin (cool/wet)	N	S	S	Yes[‡]	(Morales et al., 2015)
Karoo scrub-robin (hot/dry), Karoo scrub-robin (cool/wet)	M	S	S	Yes[‡]	(Ribeiro et al., 2011)
Willow warbler, chiffchaff	N	–	S	Yes	(Bensch et al., 2006)
Great tit, Japanese tit	M	–	S	Yes	(Kvist and Rytkonen, 2006)
Spotted towhee, collared towhee	N	–	S	Yes	(Kingston et al., 2012)
Icterine warber, melodious warbler	N	–	S	Yes	(Secondi et al., 2006)
~20 species gulls in genus *Larus*	M	–	S	Yes	(Crochet et al., 2003, 2018)
6 species holactic ducks	M	–	S	Yes*	(Peters et al., 2014)
5 species Locustella warblers	N/S*	–	S	Yes[†]	(Drovetski et al., 2015)
Western scrub jay, Woodhouse's scrub jay	N	–	S	Yes	(Gowen et al., 2014)
Common raven California type, common raven Old World	S	–	S	No	(Omland et al., 2000; Webb et al., 2011)

* Four of six comparisons showed the predicted pattern.

† There was no mt introgression between any species pair except one complete replacement of mt genotype between an invading and resident species.

‡ mt genes introgressed much less than N genes as predicted but Z-linked genes were not different than other N genes.

§ Differentiation between mt and autosomal N genes similar; Z-linked genes more divergent.

N = little or no gene flow between populations; M = modest gene flow between populations; S = substantial gene flow between populations.

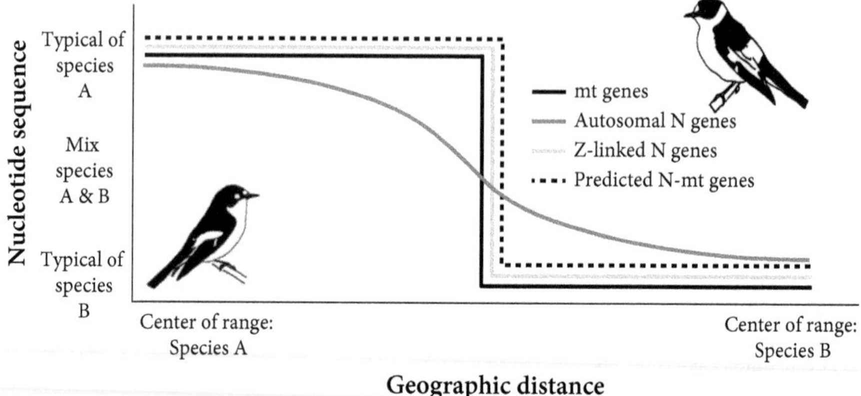

Figure 7.8 A simple graphical sketch of the common pattern of introgression and divergence of mt genes, Z-linked N genes, and autosomal N genes emerging in studies of the population structure of birds (Table 7.1). Z-linked and mt genotypes (two middle lines in stacked set) typically show an abrupt transition at species boundaries, while autosomal N genes show introgression across the boundary. The pattern that is predicted, but as yet untested, for N-mt genes that create co-function with mt genes is shown as a dashed line.

chromosome. These patterns of gene exchange are consistent with the hypothesis that selection is preventing the exchange of mt genotypes between the two species. I propose that the two populations have mt and N-mt genotypes that are incompatible. N-mt genes that are coadapted with particular mt genotypes segregate with plumage color, song type, and mate preference genes. The result is abrupt transitions in color, song, and mt genotype at a species boundary (Hill, 2017). It should be straightforward to test the prediction that there will be N-mt genes among the relatively few N genes that show fixed difference between golden-winged and blue-winged warblers.

The details of the golden-winged/blue-winged warbler story are unique to that pair of species, but the observation of little or no introgression of mt genes but substantial introgression of N genes has now been documented in over two dozen species of birds (Table 7.1). In addition, while the N genes on most chromosomes show a tendency to introgress across avian species boundaries, Z-linked genes seem nearly as resistant to introgression as mt genes (Carling and Brumfield, 2008; Rheindt and Edwards, 2011; Hill, 2017). To explain a lack of introgression of Z-linked genes, we have to consider the effect of sex linkage on the interaction of mt and N-mt genes.

Sex linkage and speciation

Effects of sex linkage on mitonuclear gene interactions

Genes on sex chromosomes evolve faster than genes on autosomes (Charlesworth et al., 1987). Faster evolution of sex-linked genes, which was first observed for

X-linked genes and dubbed the "fast X effect," was subsequently documented also for Z-linked genes called "fast Z effect" (Mank et al., 2007). The basis for this faster evolution has been attributed to differences in recombination rates of sex-linked genes and especially to loci in the heterogametic sex existing in a hemizygous state (one unpaired allele) in half of the population (Coyne, 1985). These characteristics of sex-linked loci would expose alleles at such loci to different intensities of selection and permit different rates of evolution (Ellegren, 2009a; Mank et al., 2014). Key to the topic of this chapter is that biologists have long recognized that loci on sex chromosomes also commonly play an important role in adaptive evolution and particularly in maintaining species boundaries (Coyne, 1985; Charlesworth et al., 1987). Effects of X-linkage have received most of the attention in the speciation literature, but effects of Z-linkage on speciation seem even more pronounced than effects of X-linkage (Ellegren, 2009b; Petit and Excoffier, 2009; Rheindt and Edwards, 2011; Irwin, 2018).

Sex linkage can significantly change the cost of mating across species boundaries if species carry uniquely coadapted sets of mt and N-mt genes (Hill, 2017). If there is XY sex determination or no sex chromosomes, then in the F1 generation there is never exclusive pairing of a paternally derived allele with maternal mitochondrial alleles (Hill, 2014a) (Figure 4.4). If N-mt genes are X-linked, then males in the F1 generation receive all of their N-mt alleles from their mother, such that there can be no effects of mitonuclear incompatibilities in F1 males (Figure 4.4). Females in XY systems receive one X chromosome from their mother and one from their father, so if N-mt genes are X-linked, the effects of mitonuclear incompatibilities will be the same as if the N-mt genes were on an autosome. The take home message from these patterns is that X-linkage of N-mt genes lessens the negative effects of hybridization between species with incompatible mt and N-mt genes because, in the F1 generation, paternally derived alleles are never the exclusive source of N-mt genes. When there is X-linkage of the genes that co-function with mt genes, paternally derived alleles are present only in combination with maternally derived alleles (Hill, 2014a).

On the other hand, the different pairings of maternal and paternal alleles in a ZW system (Figure 4.4) mean that the full costs of hybridization are manifest in the F1 generation if N-mt genes are Z-linked rather than autosomal or X-linked (Hill, 2014a). When there is ZW sex determination, females derive all of their Z-linked genes from their father (Figure 4.4). As a consequence, if the father's N-mt genes are Z-linked and not compatible with the mother's mt genes, the full negative consequences of the mitonuclear incompatibility will be experienced in the F1 generation by females (Figure 4.4). Male offspring, in contrast, receive one Z chromosome from their mother and one from their father, and negative consequences of hybrid pairing will depend on the penetrance of the incompatible paternal gene. Supporting the prediction that ZW females will show greater effects of genetic incompatibilities than ZZ males, in a study of hybridizing herring (*Larus argentatus*) and Caspian (*L. cachinnans*) gulls, the survival of F1 hybrid males was not different than that of males in either parental population, but the survival of F1 hybrid females was significantly reduced (Neubauer et al., 2014). In a comparative study in which all available data on hybridizing birds were compiled, 215 of 217 species of birds showed more hybrid inviability of females than males (Schilthuizen et al., 2011). This strong pattern of loss of viability of female but not male F1 hybrids also held for ZW lepidopteran taxa (Schilthuizen et al., 2011). The dilution of paternal gene effects in ZZ males but not in ZW females in birds and other eukaryotes with ZW sex determination is a potential explanation for Haldane's rule, which is the pattern across many eukaryotic lineages that the heterogametic sex suffers more in hybrid crosses than the homogametic sex (Schilthuizen et al., 2011; Hill and Johnson, 2013; Irwin, 2018) (Box 4.1).

In Chapter 4, I discussed the advantages of N-mt genes evolving to be on the Z chromosome in terms of enabling independent evolution of mt and N-mt genes. Because the Z chromosome is only 33 percent maternally transmitted (two-thirds of Z genes in a population are carried by males), its rate of co-transmission with the mt genome is lower than that of autosomal N genes (50 percent) or X-linked genes (67 percent). Independent transmission promotes independent evolution of mt and N genes (Hill, 2014a). Paradoxically, positioning N-mt genes on Z chromosomes could also be advantageous because it *promotes* the maintenance of coadapted mitonuclear gene complexes (Hill and Johnson, 2013). For Z-linkage to promote coadaptation, then both N-mt genes that co-function with mt genes and the genes for a phenotype that is the object of mate choice should be positioned on the Z chromosome such that genetically compatible mate can be recognized (Hill, 2017). Mate choice for compatible N-mt genes will be the focus of the discussion of mitonuclear mate choice in Chapter 8. Here, I make the point that positioning N-mt genes on the Z chromosome promotes both pre- and post-zygotic sorting of compatible mt and N-mt genes and hence promotes mitonuclear coadaptation. Because coadaptation is a key to the fitness of future generations, lineages with Z-linkage of N-mt that produced offspring with better mitonuclear coadaptation would produce offspring with higher fitness.

The hypothesis that key N-mt genes that co-function with mt genes should be Z-linked can be tested, but this is not a statistical prediction. It cannot be properly tested by simply looking at the total proportion of all N-mt genes that are sex-linked. For instance, the pattern of sex linkage of N-mt genes shown in Figure 4.5 is for all N-mt genes, of which only about 10 percent are expected to engage in key co-function with mt genes. The prediction is that those specific N-mt genes that differ between potentially hybridizing species and that create dysfunction when matched to the mt genes of the other species should be Z-linked. It only takes one locus with alleles with significant effects for Z-linkage to significantly change the dynamics of gene flow between species. An important caveat is that organismal biologists tend to be drawn to exceptional species pairs when studying speciation, such as species pairs that show higher rates of hybridization than typical species pairs in their clade. It is likely that such species pairs are exceptional for a reason, such as having atypical linkage of N-mt genes, and thus these sorts of exceptional species should not necessarily be taken as the models for the chromosomal position of N-mt genes. Rather, boring and typical sister species with low rates of hybridization should be studied to deduce the common pattern of linkage of N-mt genes.

Darwin's corollary to Haldane's rule

As presented in Box 4.1, Haldane's rule recognizes that the negative effects of hybrid pairings are not borne equally by the two sexes: the heterogametic sex pays a much steeper price. Another interesting outcome of hybrid pairings that has long been noted by animal and plant breeders is that hybrid incompatibilities are often not symmetrical: crossing males of species A with females of species B may not yield the same outcomes as crossing males of species B with females of species A (Figure 7.9). Because Darwin (1859) wrote about this phenomenon (chapter 8 in *On the Origin of Species*), it has been dubbed Darwin's corollary to Haldane's rule (Turelli and Moyle, 2007). Such asymmetries of hybrid pairings would not occur if the incompatibility factors causing hybrid dysfunction were located on autosomes; heterozygous gene combinations function the same regardless of the parental origin of each allele. Rather, differing viabilities in reciprocal crosses between species are necessarily due to uniparentally inherited loci. Such loci could be on sex chromosomes or they could be mt or chloroplast genes.

The strongest evidence that asymmetries in reciprocal crosses can be due to mitonuclear interactions comes from a comparative study of freshwater fish in

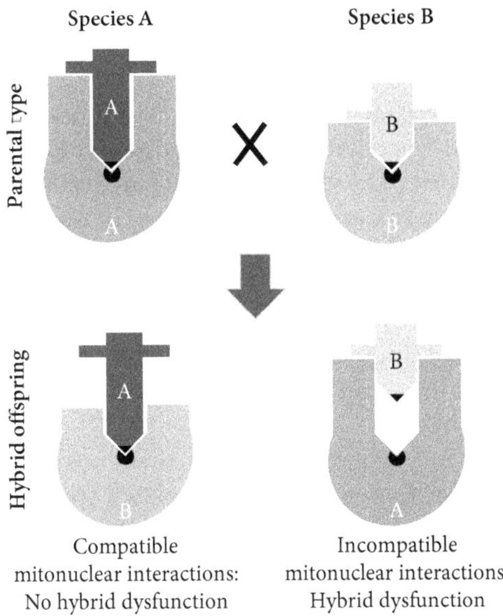

Figure 7.9 Simple pictorial illustration of asymmetrical mitonuclear effects on hybrid viability. Mitochondrial gene products are shaped like pointed bars and N gene products are cup-shaped. The key point of functional interaction is shaded in black. The mt gene product of Species A can co-function with the N gene product of Species B, but the mt gene product of B is conformationally incompatible with the N gene product of A resulting in reduced function and fitness.

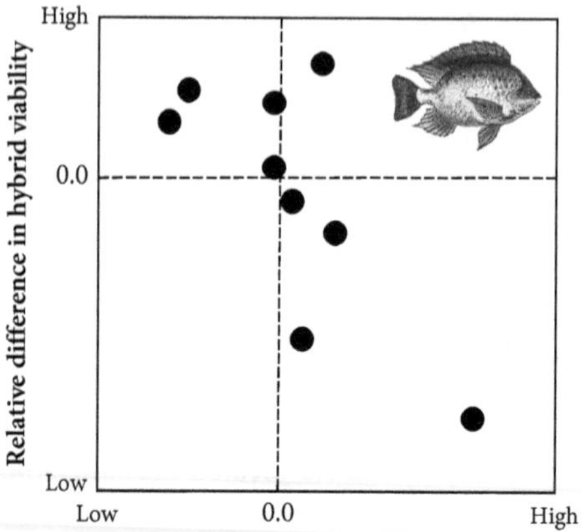

Figure 7.10 The relationship between rate of evolution of mt DNA (nucleotide substitution rate) and magnitude of asymmetry in hybrid viability in comparisons of Centrarchid fishes. The latter variable was a measure of how much the mt genotype of the taxon in question contributed to negative effects of hybrid crosses compared with the mt of a taxon with less mitochondrial evolution. Each plotted point is a comparison between two clades. This comparative analysis indicates that, for these fish, asymmetries in reciprocal crosses are due to differences in the function created by different crosses of mt and N genotypes. Figure adapted from Bolnick et al. (2008), which has details of the analysis.

family Centrarchidae, a taxon lacking sex chromosomes. Bolnick et al. (2008) found that the species with accelerated mitochondrial evolution, presumably due to mutational erosion, tended to be worse maternal parents in terms of viability of F1 hybrids than species with less accumulated change in mt genotype (Figure 7.10). These observations suggest that loss of fitness in F1 hybrids in these fish is due to mitonuclear incompatibilities and that hybrid dysfunction in the F1 generation is dependent on both mt and N genotype. This study supports mitonuclear interactions as a basis for hybrid dysfunction. It also demonstrates that even with no sex linkage of N-mt genes, there can be negative fitness consequences from mitonuclear interactions in F1 hybrids.

Earlier in this chapter, I advocated the idea that a species can be defined as a population of eukaryotes that is genetically isolated from other populations by incompatibilities in mt and N-mt genes (Hill, 2016, 2017). What, then, is the status of two populations that are incompatible with one male/female combination, but fully compatible with the remaining male/female combination? I propose that such populations are not species. Gene flow is only partially disrupted, and speciation is not complete until gene flow is disrupted through mating by any male/female combination. On the other hand, species with incompatibilities in both directions but with varying levels of severity would be recognized as good species.

What does mitochondrial introgression mean for speciation?

In the preceding sections, I presented data and explained why N genes introgress across species boundaries more readily and more often than mt genes. In this discussion, I focused particularly on birds, where the pattern is clear and well supported. Greater introgression of N than mt genes is not, however, the universal among eukaryotes. Indeed, there are numerous examples of complete or nearly complete (often referred to as "rampant") capture by introgression of mt genotype from one species into another, and in some eukaryotic lineages in particular, mitochondrial introgression appears to be, if not a common phenomenon, at least not a rare phenomenon among closely related species (Funk and Omland, 2003; Toews and Brelsford, 2012; Sloan et al., 2017).

Mitochondrial introgression is typically recognized when there is discordance in the boundaries between species that are identified by mt versus N genes. Toews and Brelsford (2012, p. 3908) defined genetic discordance as "a significant difference in the patterns of differentiation between [mt and N] marker types." In some cases, claims of introgression of mt genes are simply mistakes. The boundaries of a species might be demarcated so that it includes two populations with divergent mitonuclear genotypes when, under any species concept, proper analysis of the organisms would divide individuals into two species (McKay and Zink, 2010). Subsequent attempts to explain two mitochondrial types within one species are then founded on an assumption of neutral mitochondrial evolution, and introgression is invoked (Toews and Brelsford, 2012). There are, however, many well-documented cases of rampant introgression of mitochondria from one species into another species and hence of the movement of mitochondria from one nuclear background to another (Funk and Omland, 2003; Toews and Brelsford, 2012; Sloan et al., 2017). Such introgression of mt genotypes is predicted *not* to occur under the mitonuclear compatibility species concept (Hill, 2016, 2017) because it would lead to combinations of mt and N-mt genes that are not coadapted and hence to mitonuclear incompatibilities and loss of function. Indeed, introgression of mt genomes has been identified as the biggest challenge to the mitonuclear compatibility species concept (Sloan et al., 2017).

In an insightful perspective, Sloan et al. (2017) proposed that the selection to overcome the fitness costs of mutation erosion of mt genotypes might explain *both* the phenomenon of divergent coadaptation of co-functioning mt and N-mt genes leading to barriers to mitochondrial introgression (mitonuclear speciation) and the occasional replacement of one mitochondrial type by another through mitochondrial introgression. They proposed that there are two plausible evolutionary solutions to the problem of loss of mitochondrial function via mutational erosion: (1) compensatory coevolution (which is the foundation of the mitonuclear compatibility species concept; Chapter 3) and (2) wholesale replacement of less fit with more fit mt genotypes. If the explanations of Sloan et al. (2017) are correct, then mitochondrial introgression can potentially erase species, but the replacement of a more corrupted for a less corrupted mitochondrial type is driven by natural selection for better mitochondrial function. By this line of thinking, introgression of mt genomes poses no fundamental challenge to the mitochondrial compatibility species concept.

As I outlined in Chapter 3 (see especially Figure 3.11), while there will always be a fitness cost for matching sets mt and N-mt genes that are not coadapted, there can also be a fitness gain if a mt genotype that has been corrupted via mutational erosion is replaced by a mt genome with fewer deleterious mutations (Hulsey et al., 2016; Sloan et al., 2017). In addition, a foreign and non-coadapted mt genome might carry a genotype that enables better mitochondrial function in current environmental conditions providing another positive fitness benefit to counterbalance a loss of coadaptation (Boratyński et al., 2011, 2014; Melo-Ferreira et al., 2014, Hulsey et al., 2016) (see Chapter 9 for a complete discussion of adaptation related to mt genotype). Consequently, when the mt genotype is introgressed into the novel nuclear background of a closely related species, there are at least three measures of mitochondrial fitness to be considered: (1) How well do the mt genes co-function with the new N-mt genes? (2) How many deleterious alleles are lost or gained in the exchange? And (3) how well does the novel mt genotype perform under current environmental conditions? When the benefits of replacement exceed the costs of incompatibility, then introgression is predicted (Figure 3.11). In support of these ideas, Bonnet et al. (2017) used individual-based computer simulations of hybridization and gene flow to study the conditions under which mt genomes would introgress across species boundaries. They concluded that neither drift changing the frequencies of mt genotypes nor selection on N genes changing the frequencies of mt genotypes could explain observed patterns of mitochondrial introgression. On the other hand, natural selection on mt genotypes could generate the observed patterns.

The hypothesis that mitochondrial introgression can result from the replacement of more corrupted mt genomes with less corrupted genomes leads to specific predictions about the direction and circumstances of mitochondrial introgression. If genotypes can be assessed directly, then the prediction is that mitochondria with higher mutational loads should be replaced by mitochondria with lower mutational loads. Moreover, because drift and the accumulation of slightly deleterious genes should be greater in small versus large populations, flow of mt genes is predicted to be from large populations to small populations (Hulsey et al., 2016). If populations vary in mutation rate, for instance genetic bottlenecks might affect the mutation rate of populations (Lynch and Hagner, 2015), then the flow of mt DNA should be from the population with lower mutation rate to the population with higher mutation rates.

Introgression of mt genomes should also follow patterns of environmental change. Consider a simple example. If sister taxa carry mitochondria with different thermal optima and the environment of the cool-adapted species is undergoing rapid increase in mean temperature, then, all else being equal, the warm-adapted mt type is predicted to replace the cool-adapted type. Several potential examples have been published that suggest that mt genotypes have introgressed from one species to another under directional selection for better respiratory performance in the local environment that is enabled by the introgressing mt genotype (Boratyński et al., 2011, 2014; Melo-Ferreira et al., 2014; Hulsey et al., 2016). No systematic assessment has been published regarding the direction of mitochondrial introgression in relation to mutation load or adaptation, but anecdotal observations support predictions regarding mutational erosion (Figure 3.13).

Other potential drivers of mt introgression

Cytoplasmically inherited bacteria

In the previous sections, I focused on selection as the primary force both preventing the introgression of mt DNA and leading to rapid introgression from one species to another. However, in the current literature, non-adaptive mechanisms for introgression of mt genes are invoked much more commonly than adaptive explanations (Rheindt and Edwards, 2011). Because the mitochondrion is maternally inherited in most eukaryotes, mitonuclear discordance and introgression of mt genes across species boundaries could result from a range of female-specific effects. The most commonly invoked mechanism for discordance in the introgression of mt and N genes is female-biased dispersal. If females disperse farther and at a greater rate than males and if there is no selection on the mt genotype, then they will move mt genes across species boundaries faster than N genes (because the N genotype of offspring is diluted by half). In simulation models, Bonnet et al. (2017) found that female-biased dispersal could be an important mechanism for introgression of mt genomes and mitonuclear discordance.

A better documented, and perhaps more insidious, factor in the introgression of mitochondria is cytoplasmically inherited bacteria, which are obligate associates of a particular host and are variously called symbionts or parasites (Hurst and Frost, 2015). Here I will focus on the best studied of the bacterial associates, *Wolbachia*, which commonly infect arthropods and some other invertebrates (Werren, 1997). Because these microbes are strictly maternally inherited, they co-transmit with mitochondria (Perlman et al., 2015). Two important consequences of this mode of transmission are that (1) the microbes, like mitochondria, gain no fitness advantage from the production of males and thus they have evolved mechanisms for suppressing the production of males, and (2) the success or failure of a *Wolbachia* lineage is the same as the success or failure of the mt lineage with which it is associated (Perlman et al., 2015). Thus, if a hybridization event transfers both a novel symbiont and a novel mt genotype across a species boundary, then the success of the *Wolbachia* infection can drive the spread of the novel mt genotype, even if mitonuclear incompatibilities lead to loss of fitness. In other words, the mt genotype can hitchhike to fixation if the symbiont spreads to infect all individuals in the new population (Charlat et al., 2009; Klopfstein et al., 2016). Few examples of mitochondrial introgression driven by *Wolbachia* infection have been documented (Klopfstein et al., 2016), but it is a potentially important mechanism for mitochondrial introgression in some animal taxa.

Co-introgression of coadapted mt and N-mt genes

If coadapted sets of mt and N-mt genes move *together* across species boundaries, then there is potential to avoid fitness costs of introgression of a mt genome into an otherwise non-coadapted and incompatible N background (Burton and Barreto, 2012; Sloan et al., 2017). However, co-introgression of coadapted mt and N-mt genes is far from a straightforward process. As discussed in detail in Chapter 4, mt genes and N-mt genes are not physically linked, so tight co-transmission of mt and N-mt genes is not possible. Without sex linkage, mt and N-mt genes will transmit together only 50 percent of the time; stated another way, 50 percent of the linkage disequilibrium between mt and N-mt genes that exists in a parent will be lost in every subsequent generation of outbreeding (Lewontin, 1975). Rapid loss of association with the mt genome is trouble for introgressed N-mt genes trying to promote respiratory function: the benefits of the occasions when the N-mt genes are matched to mt genes will have to balance against the costs when they are expressed with a foreign mitochondrial type. For these reasons, co-introgression is predicted to actually work as sequential introgression: mitochondrial type A replaces mitochondrial type B in a sister taxon, for the reasons discussed in detail in the previous sections. Once mitochondrial introgression is complete, then there should be strong selection to secondarily introgress coadapted type A N-mt genes to restore full function of the introgressed mt genes by restoring mitonuclear coadaptation. It should be noted that, by the mitonuclear compatibility species concept, introgression of coadapted sets of mt and N-mt genes is an erasure of a species or "speciation in reverse" as mitochondrial introgression was recently called by Webb et al. (2011).

To date, there have been few well-documented cases of co-introgression of mt and N-mt genes. Among the best examples involves two species of fruit fly, *Drosophila yakuba*, a species that is widespread across central Africa, and *D. santomea*, a species that is endemic to the island São Tomé. *Drosophila yakaba* is found in the lowlands of São Tomé, while *D. santomea* is restricted to wetter uplands. The two species of fruit fly hybridize when they come into contact, and the mt genome of *D. yakuba* has been replacing the mt genome of *D. santomea* in *D. santomea* populations. Beck et al. (2015) predicted that N-mt genes might be co-introgressing with mt genes, and they found evidence that three N-encoded Complex IV proteins that have direct interaction with core, mt-encoded Complex IV gene products, particularly during complex assembly, were also introgressing from *D. yakuba* into *D. santomea* (Figure 7.11). It seems that the wave of introgressing mt genomes was being followed by introgression of coadapted N-mt genes, so this is potentially an example of sequential co-introgression but with little delay between the movement of the mt genome and the subsequent introgression of key N-mt genes.

Another study providing direct evidence for co-introgression of mt and N-mt genes concerns population of the Australian songbird, eastern yellow robin (*Eopsaltria australis*). Pavlova et al. (2013) found two highly divergent mt genotypes within what had long been considered one songbird species, and they documented an abrupt transition between mt genotypes across eastern Australia (Pavlova et al., 2013; Morales et al., 2015, 2017) (Figure 7.12). In the same comparison of individuals, however, most N genes showed no change in allele frequency across this same divide, showing instead a north to south cline perpendicular to that shown by mt DNA (Figure 7.12). Further analysis showed that on Chromosome 1A there was a cluster of genes that was highly enriched in N-mt genes. The Chromosome 1A genotypes of

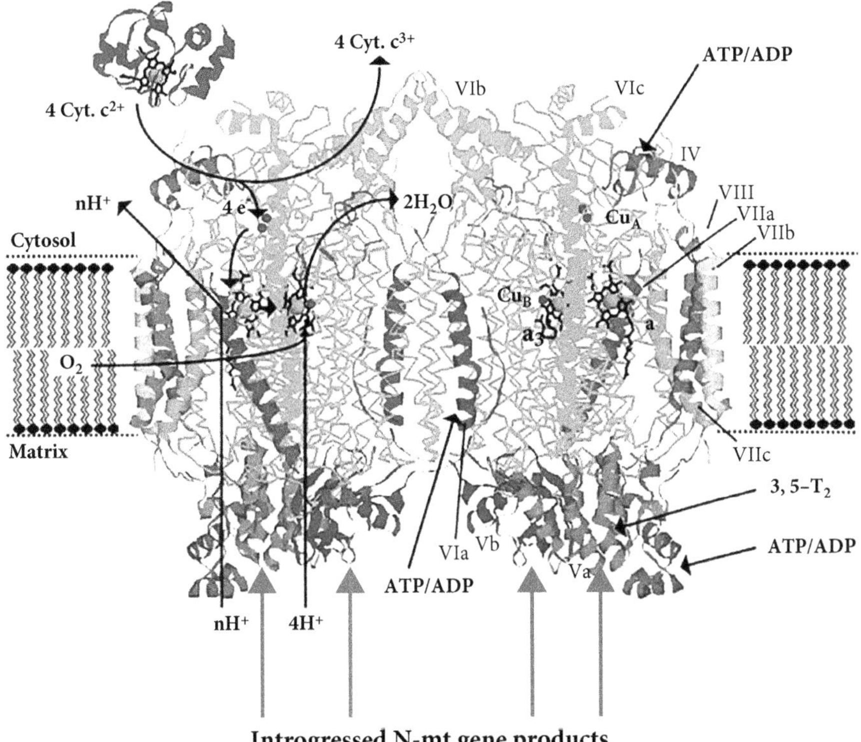

Figure 7.11 A structural model of Complex IV of a fruit fly. Genetic analysis indicates that there has been rampant introgression of the *Drosophila santomea* mt genotype into populations of *D. yakuba*. In addition, it appears that the N-mt genes that code for subunits Va and Vb (the bottom protein elements indicated by arrows) in Complex IV of the electron transport system have co-introgressed with the mt genes (Beck et al., 2016). Subunits Va and Vb interact directly with mt-encoded core subunits I, II, and III shown. Complex IV model reprinted from Kadenbach et al. (2000).

birds fell into two highly differentiated groups, the boundaries of which were coincident with the boundary between mitochondrial types (Morales et al., 2018) (Figure 7.12). One explanation for this pattern is that N-mt genes on Chromosome 1A co-introgressed with mt genotypes (Morales et al., 2018). The two divergent mt genotypes in eastern yellow robin are also linked to thermal adaptation (Box 9.3).

Another possible case of N-mt genes co-introgressing with mt genes concerns two sister species of bank voles in the genus *Myodes*. There has been rampant introgression of the mt genome of the northern/western species into the southern/eastern species that appears to be driven by climate adaptation (Boratyński et al., 2014). To test for evidence for co-introgression of N-mt genes along with the mt genome, Boratyński et al. (2011, 2016) conducted experimental crosses of the two vole species and documented sex-specific effects on physiology including core respiratory performance of hybrid voles. They concluded that the observed patterns were

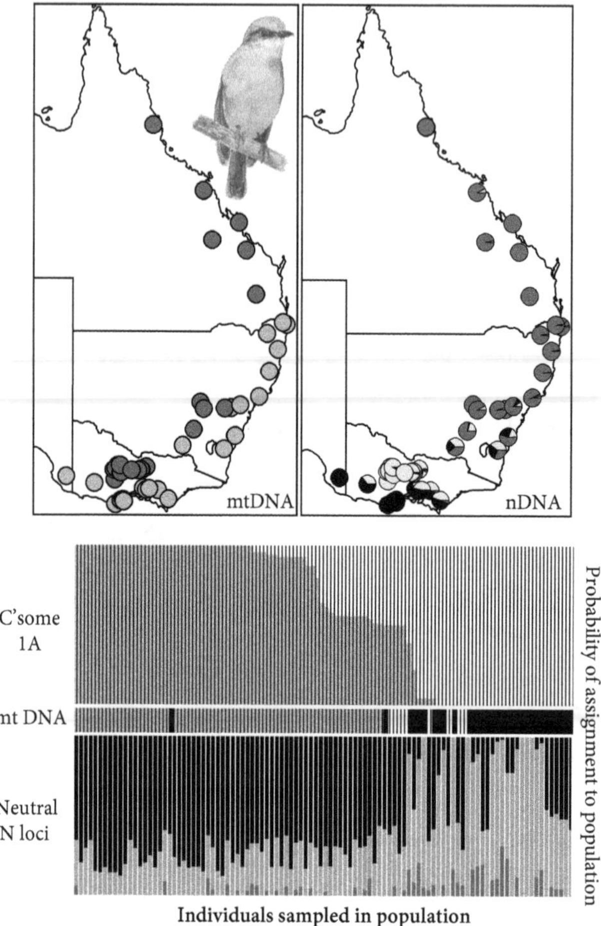

Figure 7.12 Evidence for co-introgression between two divergent populations of eastern yellow robin (*Eopsaltria australis*) of mt genotype and the N-mt genes that co-function with mt genes. Maps at top: the distribution of mt (left) and N genotypes (right), coded as dark or light shading, across eastern Australia. Bottom panels: Bar plots showing probability, for each sampled robin, of mt alleles, neutral N alleles, and N alleles on 1A chromosome being assigned to a specific population (shaded dark or light). Chromosome 1A holds a high number of N-mt genes and it shows a transition in genotype in concert with a transition in mt DNA haplotype. Most N loci, in contrast, show no change in allele frequency associated across the same boundary (Pavlova et al., 2013; Morales et al., 2015; Morales et al., 2017; Morales et al., 2018). Eastern yellow robin image by Paul Sunnucks used with permission. Maps adapted from Pavlova et al., 2013; Structure image provided by Hernan Morales.

consistent with sex-linked N-mt genes co-introgressing with mt genes between the two species.

Perhaps the most dramatic demonstration of the potential for co-introgression of mt and N-mt genes comes from an experimental hybridization study with *T. californicus* copepods. As described previously, populations of *T. californicus* copepods show mitonuclear incompatibilities that impact hybrid fitness in F2 and later generations (Burton et al., 2013). Pritchard and Edmands (2013) observed multiple laboratory populations after they were founded by mixing divergent wild populations of *T. californicus* copepods. The fitness and genetic outcomes of these mixed-source populations was highly dependent on the starting populations and stochastic factors in the first few generations, but one emerging pattern was a decline in mitonuclear incompatibilities with generational time. This pattern supports the hypothesis that during hybridization events, natural selection removes individuals carrying incompatible combinations of mt and N-mt genes because there are fitness advantages for maintaining coadapted mitonuclear genotypes (Pritchard and Edmands, 2013). Such selection would lead to the co-introgression of mt and N-mt genes.

A unified concept of species

Butlin and Smadja (2017) argued that mechanisms for isolation of gene pools tend to build upon each other to an endpoint at which many mechanisms halt the flow of genes, and speciation has been achieved. Orr (1995) had previously calculated that, once gene flow between species is halted by a single incompatibility factor, the evolution of secondary incompatibility factors should grow at a rate that is faster than linear. This hypothetical rapid accumulation of incompatibility factors has been dubbed "the snowball effect" (Presgraves, 2010). Such an endpoint with a multitude of mechanisms disrupting gene flow makes reconstruction of the path to speciation obscure and open to debate (Ortiz-Barrientos et al., 2004; Garner et al., 2018).

Given the obfuscation of the speciation process by the piling on of secondary incompatibility factors, it is hardly surprising that the past decades of speciation research have been characterized by consternation and endless debate about which particular processes are likely to drive speciation in different taxa (Wolf et al., 2010; Nosil and Schluter, 2011; Butlin et al., 2012). The conundrum is that well-bounded populations that are recognized as species exist in all eukaryotic lineages, and such a universal and defining characteristic of eukaryotes strongly suggests that there should be a common mechanism for such a ubiquitous endpoint. Paradoxically, however, we have literally dozens of proposed mechanisms to halt gene flow, each presented as an important mechanism of speciation, and many leading to new definitions of species. Divergence in coadapted sets of mt and N-mt genes is a process that may be common to all eukaryotes and could give rise to a single definition of species for all eukaryotes (Box 7.3).

Summary

Current models of speciation propose that species arise from disruption of gene flow between populations leading to divergence in N genotypes. This proposed process of speciation predicts the existence of nuclear incompatibility genes that stop the flow of genes between populations. Such nuclear incompatibility genes, however, have rarely been identified. Rather, it is much more common to find cytonuclear incompatibilities between divergent populations.

The new concept of speciation that I advocate in this chapter is that species evolve through the divergence in coadapted mitonuclear gene complexes with most N genes playing little or no role in most speciation events. According to the mitonuclear compatibility concept of species, mitonuclear coevolution in isolated populations leads to speciation because population-specific mitonuclear coadaptations create between-population mitonuclear incompatibilities and hence barriers to gene flow between populations. Elevating the interaction of mt and N genes to a central process in discussions of speciation leads to novel explanations for some of the most interesting and perplexing characteristics of species including Haldane's rule, asymmetrical outcomes of hybrid crosses, and the tendency for N genes to diffuse across species boundaries while mt and Z-linked genes typically show discrete transitions. Paradoxically, the rampant replacement of one mitochondrial type with another, which seems to pose a significant challenge to the mitonuclear compatibility species concept, might also be best understood through consideration of the role of mitochondrial function in species evolution. Sexual selection, speciation, and adaptation are intimately interconnected phenomena, and I will build from considerations of mitonuclear interactions in the process of speciation to show the importance of mitonuclear interactions in sexual selection and adaptation.

8

Mitonuclear mate choice

In the early chapters of this book, I emphasized the risk that is posed to eukaryotic lineages through the slow accumulation of deleterious mutations in the mt genome leading to mutational meltdown. Gradual mutational erosion, however, can be viewed as a small and manageable problem when compared with the immediate and disastrous outcome when an entire set of incompatible mt or N-mt genes is recruited for the production of offspring. The Titanic may have eventually sunk if it suffered from chronic maintenance problems that were not addressed over decades, but it sunk immediately due to catastrophic damage when it struck an iceberg. Choosing a sexual partner with incompatible mt or N-mt genes can be a mistake that is as catastrophic to the reproductive success of a eukaryote as striking an iceberg is to a ship in the Atlantic.

One way or another, selection will halt the mixing of mt and N-mt genes that are not coadapted. The brute-force solution is simply to eliminate, through natural selection, the unfit individuals produced through poor gene combinations. This is post-zygotic selection for reproductive isolation of coadapted mitonuclear gene sets, and it is the reproductive equivalent of hitting an iceberg. The result of poorly matching mt and N-mt genes is poor respiratory function, which will be revealed to natural selection by poor performance across a range key phenotypes that will result in low fitness. By this process, mating can be random with respect to the genetic suitability of potential mates, and post-zygotic selection perpetually cleans up the mess (Figure 8.1). For an individual engaged in sexual reproduction, however, there are huge advantages of pre-zygotic sorting of genotypes in the form of assessment of prospective mates for genetic suitability (Figure 8.1). Mitonuclear mate choice is the process of pre-zygotic assessment of mates for compatibility of mt and N-mt genes (Hill, 2018).

A quarter of a century ago, before the publication of thousands of additional studies, Malte Andersson required a 600-page tome to summarize the state of knowledge of sexual selection (Andersson, 1994). To make headway on this topic in a succinct chapter, I will necessarily summarize and simplify. This chapter is not presented as a comprehensive review of the literature on mate choice. Rather, my goal is to introduce the sorting of coadapted mt and N genotypes into the discussion of mate choice and the evolution of ornamentation.

Mitonuclear Ecology. Geoffrey E. Hill, Oxford University Press (2019).
DOI: 10.1093/oso/9780198818250.001.0001

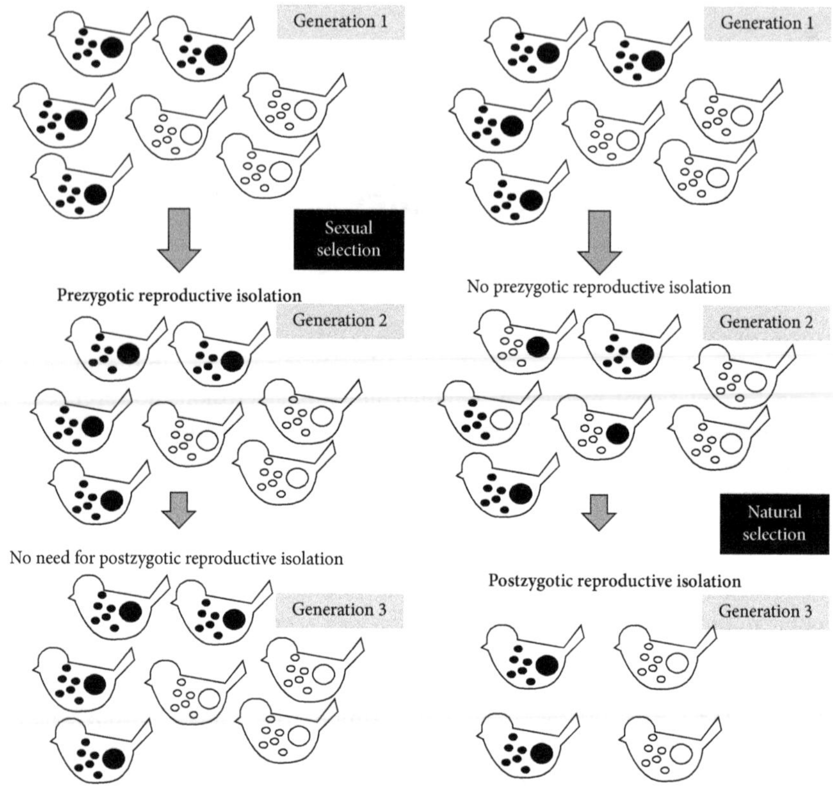

Figure 8.1 An illustration of the concepts of pre- versus post-zygotic reproductive isolation to maintain coadaptation of mitochondrial and nuclear genes. Bird shapes represent eukaryotic organisms. Arrows indicate transitions between generations. Large ovals represent nuclear genomes and small ovals represent mitochondrial genomes. Compatible genomes share dark or light shading. Pre-zygotic sorting in the form of mate choice allows for correct matching of compatible sets of mt and N genes across generations. Without mate choice, incompatible combinations are created that are subsequently removed from the population via natural selection in the process of post-zygotic reproductive isolation.

Mate choice basics

Given that two mating types comprise almost all eukaryotic populations (Chapter 5), choice for an opposite mating type is fundamental. But sorting males from females is not the interesting aspect of mate choice. Rather, the central and long-standing topic in evolutionary biology related to mate choice is whether, how, and why individuals of the opposite sex are sorted and assessed. What are the benefits of assessing potential mates (Reynolds and Gross, 1990; Hill, 2015c) (Box 8.1)?

Mate choice is often discussed strictly in the context of female mate choice, with the disclaimer that, in most species, females are choosy and males are not. Hence, discussions of mate choice are typically discussions of *female* mate choice (Andersson, 1994; Kuijper et al., 2012). But choosiness is a function of resource investment (Trivers, 1972). If an individual can mate many times per lifetime with little cost incurred in any individual mating, then mating mistakes with little chance of producing functional offspring are of no consequence and individuals will mate without discrimination (Kokko et al., 2003). As mating opportunities become limited and costly, individuals become more discriminating among potential mates (Trivers, 1972). Females often invest much more in reproduction than males, and hence females are often much choosier than males about their sexual partners (Bateman, 1948; Wirtz, 1999). In many eukaryotes, however, males invest as much or even more than females in the production of offspring and there is mutual mate choice (Burley, 1986; Hill, 1993; Johnstone et al., 1996). In a minority of species, males invest much more than females and males become much more discriminating than females (Jones et al., 2001). Thus, the emphasis on female mate choice and male display in the sexual selection literature is an emphasis on one extreme of a continuum across eukaryotes from high female investment/low male investment to high male investment/low female investment. In this chapter, I will discuss mate choice rather than female mate choice.

Box 8.1 A brief account of the long history of mate choice research

Invoking species recognition as a major component of mate choice and the evolution of ornamental traits gives new life to old and neglected ideas. Signaling species identity was the first specific function proposed for ornamental traits (Wallace, 1889). In contrast to Alfred Wallace's quest for adaptive explanations for mate choice, and hence his proposal of choice for conspecific mates, Charles Darwin (1871) simply accepted the phenomenon of female mate choice and moved on to a consideration of its consequences; Darwin never presented a coherent argument to explain *why* females assessed male ornamentation (Cronin, 1991). Wallace's species recognition hypothesis was widely embraced, and through most of the twentieth century, maintaining species boundaries was considered the primary, perhaps only, purpose of ornamentation and mate choice (e.g. Dobzhansky, 1937b and Mayr, 1963).

With the rebirth of interest in female mate choice and the development of modern sexual selection theory in the latter decades of the twentieth century, researchers tended to dismiss the importance of species recognition as an interesting part of the mate choice process. Instead of focusing on the benefits of choosing mates within species boundaries, researchers focused on two other classes of potential benefits from choosing mates among the pool of conspecifics: direct benefits in the form of good genes or resources and indirect benefits in the form of sexy sons (e.g. Weatherhead and Robertson, 1979; Maynard Smith, 1991; Andersson and Simmons, 2006). According to this latter hypothesis, nothing about the genetic or phenotypic quality of a potential mate is signaled by ornamental traits. Choosing individuals are proposed to mate with ornamented individuals because they find ornamentation aesthetically appealing (Prum, 2010). By mating with an aesthetically beautiful male, female produce offspring that are, in turn, aesthetically appealing and have high reproductive success (Weatherhead and Robertson, 1979). Presumably sexy daughters are also a benefit when there is female ornamentation and male choice, but mate choice for aesthetic beauty is always couched in terms of female mate choice (Lande, 1981; Kirkpatrick, 1987).

(Continued)

Box 8.1 *Continued*

Choice for direct benefits, on the other hand, is selection for better-than-average mates, with the assumption that degree of ornamentation is an honest signal of either good genes or resources (Albon and Clutton-Brock, 1979; Kodric-Brown and Brown, 1984; Hill, 2014c). These two hypotheses: choice for direct benefits or choice for indirect "sexy son" benefits, were often proposed as competing, stand-alone hypotheses for the evolution of ornaments (Bradbury et al., 1987). A goal of many studies in the late twentieth century was to devise tests to falsify one of these two hypotheses thereby supporting the remaining hypothesis, with species recognition generally not considered interesting (e.g. Hill, 1994b). The above statements tend to oversimplify decades of complex discussions, but they capture the general tendencies in these fields. Certainly, mitonuclear compatibility was never a part of any of these discussions.

In this era of sexual selection studies, species recognition as a primary function of mate choice faded from discussions among many evolutionary biologists not because the importance of species recognition was theoretically or empirically invalidated. On the contrary, data strongly supported choice for conspecifics across a diversity of eukaryotes from baker's yeast to flycatchers (Figure 8.3). Interest in the concept of species recognition waned because it was clearly not a universal explanation for all aspects of female mate choice. As one of the most influential evolutionary biologists of the late twentieth century wrote: "Choice by females of a conspecific mate cannot account for all male characteristics: it would be absurd to suppose that a male nightingale must sing like that in order for a female to tell that he is not a willow warbler" (Maynard Smith, 1991, p. 146). Because species recognition was not a plausible explanation for elaborate ornamentation (Hill, 2015a), it tended be dismissed as an explanation for the evolution of ornamentation in general. Recasting mate choice as essential for maintaining mitonuclear coadaptation returns the species recognition hypothesis to a central position in sexual selection theory.

The *raison d'être* of both males and females is to produce functional offspring. The theme of this book is that individuals will be highly functional only if they inherit compatible mt and N-mt genes (Hill and Johnson, 2013). The challenge is that each bout of sexual reproduction poses a threat to the maintenance of coadapted mt and N-mt gene sets and hence to the maintenance of core system function in eukaryotes. The defining aspect of sexual reproduction and the likely reason it evolved and exists in nearly all eukaryotic lineages is that it enables recombination of N genes (Chapter 5). Inevitably, and perhaps unfortunately, sexual reproduction also creates new combinations of mt and N-mt genes. The dissolution of sets of coadapted genes has long been recognized as the dark side of recombination (Otto, 2009). Throughout this book, I advocate that the most vital of all coadapted gene sets are the co-functioning mt and N-mt genes that create core respiratory function (Lane, 2011c). Each generation, the mixing of mt and N-mt genes through sexual reproduction creates the potential for offspring with mitonuclear incompatibilities. Mate choice can be a crucial mechanism to hold coadapted gene sets together.

Traditionally, mate choice is discussed within the context of animals using sensory systems to assess ornamentation (Jones and Ratterman, 2009), but the scope of mate choice is potentially much broader. There should be assessment of prospective mates whenever restriction of choice of mates to a subset of available individuals provides fitness benefits. Such mate assessment does not require sensory perception or behavioral response mechanisms. It simply requires a means to sort among potential mates. In support of the universal necessity of assessment of mates in eukaryotes, mate choice appears to be common in yeast (Murphy et al., 2006) and in plants (Moore and Pannell, 2011). A general theory to explain the evolution of mate choice, therefore, must consider the core assessments of sexual partners that are relevant to all eukaryotes. Theories of mate choice founded on sensory bias (Arnqvist, 2006) or on a sense of aesthetic beauty (Prum, 2012) cannot accommodate choice without sensory perception, and most mate choice across eukaryotes proceeds without a nervous system. Mate choice for mitonuclear compatibility is fundamentally mate choice for genomic interaction to achieve biochemical function and works at least as well with biochemical assessment as sensory assessment (Lymbery et al., 2017).

Because of the mode of transmission of mt genomes, males and females are predicted to make different assessments of the genotypes of prospective mates during mate choice (Figure 8.2). With few exceptions, males transmit no mt genes to offspring. For most eukaryotes, therefore, the male mt genotype cannot directly affect the genetic makeup of offspring or mitonuclear compatibility (Figure 8.2). On first consideration, it might seem that a choosing female should be unconcerned about the type or quality of mt genes carried by a prospective mate. However, even though males do not transmit mt genes to offspring, they do transmit N-mt genes, and the

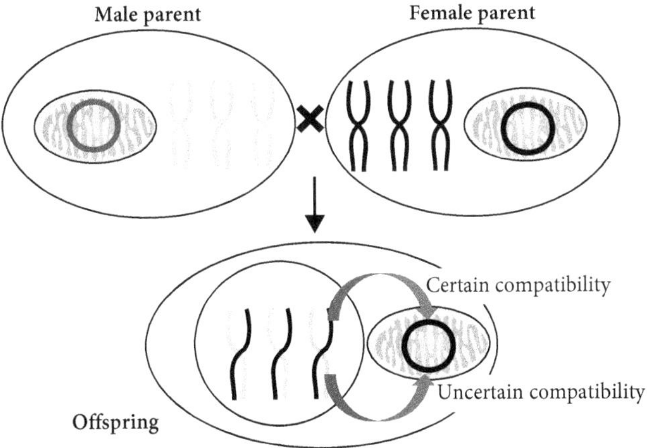

Figure 8.2 An illustration of the effect of sexual reproduction on combinations of mt and N genes. Nuclear genes are represented as paired chromosomes and mitochondrial genes as a circular haploid genome. Compatibility of genes is indicated by shading. Only autosomes are considered here. A female transmits one set of N genes that is compatible with the mt genes that she also transmits. The male contributes N genes that could be incompatible with the female's mt genes. Sexual selection is the mechanism for sorting N and mt genotypes for compatibility.

paternal N-mt genes must co-function with the mt genes of the female to enable cellular respiration in offspring (Figures 4.2 and 4.3). Thus, it is critical for females to know the mt genotype of males, even if those mt genes are not passed to offspring, in order to assess the potential compatibility of male N-mt genes. Males, on the other hand, need to worry about the compatibility, relative to their N-mt genes, of both the N-mt and mt genes contributed to offspring by a female mate. Accordingly, it is essential that both males and females choose a mate from within the population of individuals that share a common set of coadapted mt and N-mt genotypes in order to have offspring with fully compatible combinations of mt and N-mt genes (Figure 8.2).

Choice for shared mt genotype

The mitonuclear compatibility hypothesis of sexual selection

When I declare that choosing individuals should select a mate with shared coadapted mitonuclear genotype, what, specifically, is the hypothesis? Where are the boundaries between mitonuclear gene sets that have coevolved to be coadapted? As I argued in Chapter 7, the boundaries between coadapted sets of mt and N genes are species boundaries (Hill, 2016, 2017). Hence, choice for mates with shared coadapted mitonuclear genotype is choice for conspecifics (Hill and Johnson, 2013) (Figure 8.3). According to this line of reasoning, species recognition is the most fundamental aspect of mate choice because it is the key to preventing grossly incompatible mt and N-mt gene combinations (Hill, 2018). In recent discussions of the evolution of female mate choice, however, species recognition is rarely presented as a key mechanism that shapes patterns of display and mate choice.

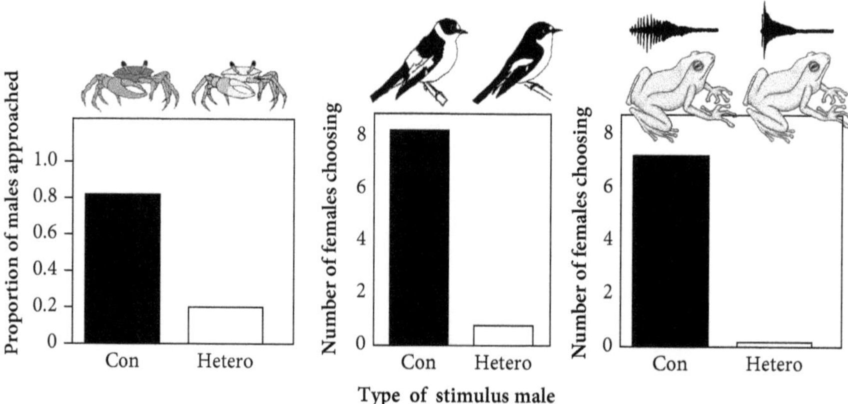

Figure 8.3 Choice for conspecific mate over heterospecific mate has been documented in a wide range of animals including (left) fiddler crabs (Detto et al., 2006), (middle) songbirds (Saetre et al., 1997), and(right) frogs (Rand and Ryan, 1993). Display traits, including color and pattern (left and middle), sounds (right), and a host of chemical, vibrational, electrical, or other signals (not illustrated) are used by choosing males and females to recognize conspecifics.

Box 8.2 Loss of ornaments on islands

When I was a student in the Bird Division of the Museum of Zoology at the University of Michigan, I took "Birds of the World" from the curator of birds at the time, Dr. Robert Payne. The class was focused on learning the distinguishing characteristics of all of the orders and families of birds in the world, and for our final exam we were informed that we would be given the study skins of twenty species of birds to identify to order and family. Upon entering the exam room, however, my classmates and I were horrified to see that all of the birds on our exam were uniformly black. These were not ravens or other species that are typically black. These were unique island taxa that had black feathers in place of their species-typical color pattern. For instance, the Bananaquit (*Coereba flaveola*) is almost always a bright yellow, white, and black bird, but the study skin in front of us on the exam table was from an all-black population of Bananaquits from the small island of Grenada.

At the time, I had nothing constructive to say about this difficult exam, but it did drive home an important point: island birds are unusual in their ornamentation. Many island taxa have either lost all species-typical plumage coloration and are plain, or they have gone melanistic and are entirely black (Peterson, 1996). From the perspective of the signal function of plumage coloration, these two endpoints might be the same—these island taxa have lost their species-typical plumage pattern.

In 1942, Mayr (1942, pp. 49–50) proposed that loss of species-typical plumage "seems to occur only in localities where no other similar species exist, i.e. where a highly specific male plumage is not needed as a biological isolating mechanism between two similar species." This remains the only viable explanation for loss of coloration in island birds (Badyaev and Hill, 2003). Genetic drift is sometimes held up as an alternative explanation, but I view drift as an extension of loss of selection resulting from the need for signaling species identity. Once the selective advantage of choosing species-typical plumage coloration is lifted, genetic drift can cause the loss of the trait. Drift will not be effective so long as there is selection for the trait.

I return to the long-standing hypothesis that choice for a conspecific mate is a fundamental component of mate choice that plays a central role in shaping sexual displays and ornamentation (Wallace, 1889) (Box 8.2). However, I recast this old hypothesis in the new framework of the necessity of mitonuclear coadaptation. In line with this argument, choice for conspecific mate is specifically choice for mt and N-mt genes that are coadapted with the genotypes of the choosing individual and that will create offspring with mitonuclear compatibility and full mitochondrial function.

Ornamentation gaps coincide with barcode gaps

The handful of behavioral studies directly demonstrating choice for conspecific mates over heterospecific mates is strong evidence for reinforcement in the process of speciation (Servedio and Noor, 2003) (Figure 8.3), but what is the evidence that boundaries of mating preferences and ornamentation coincide with boundaries between populations that have uniquely coadapted mitonuclear genes? In other words, what is the evidence for mitonuclear mate choice? I think that the most compelling evidence comes from analyses of avian mt genotypes for the purposes of establishing DNA barcodes (Hebert et al., 2004; Kerr et al., 2007; Barreira et al., 2016). The Barcode of

Life initiative was an effort to simply improve taxonomy (see Chapter 6); its progress should have been unrelated to sexual selection theory. The patterns that DNA barcoding revealed, however, were breathtaking when considered in light of the role of ornamentation in signaling mitonuclear genotype.

Across Class Aves, barcode gaps coincide with ornamentation gaps (Figure 8.4). To be more precise, clustering birds by ornamentation (song and plumage pattern), which is how taxonomists estimated the boundaries of bird species in the eighteenth and nineteenth centuries (Mayr, 1963; Price, 2015), and clustering birds based on mt

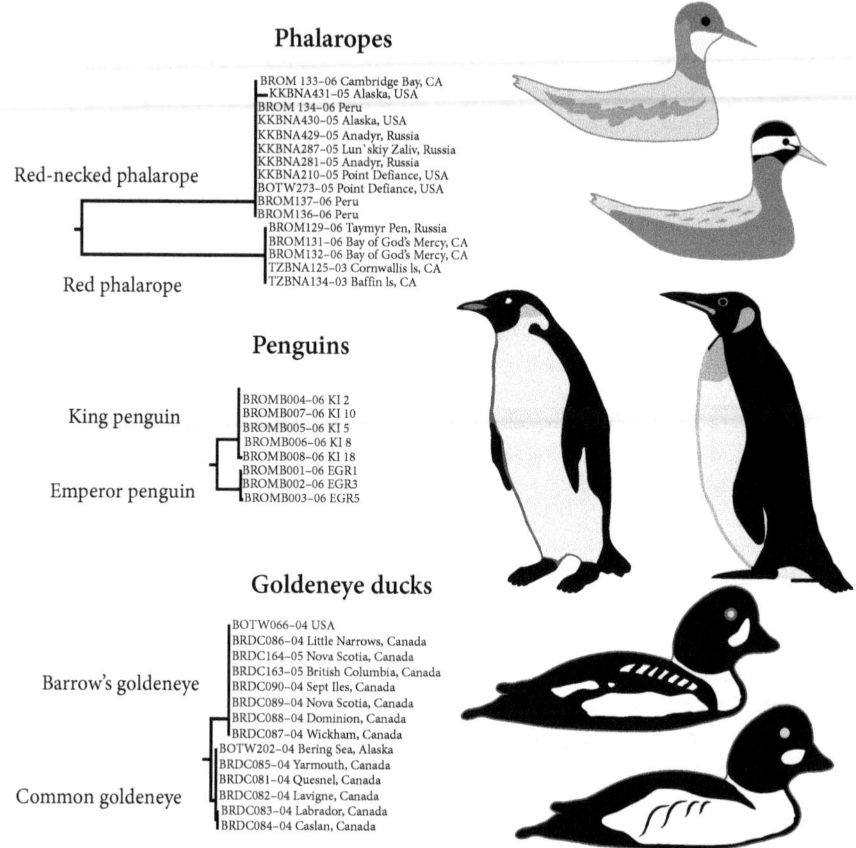

Figure 8.4 Among birds, major transitions in mitochondrial genotype correspond closely with transitions in ornamentation (either song or plumage pattern). Shown are analyses of the nucleotide sequences of COX1 genes compared between sister taxa of birds using neighbor-joining trees. Species-typical plumage patterns are illustrated next to phylogenies. In these three examples and in fifty-seven other comparisons of sister taxa of birds, both mt genotype and ornamentation showed abrupt transitions at species boundaries. In other words, a barcode gap matched an ornamentation gap. Data and explanation of genetic analysis in Tavares and Baker (2008). Figure based on Hill (2018).

genotype give almost the exact same result. The reason that bird watching is so popular and bird identification from field guides works so well around the globe is that there are very few populations of birds that cannot be quickly and unambiguously identified to species by a human observer based on plumage coloration or song. These same species-typical plumage patterns and songs are used by birds to sort conspecifics from heterospecifics during mate choice (e.g. Hill and Mcgraw, 2004; Saether et al., 2007; Uy et al., 2009). Twenty-first-century avian taxonomists then independently showed that, with remarkably few exceptions, these same species boundaries coincide with major transitions in mt genotype (Hebert et al., 2004; Tavares and Baker, 2008; Baker et al., 2009; Kerr, 2011). For instance, in goldeneye ducks, there are two distinct populations based on plumage pattern: one population that is mostly white with some black feathering and a second population that is mostly black with some white intermixed. Nineteenth-century taxonomists drew a species boundary between these two populations of ducks naming the lighter species common goldeneyes (*Bucephala clangula*) and the darker species Barrow's goldeneyes (*B. islandica*). When DNA barcoding was applied to these goldeneye duck populations, molecular taxonomists found a transition in the nucleotide sequence of mt genes that divided populations in precisely the same manner as did transitions in feather coloration (Tavares and Baker, 2008) (Figure 8.4). Moreover, for neither plumage nor genotype are these transitions gradual (or clinal as population geneticists would say). They are abrupt. This pattern of congruence between mt DNA barcode gaps and transitions in song or coloration is so universal and so conspicuous in birds that it has simply been taken for granted (Hill, 2018). I argue that the coincidence between transitions in ornamentation and transitions in mt genotype is striking support for the hypothesis that mitonuclear genotypes are sorted during mate choice in birds. The obvious and untested prediction is that transitions in mt genotype and ornamentation should also coincide with transition in the N-mt genes that co-function with mt genes. This prediction can be tested with existing data sets.

The hypothesis that changes in the plumage pattern of birds occur during speciation events as signals of divergent genotypes was further supported in a study of New World orioles. Matysioková et al. (2017) mapped song and color pattern onto a well-supported phylogeny of orioles. They found that most of the transitions in color pattern were concentrated at speciation nodes, as predicted if plumage was signaling species-typical genotype. Future studies that consider the evolution of mt genes, N-mt genes, and ornamentation should provide more direct tests of the hypothesis that species-typical ornamentation evolves as signal of mitonuclear genotype.

Sex linkage and sexual selection

ZW sex determination and ornamentation

As I introduced in the last chapter, the fitness consequences of hybrid pairing and the mixing of sets of mitonuclear genes that are not coadapted are strongly affected by the system of sex determination and the position of N-mt genes on chromosomes. If N-mt genes are positioned on autosomes or if they are X-linked, then the negative effects of mitonuclear incompatibility may be delayed until the F2 and later generations because females provide to offspring both the entire mt genome and a complete set of N-mt genes that is coadapted with her mt genotype (Burton et al., 2006) (Figure 8.2). If, on the other hand, the N-mt genes that interact with mt genes are positioned on the Z chromosome, then Fl female offspring receive all of their Z chromosomes from their father, and they will experience the full negative effects of pairing non-coadapted paternal N-mt genes with maternal mt genes (Figures 4.2 and 4.3). As a consequence, hybrid pairing could result in the loss of viable female offspring in the Fl generation, which creates a huge cost to hybrid pairings (see section on Haldane's rule in Chapter 4) and should lead to stronger selection for species recognition and choice of conspecific mate.

A conclusion from these considerations of sex linkage of N-mt genes is that Z linkage will lead to stronger selection for mate choice for mitonuclear compatibility than X linkage or autosomal positioning of N-mt genes (Hill and Johnson, 2013; Hill, 2018). Stronger selection for mating with individuals with compatible mt and N-mt genes should, in turn, lead to stronger selection for ornamentation that signals mitonuclear genotype. Putting this together, mitonuclear mate choice predicts that organisms that have ZW sex determination will be more ornamented than taxa that lack sex chromosomes or that have XY sex determination.

Long before there was consideration of a role for sorting coadapted mt and N-mt genes in mate choice, Reeve and Pfennig (2003) conducted an expansive comparative analysis that assessed the effects of chromosomal sex determination on ornamentation. The patterns that they observed were clear and striking. In their study, they included birds, lizards, snakes, frogs, salamanders, fish, and insects, and they found that for all of the taxa examined there was a significant positive relationship between ornamentation and whether or not sex determination was linked to the Z chromosome. They measured the "Zness" of lineages with a chromosome score in which 0 was assigned to species with no sex chromosomes—there were only paired autosomes with the same gene content. Negative scores were assigned when there were sex chromosomes and males were the heterogametic sex. The lowest, most negative values were assigned to species with the greatest disparity in size (gene content) of the X and Y chromosomes. Positive scores were assigned to species with sex chromosomes in which the female was the heterogametic sex with higher scores for greater disparity in Z and W chromosomes. The result was a "Zness Score" ranging from low negative numbers for taxa with diminutive Y chromosomes to high positive numbers for taxa with diminutive W chromosomes.

When Reeve and Pfennig (2003) crunched the numbers, they found that ZW taxa were consistently more ornamented than XY taxa (Figure 8.5). This stunning pattern held up across all of the taxa examined. Even in birds, which have exclusively ZW sex determination, Zness predicted ornamentation. This comparative analysis supports the prediction that species with ZW sex determination will be subject to stronger selection for signals of species identity than species with no sex determination or with XY sex determination.

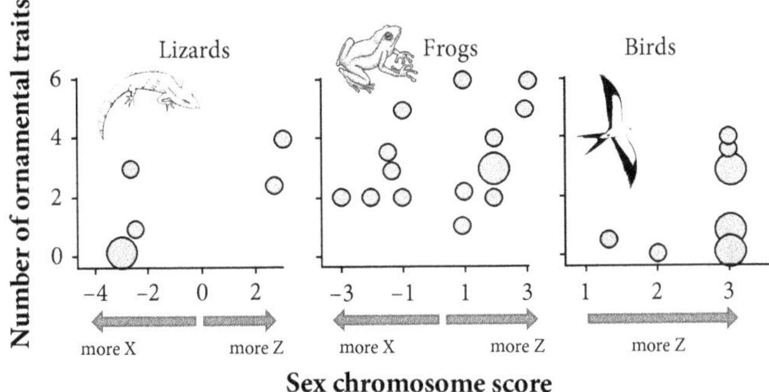

Figure 8.5 Relationships between sexual ornamentation and system of sex determination for three classes of vertebrates. The sex chromosome score is based on the degree of size disparity between sex chromosome with negative values used when males are the heterogametic sex (XY) and positive values used when females are the heterogametic sex (ZW). Ornamentation was measured as the number of ornaments. Points are average values for species within an order and larger points are overlapping data for two to five orders. For these three comparisons, as well as for comparisons of fish, insects, snakes, and salamanders (not shown), there was a significant relationship between sex determination and ornamentation. Redrawn from Reeve and Pfennig (2003).

There are certainly other explanations for this association between sex determination and ornamentation (Hastings, 1994; Reinhold, 1998). Reeve and Pfennig conducted their analysis against the backdrop of the hypothesis that rare alleles are better protected from loss due to drift when they are Z-linked and hence novel ornamental traits are more likely to evolve in ZW systems. Kirkpatrick and Hall (2004) used simulation models to assess how sex linkage could affect ornament evolution and they concluded that Z linkage increases the correlation between preference and ornamentation and makes a runaway process much more likely. Note that a stronger link between preference and ornamentation also accommodates mitonuclear mate choice (see the next section) and that mitonuclear mate choice is not presented as a hypothesis for the evolution of highly elaborate ornaments (Hill, 2018). Thus, the Kirkpatrick and Hall (2004) explanation and mitonuclear mate choice are entirely compatible as explanations for the association between sex determination and sexual selection revealed in the Reeve and Pfennig study (Reeve and Pfennig, 2003).

Linkage of ornamental traits

For species-typical ornamental traits to serve as reliable signals of mt and N-mt genotype, there should be genetic linkages among genes for species-typical ornamentation and N-mt genes (Figure 8.6). Such linkages ensure that when a mate is chosen based on species-typical ornamentation, the correct N-mt genes will accompany the ornament genes (Hill, 2017, 2018). Without such linkages, species-typical ornamentation and species-typical N-mt genes could be inherited independently such that the wrong set of N-mt genes might be associated with a particular species-typical ornamentation.

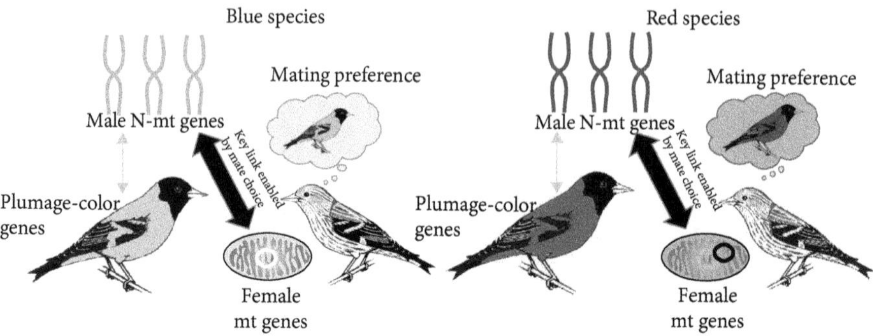

Figure 8.6 A diagram illustrating how sexual selection can sort for coadapted mt and N-mt genes via female mate choice. Genes for traits that are used to sort conspecifics from heterospecifics, in this example dark feathers versus light feathers, are genetically linked to N-mt genes that are coadapted with mt genes. By choosing the correct, species-typical ornament type, females choose N-mt genes that are compatible with their mt genes. Mating preferences could arise from behavioral imprinting on the ornament type of the father. If mating preferences are genetically determined, mate-choice genes should also be linked to ornamentation genes and N-mt genes.

As the reliability of the species-specific ornamentation decreased, the value of choosing species-typical ornamentation would be lost. Physically linking N-mt genes and genes for species-typical ornamentation used in mate choice would ensure the value of choosing based on species-typical ornamentation. In addition, if mating preference is genetically determined, then genes for mating preferences should also be part of this linkage group (Hill, 2017, 2018). Alternatively, mating preferences might be determined by imprinting (Verzijden et al., 2005), which would also maintain links between N-mt genes, ornamentation, mt genes, and mate choice.

Linkage on any chromosome would promote co-transmission of N-mt genes and ornamentation genes, but linkage on the Z chromosome places ornament genes and N-mt genes at the position where selection against hybrids will provide the strongest purifying selection to maintain the association between N-mt genes and ornament genes (Hill and Johnson, 2013; Hill, 2017, 2018). The strong purifying selection for genes that are Z-linked arises because females inherit all of their Z genes from their father and thus incompatibilities between paternal N-mt genes and maternal mt genes are fully revealed in F1 females. A lineage with Z linkage of N-mt genes would maintain tighter mitonuclear coadaptation across generations than a lineage in which N-mt genes were on autosomes, for which there may be no fitness loss in the F1 generation even if paternal N-mt genes are poorly coadapted with maternal mt genes. This should lead to Z linkage of N-mt genes.

Given the prediction that N-mt genes, genes for ornamental traits, and genes for mating preferences should be Z-linked, it is especially intriguing that genes for plumage coloration in birds frequently map to the Z chromosome (Qvarnstrom and Bailey, 2009) and that in *Ficedula* flycatchers the genes for both species-typical plumage pattern and female preferences for conspecific plumage map to the Z chromosome

(Saether et al., 2007). The prediction from the mitonuclear mate choice hypothesis is that N-mt genes that interact with mt genes will also map to the Z chromosome, but this prediction has yet to be tested, let alone confirmed.

One of the most interesting patterns emerging from genomic analysis of closely related bird species is that, even as there are abrupt and coordinated transitions in mt genotypes and species-typical ornamentation at species boundaries, sets of N genes typically show much less abrupt transitions at species boundaries (e.g. Carling and Brumfield, 2008; Gowen et al., 2014; Toews et al., 2016b; Walsh et al., 2016; Yang et al., 2017) (Figure 7.5). Moreover, the introgression of N genes in birds is not uniform. The common pattern that is emerging from genomic analyses is that Z-linked genes are much less likely to introgress across species boundaries than autosomal N genes. Genes that code for species-typical ornamentation also tend not to introgress across species boundaries. This pattern suggests that there is active selection against the introgression of mt genotypes, Z-linked genes, and ornaments—but not N genotypes—across species boundaries. I propose that the selection that generates these patterns is provided by mate choice for species-typical ornamentation linked to N-mt genes (Hill and Johnson, 2013; Hill, 2018).

Assessment within species

Avoiding the mistake of mating outside of species boundaries circumvents the genetic catastrophe of producing offspring with grossly incompatible mt and N-mt genes, but from the perspective of mitonuclear compatibility and mitochondrial function, there should still be substantial benefits to assessing the quality of prospective mates from within a pool of conspecifics (Hill and Johnson, 2013; Hill, 2018). Even if the overall genotype of a prospective mate is part of the same lineage of coadapted mt and N-mt genotypes, variation in genotype that can affect mitochondrial function may still be present because of mutation, gene flow, and adaptive variation (Picard et al., 2016). The hypothesis that there might be good genes benefits for mate choice is decades old and remains controversial (Box 8.3). Here, I focus on mitonuclear gene interactions and mitochondrial function.

Avoiding potential mates with deleterious mutations in either N-mt and mt genes that reduce mitochondrial function is likely to be an important benefit of mate choice (Hill and Johnson, 2013; Hill, 2018). As reviewed in Chapter 2, among bilaterian animals, there are thirty-seven mt genes, all of which have close functional interactions with N genes, and there are about 180 N-mt genes that produce products that are closely associated with mt gene products. Promoters and transcriptional start sites must also be coadapted (Ellison and Burton, 2010), and there may be other interacting molecules encoded by mt genomes (Breton et al., 2014; Pozzi et al., 2017). Despite what appears to be a small number of genes in play, many inherited diseases in humans involve either mt or N-mt genes, and these inherited diseases affect mitochondrial function (Lightowlers et al., 2015; Picard et al., 2016). Poulton et al. (2010) estimated that one in 400 people carry at least one deleterious mtDNA mutation that

Box 8.3 Benefits for mate choice

For more than four decades, a major focus in evolutionary and behavioral ecology has been understanding the good gene, resource, and sexy son benefits that underlie mate choice for ornamentation (reviewed in Andersson, 1994). In this chapter, however, it is as if the thousand papers that were generated through this effort do not exist, and I'm sure that my failure to mention this literature on the benefits to mate choice will be maddening to many of my colleagues. I am, of course, well aware of the mate choice literature, having spent a majority of my career adding a few bricks to that foundation. My point in writing this chapter is not to deny the importance of assessment of social status, territory quality, parental investment, parasite load, immune function, major histocompatibility complex (MHC) genotype, stress hormone levels, or any of the hundred other aspects of an individual that have been linked to ornamentation and mate choice. Rather, my goal is to lay out the hypothesis that the most fundamental aspect of mate choice is assessment of N-mt and mt genotypes toward the production of offspring with mitonuclear compatibility and high respiratory function. Such an assessment of mitonuclear genotype does not mean that the choosing individual is not also interested in, for instance, the differences in resources that two potential mates might contribute. It simply puts an emphasis on what I argue is the most fundamental aspect of mate choice. In addition, many aspects of individual condition that are linked to ornamentation such as immunocompetence, oxidative stress, cognitive ability, and motor coordination are likely to be mediated through mitochondrial function (Hill, 2014b; Koch et al., 2017; Koch and Hill, 2018).

causes a serious disease. Gorman et al. (2015) estimated that, in an English population, a minimum rate of adults expressing clinically overt mitochondrial disease related to neurological function was 20 per 100,000 for disease resulting from mt DNA mutation and 2.9 per 100,000 for disease resulting from N-mt DNA mutation. Other forms of mitochondrial disease, for instance fatigue syndromes (Myhill et al., 2009), were not included in these calculations, nor were subclinical dysfunctions. Rates of subtle mitochondrial dysfunction in the human population are unknown but undoubtedly high (Tuppen et al., 2010; Wallace et al., 2015). This load of deleterious alleles in the human mt and N-mt genes is partly the result of deleterious mutations, and purifying selection seems to be relatively ineffective at purging mildly deleterious alleles from vertebrate populations (Fan et al., 2008). Avoiding prospective mates carrying such genes would certainly be a tangible benefit to mate choice.

Gene flow and adaptive variation are related mechanisms for the generation of genetic variation that can affect individual fitness within a population (a central topic of Chapter 9). In the context of this discussion, adaptive variation refers to standing genetic variation in mt or N-mt genotypes within a species that is related to a functional adaptation. Only a few years ago, the existence of functional variation in mt genotypes within a species was widely dismissed (reviewed in Ballard and Kreitman, 1995), but such functional variation in mt genotypes within populations is now well established (Dowling et al., 2008; Kazancioğlu and Arnqvist, 2014). Here, we consider the challenge confronted by an individual choosing a mating partner when mt and N-mt genotypes vary in fitness in a particular environment.

Some of the best evidence for standing variation in mt genotype that has an effect on fitness continues to come from studies of human populations. The "deleterious" alleles that are the focus of clinical studies may actually be alleles that are adaptive in some environments but maladaptive in other environments. For instance, Leber's hereditary optic neuropathy (LHON) disease is a malady in humans caused by a nucleotide change in the mt genome (Brown et al., 2002). Because it can cause blindness in some individuals, the LHON allele is typically classed as a deleterious mutation carried in the human population (Chinnery, 2014). However, whether or not the LHON allele actually causes blindness is dependent on the N-mt genes with which it is matched and the environment in which a phenotype is expressed (Cock et al., 1998; Hudson et al., 2005; Tońska et al. 2010). Potential LHON-causing alleles occur at high frequency in human populations living at high altitudes in Asia with no reports of LHON disease in those high-altitude populations (Ji et al., 2012). The implication is that there are mt genotypes that are adaptive at high altitudes, but when gene flow moves these high-altitude alleles to individuals living in lowland environments, with different mt/N-mt gene pairings and a different hypoxic environment, they cause disease and loss of fitness. This example illustrates the interaction of adaptive variation and gene flow as mechanisms for generating standing variation in mt and N-mt genotype that might be the focus of mate choice. For such low-fitness alleles to be avoided in mate choice, choosing individuals need to be able to assess mitochondrial function.

Signals of mitochondrial function

Because it is likely that within the population of potential conspecific mates there will be deleterious alleles that need to be avoided as well as alleles that enable better or worse adaptation to the current environment, a choosing individual should benefit by assessing the genetic quality among potential conspecific mating partners specifically related to mitochondrial function (Hill and Johnson, 2013; Hill, 2018). But how can there be assessment of the functionality of N-mt and mt genes carried by prospective mates? How does a mate-seeking individual assess the genotype of a potential sexual partner? Obviously, genotype is not assessed directly. Rather, I propose that *mitochondrial function* should be the focus of assessment in mate choice (Hill and Johnson, 2013; Hill, 2018) (Figure 8.7). Even with assessment moved away from direct assessment of genotype to assessment of core physiological function, the choosing individual is still faced with a considerable challenge. There is no way to *directly* assess the membrane potential or production of free radicals in the mitochondria of a potential mate. Choosing individuals need indicators of mitochondrial function that can be reliably and easily assessed. I propose that many ornamental traits in animals have evolved specifically to serve as signals of mitochondrial function (Hill, 2014b, 2018).

In discussions of ornamentation as a signal of mitochondrial function, I am not focused on the elaborate and fantastic ornamentation of a few animal taxa, like the train of a peafowl or even the full pheromone plume of a house mouse. The most elaborate ornamentation in animals, which adorn only a tiny fraction of all animals, is not condition-dependent and requires special explanations (Prum, 1997; Hill, 2015c).

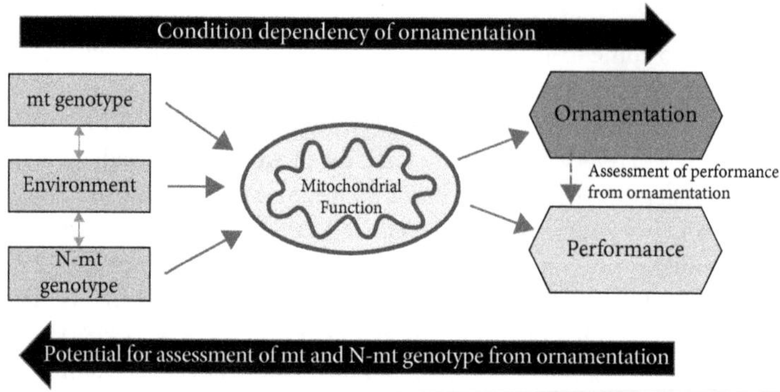

Figure 8.7 An illustration of the hypothesis that condition-dependent ornaments are signals of mitochondrial function, linking mitonuclear coadaptation to ornamentation and performance. According to this hypothesis, mitochondrial function determines both functional performance such as developmental rate, agility, cognition, and immunocompetence, and ornamentation such as song quality or carotenoid coloration. Adapted from Hill (2011).

I propose that the types of ornamental traits in animals that have been repeatedly linked to individual condition, like bursts of sounds, displays of cognitive function and motor coordination, modest patches of coloration, and chemical displays, are signals of mitochondrial function that enable choosing individuals to select mates with more fit mt and N-mt genotypes (Hill and Johnson, 2013; Hill, 2014b, 2018; Koch and Hill, 2018). The hypothesis that condition-dependent ornamental traits are signals of respiratory function is still largely speculative, but new studies on carotenoid coloration in birds provide the first evidence in support of the idea.

Species-typical vs condition-dependent ornamentation

I draw a key distinction between species-typical ornamentation and condition-dependent ornamentation (Figure 8.8). To be effective as signals of species identity, ornaments should be simple, easily assessed, and show little variation within a population (Dale, 2006). Signals of species identity should also change abruptly at species boundaries (Figure 8.4). The function of such traits is to unambiguously signal the identity of species, so having such traits vary in expression depending on the condition of an individual would lead to signal ambiguity and mistakes in distinguishing conspecifics from heterospecifics (Hill, 2015a). Signals of species identity are also predicted not to be extremely elaborate or complex, such as the plumage of birds-of-paradise, because complexity will also lead to more mistakes in species recognition (Hudson and Price, 2014; Hill, 2015a; Martin, 2015). Most bird species have a pattern of plumage colors or song that is distinctly different than the plumage pattern or song of any other species but that is not extremely elaborate.

Signals of individual condition, on the other hand, are predicted to vary in expression in relation to the condition of an individual. These types of ornaments are predicted to

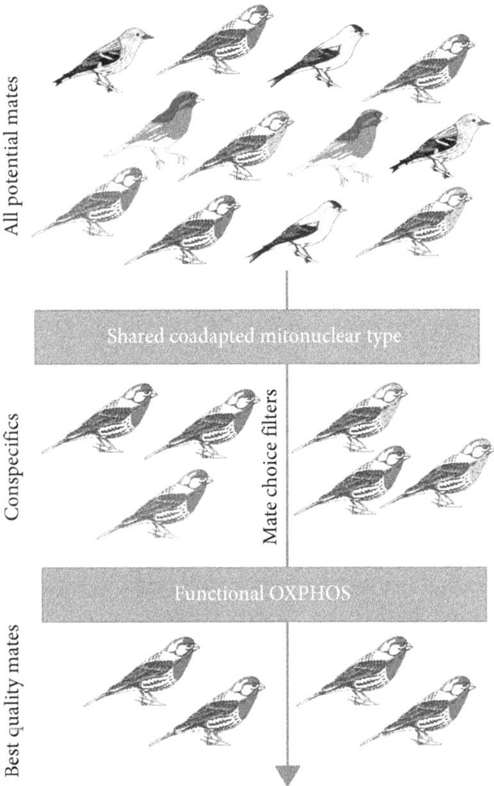

Figure 8.8 A pictorial representation of the hierarchy of mate assessment that would lead to correct sorting of mt and N-mt genotypes. In this example, choice is from the perspective of a female house finch. The first critical sorting is between conspecifics and heterospecifics. This level of mate assessment matches species-typical mt and N-mt genes in offspring. The second tier of choice involves assessment of condition-dependent ornamentation, in this case carotenoid pigmentation of feathers. If feather coloration is dependent on mitochondrial function, as hypothesized, then choice for a particular coloration enables the female to avoid genes that underlie poor mitochondrial function. From Hill (2018).

be highly variable within a population such that a choosing individual can assess the quality of prospective mates (Kodric-Brown and Brown, 1984; Dale, 2006; Hill, 2015c). Because there exist only a finite set of traits that serve as reliable signals of quality, it is expected that many taxa will converge on the same signals of individual condition (Hill, 1994a; Hagman and Ord, 2016; Badyaev et al., 2017). Red carotenoid coloration, displays of rapid and complex movement, and chemiluminescence may be examples of condition-dependent signals that have evolved repeatedly in diverse groups of animals.

Importantly, genes for species-typical ornamentation are predicted to be on the Z chromosome, while genes for condition-dependent ornamentation traits are not. Genes for species-typical plumage pattern are predicted to be Z-linked because N-mt genes with which they should be co-transmitted should be Z-linked. Because condition-dependent traits signal general condition and not specific N-mt genotypes, there is no benefit to making such genes Z-linked. For instance, the gene that codes for the ketolase enzyme that catalyzes the oxidation of yellow dietary pigments to red pigments, which is the basis for most of the red coloration in birds and turtles, was recently discovered and it is not positioned on the Z chromosome in the taxa so far examined (Lopes et al., 2016). This observation is not surprising because red carotenoid coloration is more important in signaling individual condition than signaling species identity. In contrast, genes for species-typical plumage pattern do appear to be Z-linked (Qvarnstrom and Bailey, 2009). An assessment of the genetic basis for different classes of ornamental traits will enable clear tests of the mitonuclear mate choice hypothesis.

Carotenoid coloration in birds as a signal of mitochondrial function

Most of the red, orange, and yellow coloration of birds results from carotenoid pigments. Carotenoids are especially interesting animal colorants because these pigment molecules cannot be synthesized by animals; they have to be ingested (Goodwin, 1984). Even though animals that use carotenoids as colorants, like birds, cannot make their own pigments from scratch, they can modify the pigments they eat, and indeed, most of the red carotenoid coloration used in animal ornaments is produced by biochemically modifying dietary yellow pigments to red (Mcgraw, 2006; Lopes et al., 2016; Weaver et al., 2018a). It was recently shown that pigmentation derived through modification of dietary carotenoids is more tightly linked to individual condition in birds than is pigmentation derived through the deposition of dietary pigments used unchanged (Weaver et al., 2018b) (Figure 8.9). The condition dependency of the red coloration of birds has been well established (Hill, 2006a), and females of many bird species show a mating preference for males with redder carotenoid-based coloration (Hill, 2006b). Thus, carotenoid coloration is a condition-dependent signal of individual condition in birds. In the context of the current discussion, the key question is: does ornamentation arising from deposition of red carotenoid pigments signal mitochondrial function? New studies suggest that red coloration could have evolved as a signal of mitochondrial function.

The enzyme responsible for the conversion of yellow dietary pigments to red ornamental pigments in birds and turtles is a cytochrome P450 pigment, CYP2J17 (Lopes et al., 2016; Mundy et al., 2016). This avian ketolase belongs to a subclass of P450 enzymes that are commonly targeted to endoplasmic reticulum and mitochondria (Avadhani et al., 2011). Follow-up experiments demonstrated that house finches (*Haemorhous mexicanus*), a songbird with red head and breast coloration, have high concentrations of red carotenoids in their mitochondria when they are growing and pigmenting feathers (Ge et al., 2015). This link between red carotenoid pigments and mitochondria is predicted if the capacity of the ketolase to oxidize yellow dietary pigments to red pigments used in ornamentation is dependent on the inner mitochondrial membrane potential (Hill and Johnson, 2012; Johnson and Hill, 2013).

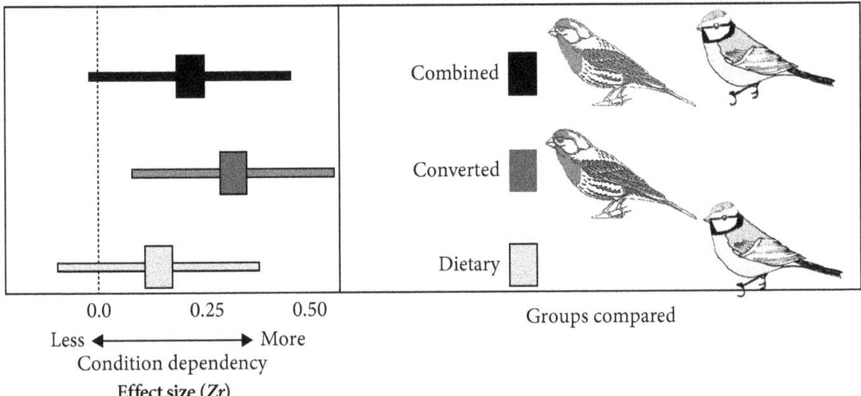

Figure 8.9 A graphical summary of the outcome of a meta-analysis assessing the relationship between individual condition and carotenoid-based plumage coloration when (top) all species are included; (middle) only species that used converted carotenoids are included; and (bottom) only species that have pigmented feathers with unmodified dietary pigments are included. The association between carotenoid coloration and individual condition arises primarily from the strong relationship with condition and coloration from modified pigments. Zr is the weighted mean correlation, with stronger effects moving away from zero. Horizontal lines represent 95 percent credible intervals. Redrawn from Weaver et al. 2018, where details of analysis can be found.

The hypothesis that red carotenoid coloration is an honest signal of mitochondrial function makes a clear prediction: there should be a positive relationship between carotenoid coloration and mitochondrial function. To test this idea, Cantarero and Alonso-Alverez (2017) conducted a study of the relationship between carotenoid-based red eye ring coloration of red-legged partridge (*Alectoris rufa*) in relation to mitochondrial function. The red pigments in the eye ring of the partridges are produced through the oxidation of yellow dietary pigments. Cantarero and Alonso-Alverez (2017) supplemented birds with a mitochondrial enhancer, the coenzyme Q10 (CoQ10), which stabilizes the inner mitochondrial membrane potential (Ng et al., 2014). They observed that this this CoQ10 treatment caused birds to become redder (Cantarero and Alonso-Alvarez, 2017). This study is consistent with the hypothesis that ketolation of carotenoids is linked to mitochondrial function, but much more testing of this hypothesis will be needed before its validity can be assessed. Testing the links between expression of ornamentation and mitochondrial function should be a major area of inquiry moving forward.

Summary

Safe to say, we all appreciate sex. But aside from its recreational value for humans, sex facilitates recombination of genetic elements enabling individuals to avoid mutational erosion through Muller's ratchet. There is, however, a dark side to sex. It inserts an element of chaos into reproduction. Sexual reproduction in species with uniparental inheritance of mitochondria creates new combinations of mt and N-mt genes, and it opens the door to the very real danger of matching sets of genes that are not coadapted and that will not function together well. Sexual selection is the primary mechanism to ensure that, each generation, sets of coadapted genes are properly matched. In line with this chain of logic, sexual selection can only be fully understood with a consideration of mitonuclear interactions.

Casting mate choice as a process for maintaining mitonuclear coadaptation presents a new perspective on the old idea of the importance of mating within species boundaries. Choosing mates with compatible mt and N-mt genes is best accomplished if species-typical ornaments are genetically linked to N-mt genes, such that choice for species-typical ornament is choice for coadapted N-mt genotype. The striking concordance between transitions in mt genotypes and species-typical ornamentation provides support for this hypothesis. Moreover, Z linkage of N-mt genes allows for assessment of mitonuclear coadaptation each generation, which is predicted to lead to better mitonuclear coadaptation over time. Within conspecific populations, mutation, gene flow, and adaptive variation in mt genotypes lead to differences in the genetic quality of prospective mates and create incentives for assessment of the quality of mt and N-mt genes in prospective mates. For assessment of the quality of genes related to electron transport system (ETS) function, traits used to choose mates should be reliable signals of mitochondrial function such that females choose mates with highly functional N-mt genes and males choose mates with both N-mt and mt genes that are functional. Mitonuclear mate choice is a new and largely untested idea, but it holds potential to explain many heretofore inexplicable patterns related to mating preferences and ornamentation.

9

Adaptation and adaptive radiation

Through eight chapters, my focus has been the functional interactions of the products of mt genes and N-mt genes to enable aerobic respiration. Only occasionally have I so much as mentioned the environment in which these genomic interactions play out. This organizational strategy was not meant to relegate environment to secondary role in mitonuclear coevolution; rather, it was the build up to a crescendo. The environment is not simply another facet of mitonuclear interactions to be considered as an afterthought; it is typically *the* determining factor in the success or failure of specific mt and N-mt gene combinations.

The role of mt and N-mt genotypes in shaping the adaptive evolution of eukaryotic organisms is a topic still in its infancy as evolutionary ecology shrugs off the last arguments for mt DNA as a strictly neutral marker (Ballard and Pichaud, 2014; Dowling, 2014; Kern and Hahn, 2018). There seems to be unlimited potential for variation in coadapted sets of mt and N-mt genes to explain patterns of adaptive evolution (Dowling et al., 2008; Breen et al., 2012; Kazancioğlu and Arnqvist, 2014). The outcome of efforts to reinterpret fundamental properties of organisms from a mitonuclear perspective holds potential not only to reshape understanding of evolutionary processes in natural systems but also to inform and redirect biomedical research (Wallace and Fan, 2010; Dowling, 2014; Picard et al., 2016; Rahman and Rahman, 2018).

In this chapter, I consider adaptive evolution of mt and N-mt genes relative to the external environment. With epistatic interactions between the two genomes and the environment, there are seven classes of interactive effects to be considered (Figure 9.1). Changes in mt genotype to improve performance in specific N backgrounds as well as changes in N-mt genotype to improve performance in association with specific mt genomes are, by definition, adaptations. I covered adaptive coevolution of mt and N-mt genes in detail in Chapter 3. Here, I focus on the function and co-function of mt and N-mt genes in relation to the *external* environment. The relationship between the external environment and mt and N-mt genotypes can be assessed at three distinct levels. First, the fitness effects of standing variation in mt genotypes *within* a population or species can be assessed relative to environmental variation (Dowling et al., 2008; Kazancioğlu and Arnqvist, 2014). Second, adaptive divergence *between* distinct species that no longer exchange genes to a significant degree can be studied with a focus on adaptation and adaptive divergence related to co-functioning mt and N-mt gene products (James et al., 2016). And finally, researchers have begun to consider the role of adaptive evolution of aerobic respiration arising from changes to mt and N-mt

Mitonuclear Ecology. Geoffrey E. Hill, Oxford University Press (2019). © Geoffrey E. Hill 2019.
DOI: 10.1093/oso/9780198818250.001.0001

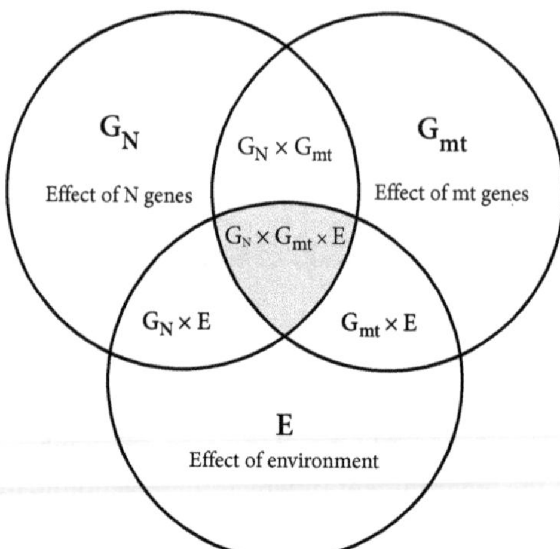

Figure 9.1 A Venn diagram summarizing the interactions that are possible when there are effects of both a mitochondrial (G_{mt}) and nuclear (G_N) genotype expressed across environments (E). The net effect of any individual interaction can range from very small to very large.

genes in the adaptive radiations of entire lineages (Hill, 2015b). I will provide an overview of how interacting mt and N-mt genes can shape adaptation at each of these levels of organization. I begin with a consideration of the mechanisms by which the products of mt and N-mt genes can enable adaptation to external environments.

Environments to which mitochondrial adaptation is responsive

The list of environmental variables to which organisms can adapt is essentially boundless, but most adaptation is the result of selection on N genes that do not cofunction with mt genes (Carneiro and Hartl, 2010; Olson-Manning et al., 2012). As I discussed in Chapter 1, in the early evolution of eukaryotes, the nucleus took on the role of information center. There are about 20,000 protein-coding genes in the N genome of a typical animal, but only thirteen protein-coding genes in the mt genome. Consequently, most adaptive evolution of a eukaryote will necessarily arise from the evolution of N genes that have no direct functional interaction with mt genes. While I will hold to my assertion that mt genes play a major role in the adaptive evolution of eukaryotes, such a statement must be qualified. Mitochondrial genes certainly do not underlie most adaptation. The key question is: in what environmental contexts are mt genes most influential?

In the literature on the function of standing variation of mt genotypes, there is a focus on temperature, oxygen pressure, and diet as the key environmental factors that shape mitochondrial evolution (Melvin and Ballard, 2017). To this list, Dowling et al. (2008) proposed that environmental salinity, via effects on the osmotic properties of the mitochondrial membrane, and exposure to hydrogen sulfide, which is a direct electron transport system (ETS) disruptor, are two other environmental factors that could shape mitochondrial evolution. Capacity to cope with desiccation is also linked to mitochondrial function and specifically to ETS properties, particularly in plants (Klein et al., 1986; Atkin and Macherel, 2009), but to my knowledge adaptation to environmental moisture has yet to be linked specifically to mt genotype. A colleague suggested that mt genotype could promote adaptation to exposure to ultraviolet radiation or other mutagens because, for instance, adjacent thymine nucleotides are more vulnerable to dimerization under exposure to UV radiation than other nucleotide pairs (Friedberg, 2008); a mt genotype with fewer adjacent thymine nucleotides might perform better in a high-radiation environment (A. Pozzi, March 2018, pers. comm.). Internal production of free radicals, which could be linked to external environment, has also been shown to shape the evolution of synonymous nucleotide positions in the mt genome (Martin, 1995). A major arena of adaptive evolution of mt genes is in response to increased or decreased energy demands related to life history changes such as exposure to specific parasites, long-distance migration, social system, or the development of energy-demanding organs such as a large brain, the musculature for flight, or a specialized digestive system (Hill, 2015b). I'll focus on mitochondrial adaptation related to life history in the section on adaptive radiation. I'll begin, however, by reviewing the mechanisms by which changes to mt and N-mt genotype might facilitate adaptation to various external environments (Figure 9.2).

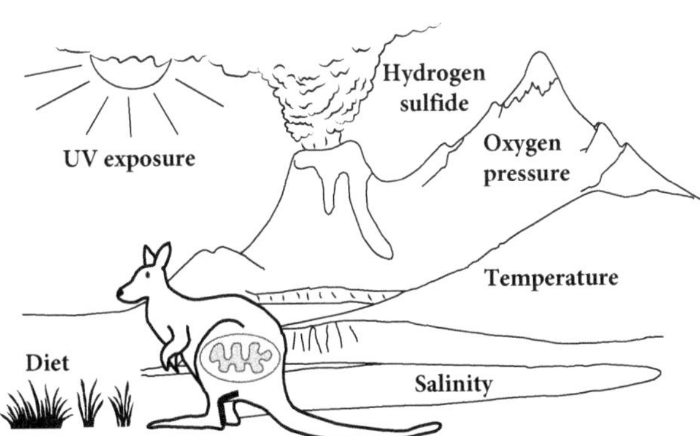

Figure 9.2 Some of the environmental variables to which adaptive evolution of mt genotypes has been linked.

Mechanisms for mitonuclear adaptation

Central to understanding how the evolution of mt genotypes can facilitate adaptation is understanding the specific mechanisms by which functional changes to the mt genome alter aerobic respiration. Studies of the mechanisms that underlie ETS function have, until recently, been largely the purvue of biochemists and cellular physiologists (e.g. Alric et al., 2006; Al-Attar and de Vries, 2013). The function of the products of mt genotypes in relation to environment has received relatively little attention by evolutionary biologists, who typically focus on whole-organism function and response (Hyatt et al., 2017; Zhang et al., 2017). It is becoming increasingly clear, however, that understanding the mechanisms by which mt and N-mt genes affect respiratory function will play a key role transitioning evolutionary ecology from a descriptive to a predictive science (Burton et al., 2013; Ballard and Pichaud, 2014; Dowling, 2014).

I should note that all references to adaptation in this chapter concern evolutionary adaptations arising from changes to genotypes. In many eukaryotic species, individuals are adept at acclimating to new environments by adjusting their physiological systems, including systems that enable aerobic respiration, but such acclimation is not the focus of this chapter. Acclimation plays a key role in determining the fitness outcome of N and mt genotypes (Seebacher et al., 2010; Schulte, 2015), but here the focus is on adaptive divergence that exceeds what can be achieved through within-system physiological adjustment and requires substitution of alleles underlying key traits.

I should also mention that the tendency for organismal biologists to typecast environments as stressful or non-stressful may hinder studies of the role of environment in the evolution of mt and N-mt genes. Environments are typically labeled as stressful simply because they lie outside the optimal range for humans and human commensals such as fruit flies and mice. It is more productive to think in terms of N-mt genes interacting with mt genes across a range of environments. Except at the very extremes of capacity to sustain any life (Dill et al., 2011), which is rarely part of the environmental range studied by evolutionary ecologists, any given environment is neither good nor bad—it is a circumstance that mitigates the outcome of the interactions of mt and N-mt genes. Removing the label "stressful" from the discussion of mitochondrial evolution eliminates prejudgements regarding expected outcomes and will lead to fundamentally better science.

Mechanisms of thermal adaptation

Seebacher et al. (2010) provide a comprehensive overview of what is known about how changes to the complexes of the ETS or to the molecules that enable transcription, translation, and replication of mt DNA can promote thermal adaptation. Among the most common mechanisms by which animals adapt to thermal environment is to change the number of mitochondria per cell; all else being equal, more mitochondria means greater capacity to produce ATP. Accordingly, a very common pattern among ectotherms is proliferation of mitochondria in cooler environments and environments with low partial pressure of oxygen (Pörtner, 2002; Lannig et al., 2005). Mitochondria can only proliferate via proper function of both mt and N-mt gene products controlling transcription, translation, and replication of mt DNA, so mitonuclear gene interactions are implicated in any adaptations emerging from changes in mitochondrial density (Taanman, 1999; Scarpulla, 2008).

Beyond simply producing more respiratory chains to produce more ATP, more subtle changes to ETS structure and function can affect how an organism responds to its thermal environment. One key response by the ETS of ectotherms is that lower temperatures result in reduced enzyme activity (Hochachka and Somero, 2002; Lannig et al., 2003). The slowing down of enzyme activity can be countered by increasing enzyme–substrate affinity or by decreasing the energy of activation (Somero, 2004; Angilletta and Angilletta, 2009), and mt genotype can play a role in both of these thermal adaptations (White et al., 2012; Schulte, 2015). The composition of the inner mitochondrial membrane, and in particular the concentration of fatty acids and phospholipids, also affects the activity of membrane-bound proteins, including ETS complexes (Hulbert and Else, 1999; Hulbert, 2008). Membrane composition also plays a critical role in the rate of passive proton leak (Brand et al., 2003), as discussed below. All of these factors contribute to direct production of heat as well as production of free radicals and ATP and could be involved in thermal adaptation.

Mechanisms that enable mitochondrial adaptation to temperature in homeotherms (organisms that maintain a constant core temperature) are not nearly as well established as the mechanisms for such adaptation in ectotherms (organisms that rely on external sources of heat and that do not maintain a constant internal temperature). The internal temperature experienced by the cells of most ectotherms will approximate ambient temperatures, so variation in ambient temperature can directly affect ETS function by affecting the temperature in which biochemical reactions occur (Simčič et al., 2014). For ectotherms, therefore, it is straightforward to comprehend how thermal environment could drive the evolution of adaptive changes to mt and N-mt genotypes—in cold environments ETS complexes must function at low temperatures and in hot environments the same reactions take place in a much warmer cellular environment. In contrast, homeotherms, such as most birds, have nearly invariant internal temperatures and the cells within the body experience little variation in the temperatures at which they operate (Whitnow, 1986). For instance, even a fever response in zebra finches changes core body temperature by less than 1°C (Skold-Chiriac et al., 2015). In such animals, regardless of whether the organism is currently in a hot or cool environment, the biochemical reactions of the ETS will operate at the same temperature. Paradoxically, however, among birds there is a strong association between basal metabolic rate (which arises primarily from aerobic respiration) and ambient temperature (White et al., 2007). With so little variation in the operational temperature of OXPHOS processes in the cells of an adult homeotherm, arguments of enzyme–substrate affinity or energy of activation make little sense. With such stasis in body temperature, how can there be adaptation to thermal environments involving the genes that code for mitochondrial function?

In birds, which develop externally as laid eggs or as embryos in nests, the body temperatures of embryos both in the egg and after hatching vary with ambient temperatures much more than the body temperatures of adults (Ricklefs, 1984; Hohtola and Visser, 1998). This is particularly true for the embryos of birds with altricial development, which hatch at an early developmental stage and have little or no capacity for maintaining internal temperature (Dawson and Hudson, 1970; Hohtola and Visser, 1998). These ontogenetic periods of poorly regulated temperature coincide with the development of major organ systems including the nervous system and could create a link between mitochondrial function and ambient temperature just as in ectotherms. (Avian embryos are ectotherms.) These basic characteristics of avian development, which will hold true to varying degrees in other homeotherms as well, underscore the importance of considering all life stages in studies of mitochondrial adaptation to temperature in homeotherms. To my knowledge, however, the paradox of constant temperature of cellular environment in light of thermal adaptation via changes to mitochondrial function has never been explicitly considered.

In adult homeotherms, generation and dissipation of heat may be the primary mitochondrial factors under selection. The importance of heat dissipation in warm environments is an old topic in physiological ecology (McNab and Morrison, 1963), but it has not been considered specifically in the context of thermal adaptation via mitochondrial function. Recently, however, the importance of heat dissipation in relation to mitochondrial function has attracted new interest in light of a study suggesting that the temperature of the inner mitochondrial membrane is much higher than had been typically assumed (Lane 2018) (Box 9.1). If the temperature of the inner mitochondrial membrane dictates an organism's capacity to function in warm

Box 9.1 Hot mitochondria

How can changes to the genes that code for the ETS enable organisms to better adapt to climates? There may be a clue in a recent paper that shows that the ETS of humans functions best at 50°C (Chrétien et al., 2018). How is it possible for mitochondria to run so hot? Animals and plants cannot survive ambient temperatures of above 45°C. Most proteins of most organisms denature at about 50°C (Dill et al., 2011).

Whether or not this reported temperature is precisely correct perhaps doesn't matter. It is hard to dispute that mitochondria run hot—really hot at the inner mitochondrial membrane (Lane, 2018). The reason that it is hotter at the inner mitochondrial membrane than the outer mitochondrial membrane, and hotter at the outer membrane than at the cell membrane, is that heat is produced at the inner mitochondrial membrane and then dissipates through the aqueous environment (Lane, 2018). The membranes that encase the ETS act like lids on pots. The rate of dissipation is roughly proportional to how well membranes serve as insulators, the difference in temperature between cellular regions, and ultimately the differences in temperatures between the inside versus the outside of the organism. The relevance of these considerations to a discussion of mitonuclear thermal adaptation is that the heat generated by an ETS will be a product of the efficiency of that ETS, and ETS efficiency is the product of mt and N-mt genes.

The challenge of dissipating mitochondrial heat has real consequences for fitness and function of organisms. Many flying insects like large beetles and locust heat up rapidly during flapping flight and have to stop flying after relatively short periods to avoid overheating (Verdú et al., 2006). Other animals seem much better at dissipating heat and maintaining activity for long periods. Bushmen are famous for running down prey animals in the heat

Box 9.1 *Continued*

of the Kalahari Desert. It has been speculated that naked skin and sweat glands are human adaptations for dissipating heat and enabling humans to run farther without overheating (Carrier et al., 1984; Liebenberg, 2008). It is easy to imagine, but entirely speculative at this point, that in addition to sweating and running naked, humans have mitochondrial adaptations for "persistence hunting," as running down prey is called.

Birds seem remarkably emancipated from the constraints imposed by heat generated through physical exertion. Consider the bar-tailed godwit (*Limosa lapponica*), a pigeon-sized bird that flies non-stop for nine continuous days from arctic Alaska to New Zealand, a one-way trip covering a staggering 11,026 km (Gill et al., 2009) (Figure Box 9.1), or the ruby-throated hummingbird (*Archilochus colubris*), a bird that weighs only about as much as the pickles on a McDonald's hamburger (3 grams) but that flies for 15 hours across the Gulf of Mexico during spring and fall migration. A migrating hummingbird flaps its wings an astounding 3 million times from takeoff to landing in crossing the Gulf of Mexico. But unlike many insects that rapidly heat up and have to stop flying in a matter of minutes (Verdú et al., 2006), migrating godwits and hummingbirds land after hours and days of flapping flight at the same temperature at which they took off (Schmaljohann et al., 2008). The biochemical basis for the ability of birds to fly without heating up is completely unknown.

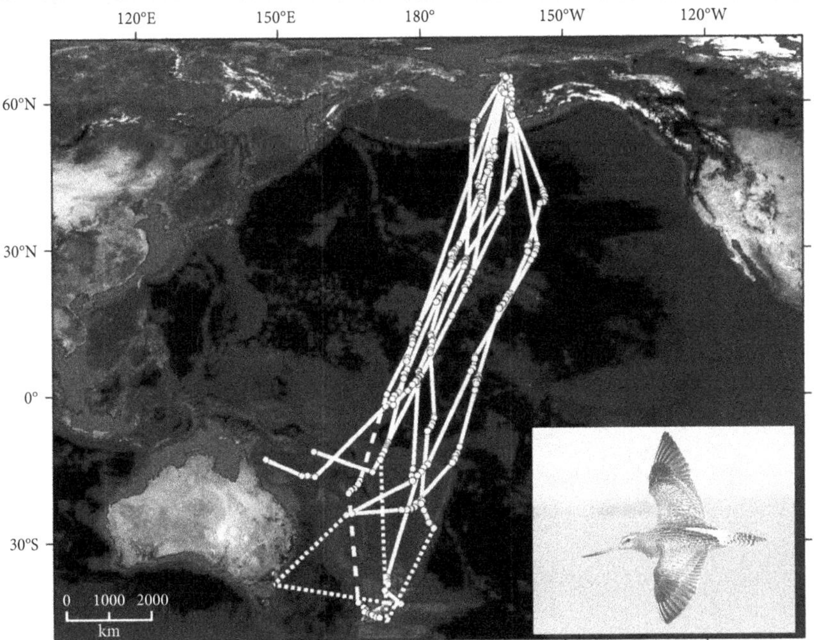

Figure Box 9.1 A map of about half of the Earth centered on the Pacific Ocean showing the track of nine southbound bar-tailed godwits (*Limosa lapponica*) launching from northern Alaska and flying non-stop to islands in the southern Pacific Ocean, including New Zealand. Inset is a photo of a flying bar-tailed godwit. Map reprinted with permission from Gill et al. (2009). Bar-tailed godwit photo by Lior Kislev, used with permission.

environments because of the necessity of dissipating mitochondrial heat, then the respiratory processes that underlie production of mitochondrial heat could certainly be under natural selection and both mt and N-mt genes could be the targets of such selection.

To date, however, it is the production of heat rather than the dissipation of heat that has been the focus of studies of mitochondrial adaptation to thermal environment in homeotherms. The proton-motive force generated by electron transfer in mitochondria can be converted to heat via three processes: (1) conversion to ATP and subsequent use of ATP to generate frictional heat (shivering thermogenesis) (Blondin et al., 2014), (2) basal proton leak, and (3) inducible proton leak (Divakaruni and Brand, 2011; Busiello et al., 2015) (Figure 9.3). I will not go into detail on shivering thermogenesis. As discussed in many sections of this book, production of ATP is intimately related to ETS function and mitonuclear coadaptation. There exist many mechanisms by which mt genotype can affect production of ATP and hence the capacity for generation of frictional heat. I will focus on the leakage of protons across the inner mitochondrial membrane.

Proton leakage can be substantial and can be a major contributor to the resting metabolic rate and heat production of a eukaryote. Basal proton leak is the unregulated movement of protons from the inner mitochondrial space to the matrix through the inner mitochondrial membrane (Jastroch et al., 2010), and it accounts for about 20–30 percent of the resting metabolic rate of rat hepatocytes and up to half of the

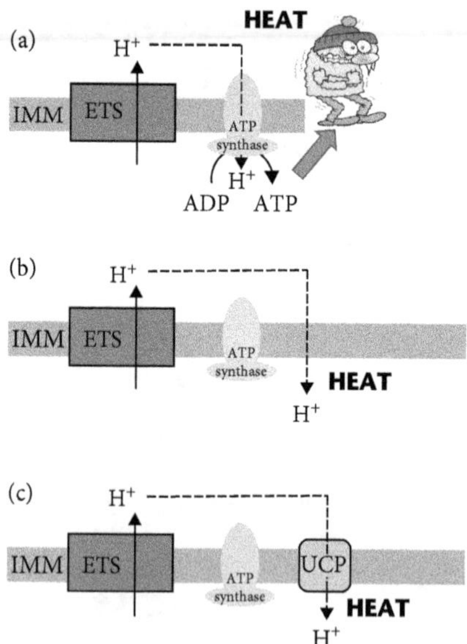

Figure 9.3 Simplistic illustrations of three means to convert proton motive force to heat: (A) the use of ATP to drive muscles to create frictional energy, (B) basal leakage of protons, (C) leakage via uncoupling proteins. All of these processes can occur simultaneously. ETC, electron transport system; UPC, uncoupling protein; IMM, inner mitochondrial membrane.

respiration of skeletal muscle of a rat (Jastroch et al., 2010). In the process of basal proton leak, the energy that was captured when a proton was pumped across the inner mitochondrial membrane by the ETS is released as heat as that proton moves back across the membrane. Basal proton leak is sometimes discussed as an imperfection in the respiratory system as though 100 percent conversion of the proton-motive force to ATP is the optimal respiratory state for all eukaryotes, but 100 percent capture of energy as ATP is impossible for any OXPHOS system. Basal proton leak is an inevitable part of mitochondrial function. A degree of proton leak can be beneficial, and the rate of basal proton leak can evolve by natural selection (Brand, 2000; Lane, 2011b, 2015a).

It is unclear if mt or N-mt genes that code for OXPHOS processes play a significant role in the evolution of basal proton leakage. I know of no studies linking genes that code for the OXPHOS system to the rate of basal proton leak. However, the production of free radicals can directly affect basal leakage of protons through feedback mechanisms controlling ETS and the inner mitochondrial membrane potential (Brookes, 2005), and production of free radicals and the mitochondrial membrane potential are certainly under control of coadapted mt and N-mt genes. There is a need for direct comparisons of basal proton leak in relation to mt genotypes linked to thermal adaptation if we are to fully understand the role of mt genotype in thermal adaptation.

The third mechanism for converting a membrane potential into heat is inducible proton leak, which is the controlled backflow of protons through proteins in the inner mitochondrial membrane. Uncoupling proteins are the best-known regulators of inducible proton leak. These N-encoded proteins have long been a focus of thermal adaptation of mitochondria and are fairly well understood, at least in model mammalian species (Criscuolo et al., 2005; Azzu and Brand, 2010; Mailloux and Harper, 2011). Another, much-less-studied mechanism for inducible proton leak is adenine nucleotide translocase, which is a N-encoded protein of the inner mitochondrial membrane that functions primarily as an ADP/ATP carrier, thus playing a critical role as a substrate recycler (Wittig and Schägger, 2008). Under certain conditions, mostly studied in the context of severe cellular perturbation leading to cell death, adenine nucleotide translocase can form what is known as a mitochondrial permeability transition pore that enables the flow of protons across the inner mitochondrial membrane (Brustovetsky and Klingenberg, 1994; Halestrap et al., 2003). In only a few studies has adenine nucleotide translocase been considered in the context of heat generation and thermal adaptation.

Both uncoupling proteins and adenine nucleotide translocase form micro-pores in the inner mitochondrial membrane through which backflow of protons, and hence heat production, can be regulated (Halestrap et al., 2003; Jastroch et al., 2010). Membrane uncoupling via uncoupling proteins is widely evoked as an adaptation to thermal environment, especially in homeotherms, because (1) it is a regulated mechanism that enables the conversion of the proton-motive force to be transferred directly into heat rather than ATP, (2) allelic variation in uncoupling proteins has been shown experimentally to affect cold tolerance (Nishimura et al., 2017), and (3) there is a tight association between the frequency of specific uncoupling protein alleles in human populations and regional temperature (Hancock et al., 2011) (Figure 9.4).

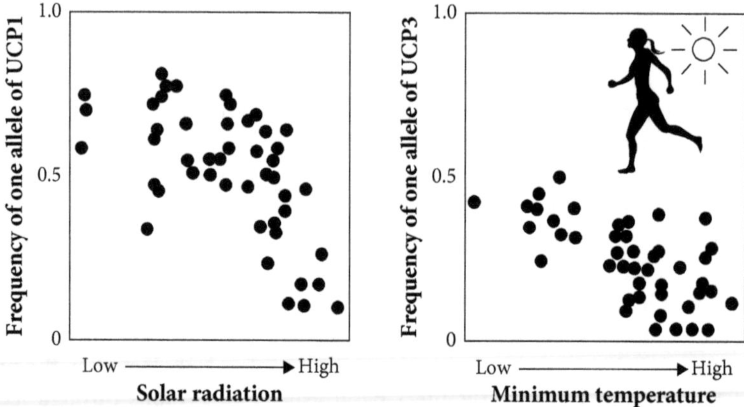

Figure 9.4 Clinal variation in the frequency of two N-encoded uncoupling protein alleles in human populations in relation to two measures of the thermal climate of the home environment of that human population. The strong positive association between allele frequency and solar radiation and minimum temperature suggests that the uncoupling protein alleles are under strong positive selection for climate adaptation. Whether mt genotype plays any role in membrane uncoupling remains unknown. Adapted from Hancock et al. (2011).

Given that uncoupling proteins are N-encoded and do not appear to engage in direct functional interaction with mt-encoded products, an effect of mt genotype on inducible proton leak is not necessarily expected and remains to be demonstrated.

The role of adenine nucleotide translocase in heat production is almost unstudied, and to my knowledge, the potential for mt genotypes to affect leakage via adenine nucleotide translocase has never been considered. In one of the few studies reporting direct links between thermal adaptation and adenine nucleotide translocase, Talbot et al. (2004) studied cold aclimation in king penguins (*Aptenodytes patagonicus*). They found that adaptive thermogenesis was associated with inducible proton leak by both uncoupling proteins and adenine nucleotide translocase. This penguin study focused on within-individual functional variation, but it showed the potential for adenine nucleotide translocase to play an important role in inducible proton leak. Adenine nucleotide translocase can significantly affect mt DNA replication (Liu and Chen, 2013), suggesting a possible connection to mt genotype through the regulation of the mt and N-mt genes that enable mt DNA replication, but this connection has yet to be explored.

The take-home message is that, despite a focus on thermal adaptation in lab studies of model organisms, the mechanisms by which mt genotype can affect respiratory performance in different thermal environments remain poorly understood. Studies to date have demonstrated the importance of proton leakage to thermal adaptation, but there is little evidence that mt genotype is involved in variation in proton leakage. The proteins involved in inducible proton leak are N-encoded as are the genes that control the fatty acid and phospholipid concentrations of the inner mitochondrial membrane. A major challenge going forward will be to test whether standing variation in mt genotypes that is linked to thermal adaptation is acting through the generation of heat from membrane uncoupling or by other mechanisms.

Mechanism of adaptation to partial pressure of oxygen

The second environmental variable that is most frequently invoked in studies of adaptation associated with mt genotype is the partial pressure of oxygen in the environment. Such studies are typically framed in terms of adaptation to hypoxia, which is defined as a deficiency in the amount of oxygen reaching cells throughout the body (e.g. Hamanaka et al., 2016 and Fuhrmann and Brüne, 2017). Because deficiency versus over-abundance is a matter of perspective arising from the evolutionary history of a particular organism, I prefer to frame discussions in terms of adaptation to different partial pressures of oxygen. The partial pressure of oxygen is the portion of atmospheric pressure contributed by oxygen or, in water, the portion of pressure from dissolved gas that comes from oxygen. In terrestrial environments, the partial pressure of oxygen decreases with distance from sea level, so there is lower partial pressure of oxygen at high altitudes than at low altitudes. For organisms adapted to sea level, areas thousands of meters above sea level are hypoxic environments (Luo et al., 2013). Conversely, for organisms adapted to environments thousands of meters above sea level, habitats at sea level are hyperoxic—they are bathed in too much oxygen relative to the needs and capacities of tissue in the body. The partial pressure of oxygen can also vary in closed environments such as burrows, where oxygen is depleted (Widmer et al., 1997), and the oxygen concentration in water varies with temperature, salt concentration, surface exposure, and agitation (Verberk et al., 2011). Cold water holds more oxygen than warm water, leading to interesting adaptations to the hyperoxic environment provided by perpetually very cold water (Box 9.2). Because oxygen is the terminal electron acceptor in the ETS, its availability can have huge effects on ETS function, and different configurations of ETS complexes achieve

Box 9.2 Life without hemoglobin

Cold water holds more oxygen than warm water, so the water in the Southern Ocean around Antarctica, which is the most consistently cold seawater in the world, is also the most oxygen-rich seawater in the world. In this unique, perpetually cold and oxygen-rich environment, the icefish evolved. And the hyperoxic (with excess oxygen) environment of icefish has led to some bizarre physiological adaptations. None of the sixteen species in family Channichthyidae has hemoglobin and all but one also lack myoglobin (Ruud, 1954). Hemoglobin and myoglobin are the oxygen transporters of vertebrates. Consequently, not only do icefish have blood that looks like water (Figure Box 9.2), they have blood with the oxygen-carrying capacity of water—the oxygen-carrying capacity of icefish blood is only 10 percent that of red-blooded species (Ruud, 1954). Each pulse of an icefish's gill exposes its blood to abundant ambient oxygen, but icefish ineffectively transport that oxygen to mitochondria. To compensate for the lack of oxygen-transporting hemoglobin and hence low oxygen pressure in blood, icefish have larger hearts, a higher blood volume, and denser capillary beds in key tissues. These are N-encoded adaptations to oxygen-poor blood. Icefish also have a higher density of mitochondria than other fish, and the mitochondria of icefish are huge (O'Brien and Mueller, 2010). How does the presence of many big mitochondria help compensate for lack of hemoglobin? Oxygen diffusion across membranes is

(Continued)

Box 9.2 *Continued*

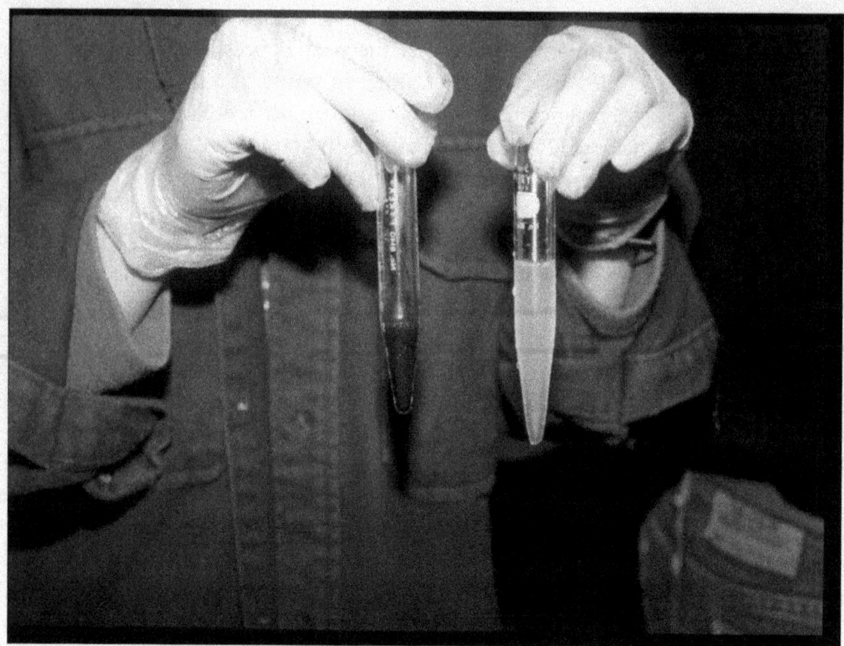

Figure Box 9.2 Icefish are unique among vertebrates in lacking hemoglobin, making their blood an icy blue (right) rather than the deep scarlet typical of boney fish (left). Photo by Kristin O'Brien, used with permission.

impeded by embedded proteins. Maximum oxygen diffusion is achieved when membranes are mostly hydrocarbons with fewer proteins. This is exactly the design of icefish mitochondria—an expanded area of membrane with an associated lower density of embedded proteins. Icefish up-regulate the biosynthesis of mitochondrial phospholipids independently of the synthesis of ETS proteins, thus the mitochondrial membranes get bigger and mitochondria balloon out—but ETS capacity remains the same (O'Brien and Mueller, 2010). These modifications seem to significantly enhance intracellular diffusion of oxygen, but it still seems that icefish would be better off if they just produced hemoglobin.

different effects in different oxygen environments. Differences in efficiency of ETS complexes in different oxygen environments create potential for adaptive evolution of the OXPHOS system in response to oxygen pressure (Fuhrmann and Brüne, 2017).

Many of the adaptations to the partial pressure of oxygen that are observed in eukaryotes are the product of N genes that have no direct functional interaction with mt genes (Figure 9.5). For instance, hemoglobin is a N-encoded gene that binds oxygen and is critical to oxygen transport in the blood. Natural selection leads to changes in the structure and oxygen affinity of hemoglobin to enable organisms to adapt to

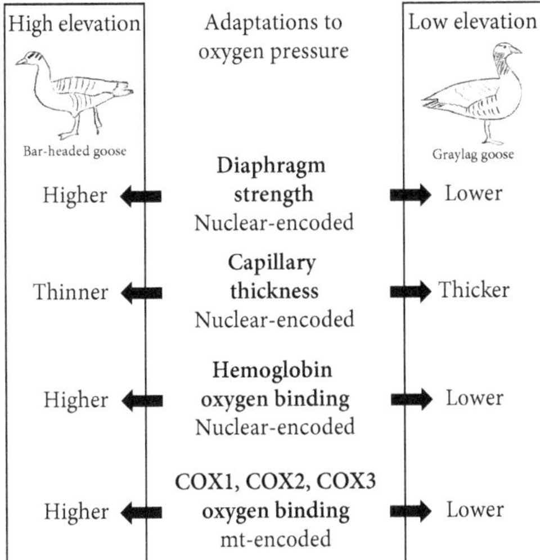

Figure 9.5 A summary of adaptations that support aerobic respiration in the high-altitude bar-headed goose versus the lowland-adapted graylag goose. Each of the changes in the bar-headed goose enables adaptation to reduced oxygen pressure by increasing oxygen tension and hence enhancing efficiency of oxygen use. In this example N-encoded gene products do not directly interact with mt gene products.

environments with different partial pressures of oxygen (Storz, 2007; Projecto-Garcia et al., 2013). Many other adaptations to high altitude and low oxygen pressure, such as increased lung capacity, enlarged heart, pulmonary vasoconstriction, and a blunted erythropoietic response (the condition wherein lowered oxygen pressure does not cause a corresponding increase in hemoglobin levels in the blood) appear to be under control of N genes with no direct functional connection to mt and N-mt genes (Storz et al., 2010). Nevertheless, there is strong evidence that functional changes to mt and N-mt genes altering ETS function can also underlie adaptations to altitude (Luo et al., 2013; Hamanaka et al., 2016).

Adaptive changes to the ETS in response to environmental oxygen concern primarily Complex I and Complex IV. Complex I adaptations are related to free radical production. As the partial pressure of oxygen drops, cells switch from production of acetyl CoA (a precursor of NADH needed for aerobic respiration) via the citric acid cycle to production by an alternative pathway involving reductive carboxylation (Kim et al., 2006). This switch to reductive carboxylation with decreasing oxygen pressure can increase free radical production, and free radical homeostasis is maintained primarily through changes to Complex I of the ETS (Fuhrmann and Brüne, 2017). The implication of this connection between partial pressure of oxygen and Complex I function is that changes to Complex I, via either mt-encoded or N-mt-encoded subunits, can potentially serve as adaptive responses to changes in oxygen pressure.

While Complex I can play a role in adaptation to oxygen pressure through effects of free radicals, Complex IV is the component of the ETS that has the most intimate assocation with oxygen and that can adapt most directly to changes in oxygen availability. Complex IV receives electrons from cytochrome *c* molecules and uses those electrons to convert molecular (atmospheric) oxygen to water. This is the outflow of electrons at the end of a chain that begins with the receipt of electrons at Complexes I and II. As emphasized in Chapter 2, Complex IV is the rate-controlling step in aerobic respiration. If the transfer of electrons to oxygen is too slow, then the flow of electrons down the respiratory chain is impeded and energy in the stalled electrons ends up being dissipated as heat and free radicals. A proper rate of transfer of electrons from Complex IV to oxygen is essential for proper ETS function.

Oxygen binding by Complex IV, which occurs deep within that portion of Complex IV that is embedded in the inner mitochondrial membrane, is mediated by the mt-encoded subunit COX1 (Little et al., 2017). Mitochondrial-encoded COX3 also plays a key role in oxygen binding by controlling the diffusion of oxygen from the intermembrane space to the binding site on COX1. And, perhaps most significantly, changes in configuration of COX subunits can change the flow of and affinity to oxygen (Little et al., 2017) thus presenting an avenue for genetically based changes to subunits of Complex IV to enable adaptation to environmental oxygen (Lau et al., 2017). For example, in a study of adaptation to high elevation environments by the migratory locust (*Locusta migratoria*), Zhang et al. (2013b) assessed the activity of Complex IV in lowland versus high elevation populations. They documented superior function of Complex IV from the high elevation population, but this improved activity was not simply a function of protein content. In other words, locust that had evolved at high elevation did not simply create more ETS units to compensate for low efficiency of individual units in a low-oxygen environment. Rather, they showed elevated catalytic efficiency of Complex IV that was a result of greater oxygen affinity by the Complex IV catalytic center (Zhang et al., 2013b). In this way, changes to mt-encoded subunits can create important functional changes to the manner in which Complex IV interacts with oxygen and hence can very plausibly underlie adaptation to different oxygen pressures.

Given the functional involvement of Complexes I and IV in mitigating the effects of different environmental oxygen pressures, it is instructive that most of the documented adaptive changes in composition of ETS complexes in response to changes in oxygen pressure occur within Complex I and especially Complex IV (Fuhrmann and Brüne, 2017). Furthermore, adaptation to reduced oxygen pressure by Complex IV most typically results from functional changes to oxygen binding by specific subunits and particularly the mt-encoded subunits, COXI, COXII, and COXIII (Scott et al., 2015; Lau et al., 2017). Thus, unlike the uncertain mechanisms by which functional changes to the OXPHOS system can mediate adaptations to thermal environment, there are clear mechanisms for adaptive evolution of the components of the ETS to changes in response to oxygen pressure.

Mechanisms of adaptation to diet

Diet is another environmental variable that has been studied rather extensively in the context of adaptative evolution of mt and N-mt genes. Ballard and Youngson (2015)

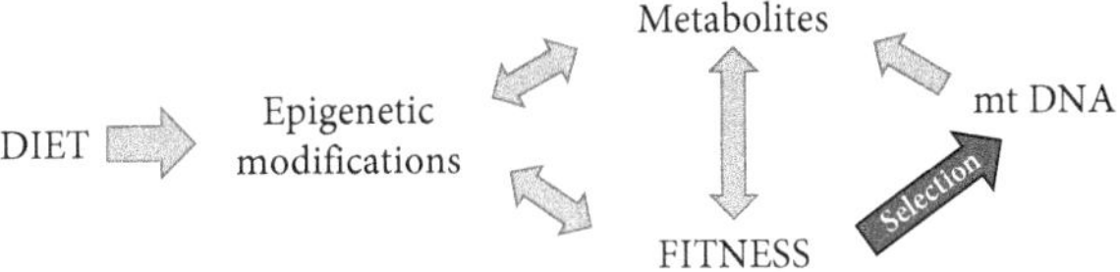

Figure 9.6 A schematic depiction of the potential role of epigenetics in the mediation of diet-induced changes in mt genotypes in populations. The metabolism of different foods produces different kinds and amounts of metabolites such as free radicals, NADH, and FADH. Mitochondrial genotype can influence the production of metabolites, and metabolites in turn can change the epigenetic state. These components of cellular biochemistry have fitness consequences and can impose selection on mt DNA. Arrows indicate connections among components with the arrow indicating natural selection labeled. Adapted from Ballard and Youngson (2015).

reviewed the evidence that evolution of mt genotypes can enable better function in the face of varying diets. They considered how different dietary components, such as proteins, carbohydrates, and fats, produce varying amounts of metabolites such as methyl-donors, reactive oxygen species (ROS), NAD+, ATP, succinate, and fumarate. Mitochondrial DNA genotypes seem to influence the relative levels of these various metabolites that are produced from various foods. These metabolites, in turn, can alter the epigenetic state of the cell (Figure 9.6). Connecting these various components, Ballard and Youngson (2015) proposed that altering mitochondrial function through the evolution of the mt genotype can lead to more adaptive metabolite production under different dietary regimens (Figure 9.6). These ideas remain largely speculative, but a growing number of studies are linking variation in mt genotype to variation in diet. Aw et al. (2016) even proposed that many mild mitochondrial dysfunctions in humans might be treated by manipulating macronutrient ratios.

Mechanisms of adaptation to salt and hydrogen sulfide

There are some data to suggest that mitochondrial function, and hence mt and N-mt genotype, might play an important role in adaptation to at least two other axes of environmental variation—exposure to salt (NaCl) and sulfides. In plants where most studies of effects of salinity on organism physiology and mitochondrial function have been conducted (Jacoby et al., 2011), mitochondria seem to be involved in salt tolerance through (1) adaptions in the tricarboxylic acid cycle, which is entirely N-encoded, (2) the transport of metabolites across the inner mitochondrial membrane, which could involve mt-encoded products (Che-Othman et al., 2017), and (3) ATP production, which could be affected by mt genotype. With regard to effects of salt exposure on the function of ETS complexes, salt promotes the destabilization of the ETS complexes affecting the flow of electrons and the leak of protons. As catalytic centers become more reduced under the influence of salt, more ROS are produced. Structural changes to ETS complexes encoded by mt and N-mt genes might stabilize the complexes and enable function in more saline environments. As an example of the potential role of mitochondrial function in salt adaptation, in a study of the root tissue of corn (*Zea mays*), a plant species that is not adapted to be tolerant of high salt exposure, salt directly negatively affected the function of both Complex I and II of the ETS (Hamilton, 2001).

Environmental gradients in hydrogen sulfide concentration might also be linked to adaptive evolution of mt genotypes. The mechanisms by which mt genotype can enable adaptation to environments rich in hydrogen sulfide is perhaps the most well documented of all mechanisms of mitochondrial adaptation. Like carbon monoxide, nitric oxide, and hydrogen cyanide, hydrogen sulfide binds to and directly blocks the function of Complex IV of the ETS of most metazoans (Cooper and Brown, 2008). However, relatively simple configurational changes in the mt-encoded subunits COX1 and COX3 can prevent hydrogen sulfide binding and allow ETS function in environments that would otherwise be unlivable (Pfenninger et al., 2014).

The next generation of studies of functional mitochondrial adaptation

There is a tendency to restrict mechanisms for the evolution of mitonuclear genotypes in response to the external environment to structural changes in the protein subunits of the ETS. It is important to remember, however, that control of aerobic respiration in mitochondria is under control of highly elaborate and interconnected regulatory pathways. Mitonuclear interactions can involve the transcription, translation, and replication of mt genes (Figure 2.2) as well as anterograde and retrograde signaling that regulates mitochondrial function. Because many of these pathways as well as mitonuclear interactions are incompletely understood, there will certainly be cases in which a specific mt DNA sequence is clearly linked to a functional response that appears to be adaptive, but for which the mechanism for the functional response is uncertain (Rand, 2017).

The potential complexity of adaptive evolution involving mt genotype was made clear in a remarkable comparative study of closely related fish species with dramatically different life histories and aging processes. Sahm et al. (2017) assessed the evolution of not only ETS proteins but a full suite of N-encoded genes that enable the coordinated synthesis and assembly of respiratory chain complexes to enable aerobic respiration (Figure 9.7). The fish clade on which these researchers focused has sister taxa that either show a normal pace of life for a small fish or that undergo their entire life cycle in less than 1 year: going from an embryo to aging old fish in the time from when it begins to rain to the time at which their shallow waterholes dry up several months later. This accelerated life history is not a plastic response to the environment; the life history is retained in a constant environment in the lab. Sahm et al. (2017) found evidence for positive selection related to an accelerated pace of life on genes in all of the arenas in which mt genes and N-mt genes co-function (Figure 2.1); specifically, they found evidence for selection on genes that are involved in the replication of mt DNA, that enable transcription and translation of mt genes, that function in the assembly of respiratory chain complexes, and that form the ETS complexes (Figure 9.7). This paper looked exclusively at N-encoded genes, and thus focused on

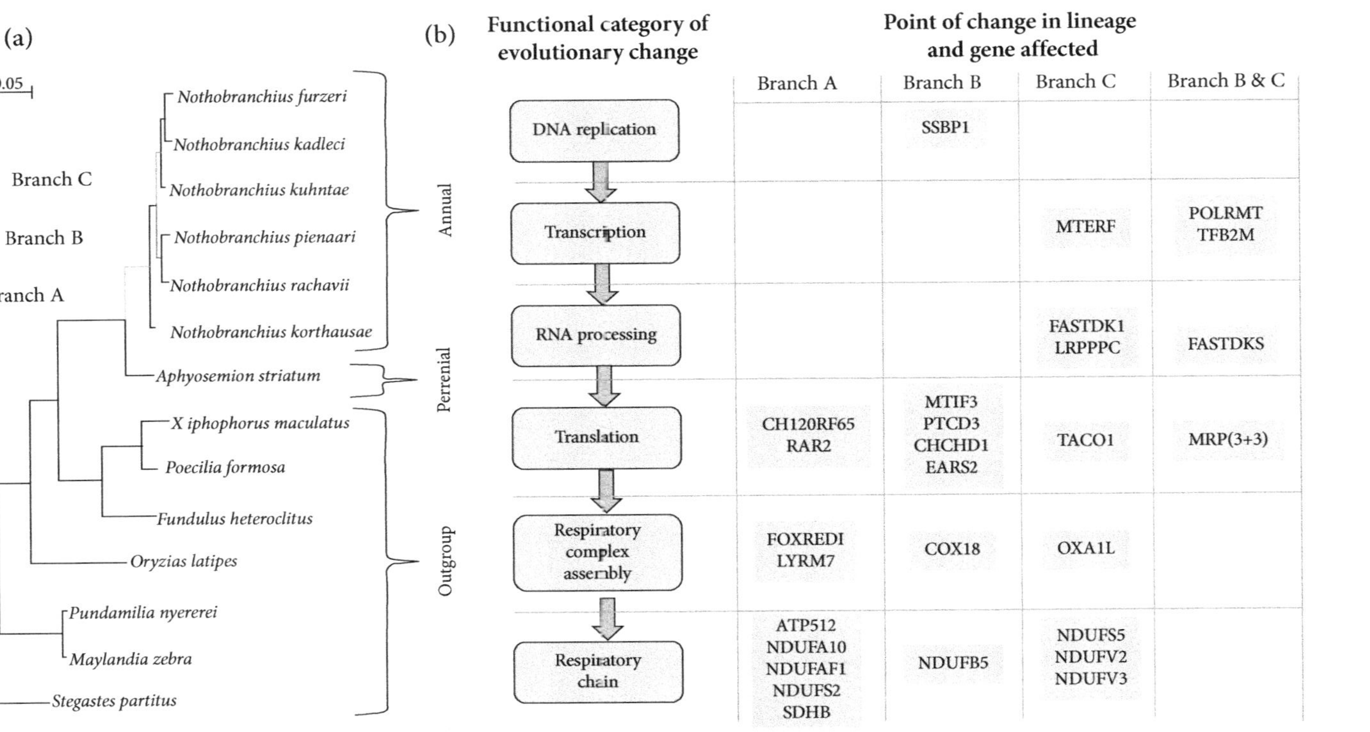

Figure 9.7 A summary of a study that assessed adaptation not just in genes that code for protein subunits of the ETS but across the entire suite of mitonuclear interactions. (A) A phylogeny of the study species of fish indicating the three points in the evolution of this taxon (Branch A, Branch B, and Branch C) that were the focus of an investigation of adaptive evolution in N-mt genes. (B) A list of the categories and the specific genes controlling mitochondrial biogenesis and mitonuclear balance that were found to be under selection in the three focal lineages. Adapted from Sahm et al. (2017).

changes to N-mt genes in the regulation of mitochondrial function. It would be most instructive to expand the analysis to include mt-encoded genes, but nevertheless, the work by Sahm et al. (2017) represents a milestone in revealing the importance of factors other than ETS protein interactions in studies of the effects of selection on interacting mt and N-mt genes.

Theoretically, one could model the fitness consequences of different ETS configurations to assess functional optima for a given environment (Schulte, 2015). In one of the first attempts at such a functional model, Simčič et al. (2014) modeled overall respiratory performance in different thermal environments in relation to ETS function in two species of crayfish. They demonstrated that ETS function directly affects thermal adaptation and that the respiratory function of the two crayfish species was shifted in predicted directions (according to the thermal environment of each species). This study looked at whole-organism and whole-mitochondrial function, not the action of any specific mt and N-mt gene on a specific component of aerobic respiration. Other studies have used models of protein structures to assess molecular interactions between N-mt and mt gene products (Grossman et al., 2004; Scott et al., 2011). Detailed, three-dimensional reconstructions are now available for all of the ETS complexes of mammals (Tsukihara et al., 1996; Iwata, 1998; Fiedorczuk et al., 2016), which provides a platform from which configurational changes to complexes can be assessed. Models of the functional interactions of variants of subunits enable assessment of how nucleotide substitutions may affect the structure and function of OXPHOS complexes (e.g. Garvin et al., 2016 and Morales et al., 2018). It seems inevitable that in the not-too-distant future the performance of the outcomes of different mt and N-mt gene combinations might be assessed across different environments entirely through simulation models.

Evidence for adaptive evolution of mt and N-mt genes

Adaptation arising from standing variation in mt genotypes

Rarely are *Homo sapiens* the model organism for the study of evolution in natural environments. For many centuries, humans have grossly modified key aspects of our environment such as hot/cold exposures, nutrition, and disease risk, and large populations of humans have trotted around the world moving, for instance, from cool temperate regions to hot tropics or from coastlines to mountaintops in a single generation. The very definition of a natural environment is an environment that has not been altered by humans. Nevertheless, before the rise and spread of western civilization, human populations evolved, largely in isolation, for tens of thousands of years in very diverse habitats. Indeed, humans may have existed during those thousands of years in more diverse thermal and altitudinal environments and with more disparate diets than any other species on Earth. Not surprisingly, therefore, the global population of humans has significant genetic structure, including significant diversity of mt genotypes among populations (Figure 9.8). Given these characteristics of the human

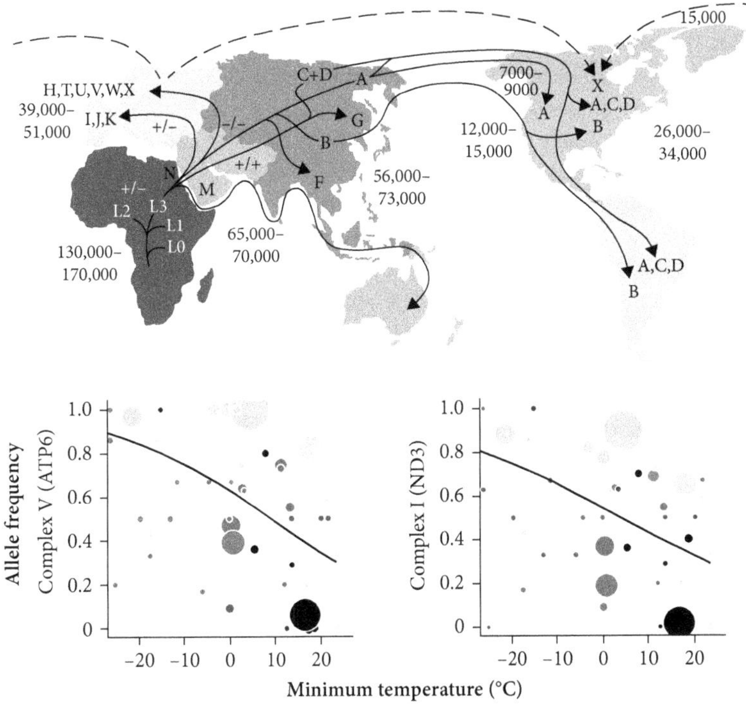

Figure 9.8 The evolution and flow of human mt genotypes across the planet. (Top) A reconstruction of the phylogenetic relationships and geographic distribution of human mt genotypes, which are designated as letters and connected phylogenetically by arrows. Dates are estimates of the age of particular genotypes based on an assumption of a mutation rate of 2.2–2.9 percent base pair substitutions per million years. (Bottom) The association between temperature of the home region for a human population and the frequency of two alleles: left, 8701A, which is a non-synonymous substitution in the sequence for ATP6, a mt-encoded subunit of Complex V, and right, 10398, which is a non-synonymous substitution in the sequence for ND3, a mt-encoded subunit of Complex I. Different human races are grouped by intensity of shading of ovals and oval size is proportional to number of overlapping observations. Top figure reproduced with permission from Lott et al. (2013). Bottom panel adapted from Balloux et al. (2009).

populations and the availability of huge genetic databases, it is really not so surprising that studies of humans provided among the first indications that standing variation in mt genotypes in a natural population could link to local adaptation.

In two seminal papers, Dan Mishmar, Douglas Wallace, and their research teams analyzed complete mt DNA sequences collected from indigenous people from around the globe (Figure 9.8) to rigorously test the widely stated hypothesis that variation in human mt DNA was functionally neutral and the result of drift (Mishmar et al., 2003; Ruiz-Pesini et al., 2004). Mishmar, Wallace, and colleagues were intrigued by several suspicious patterns in the geographic distribution of human mt genotypes including, especially, that there was a five-fold enrichment of a subset of mt genotypes in moving from warm regions of Asia to Siberia (Wallace et al., 1999). That sort of strong clinal variation in allele frequency is commonly linked to a selection gradient in non-human animals and plants (Endler, 1977). In their 2003 and 2004 papers, they laid out a detailed argument that the pattern of global diversity in mt genotypes of human populations was shaped, in part, by selection on different mt genotypes in response to climate (Mishmar et al., 2003; Ruiz-Pesini et al., 2004).

The distribution of many mt genotypes, including all of the genotypes in Africa, did not differ from the pattern predicted by a model of neutral evolution (Mishmar et al., 2003). For a significant set of genotypes in Europe, North American, and especially Asia, however, the pattern of change in mt genotype across the landscape deviated strongly from that predicted by drift (Mishmar et al., 2003; Ruiz-Pesini et al., 2004). And importantly, the mt genotypes whose distribution appeared to be shaped by natural selection coded for different amino acid sequences in proteins, including especially ATP6, which is a subunit of Complex V (ATP synthase) of the ETS. The changes in ATP6 and other mt-encoded protein products in the different human mt genotypes were associated with thermal environment (Figure 9.8). Several follow-up studies, using different approaches, failed to support the patterns reported by the Mishmar–Wallace team (Elson et al., 2004; Kivisild et al., 2006; Sun et al., 2007), but a comparison of minimum temperatures of regions in relation to the frequency of mt alleles in indigenous human populations corroborated the initial reports by Mishmar, Wallace, and their research teams that selection played an important role in shaping the evolution and distribution of standing variation in mt genotypes in the human population (e.g. Balloux et al., 2009; Figure 9.8). Even though there were major challenges to some of the claims of adaptive evolution of human mt genes, these papers on functional variation in mt genotypes in the human population were enormously influential in changing the way in which evolutionary biologists thought about standing genetic variation in the mt genome (Galtier et al., 2009b).

Perhaps the greatest wake up call to evolutionary biologists clinging to the idea that most variation in the mt genomes of most species was the outcome of neutral processes was the high-profile assessment of a central prediction of the neutral theory of mitochondrial evolution: standing variation in mt genotypes should be proportional to the population size of the eukaryotic lineage. Stated another way, large populations should hold more genetic variation among mt genotypes than small populations. The basis for this prediction is very simple: change in allele frequency resulting from drift relies on chance events—stochastic deviations from perfect random assortment of all existing allelic variants. Such chance events leading to loss of genetic diversity become more likely as populations get smaller, so genetic diversity that is the result of neutral processes should be lost more rapidly in small populations compared with large populations. Bazin et al. (2006) assessed this basic assumption of neutral theory using a data set that included more than 3000 animal species, and they concluded that the data did not support the assumption—the rates of change of mt genotypes were statistically independent of the size of the population that contained the genotype.

Animal species with small populations held, on average, as much diversity of mt genotypes as animals species with enormous populations. In a striking set of paired comparisons, they showed that insect or mollusc species did not have more or less variation in their mt genotypes than mammals or birds, despite having vastly larger populations. At the same time, variation in N DNA followed the pattern predicted by neutral theory, with genetic variation in a population proportional to the size of the population. The failure of population size to predict variation in mt DNA was a major refutation of the neutral model of mitochondrial evolution, and it left natural selection as the obvious alternative force that could work independently of population size and maintain diversity in mt genotypes within small populations (Bazin et al., 2006).

Better genomic tools and an impetus to study the importance of functional variation in mt genotypes led to increasingly better analyses of functional consequences of standing variation in mt genotypes within species. I have already focused on the research program of Ronald Burton and colleagues studying the effects of mitonuclear compatibility in the maintenance of species boundaries in *Tigriopus* copepods (see Chapters 2 and 6). This same research program also provided some of the best evidence that standing variation in mt genotype within a species can evolve in response to adaptation to thermal environments. Populations of copepods in the species *Tigriopus californicus* live in shallow tide pools from the chilly coastline of Alaska in the north to the warm rock pools of southern California in the south (Edmands, 2001). Populations show strong clinal variation in heat tolerance, with copepods from southern environments more tolerant to heat stress than animals from northern environments (Willett, 2010; Tangwancharoen and Burton, 2014), and there is a genetic basis for this variation in thermal adaptation (Pereira et al., 2017). Extensive research has documented divergence in coadapted mitonuclear genotypes in *Tigriopus* populations across this same geographic region (Burton et al., 2013) (Figure 6.9). To test whether thermal adaptation may be related to mitochondrial performance, Harada and Burton (2017) measured various aspects of mitochondrial performance in animals across the north to south cline. They found that mitochondrial performance as measured by ATP production followed the same pattern observed in studies of whole-animal performance measured in previous studies: northern populations did better in cooler environments and southern populations did better in warmer environments. In paired comparisons of northern and southern populations, measures of free radical abundance, membrane potential, and ETS activity also supported better heat tolerance in southern populations. They concluded that adaptive evolution of N and N-mt genes may have played a key role in thermal adaptation through changes in mitochondrial function.

Among the most well-controlled and compelling examples of the fitness effects of different mt genotypes within a species are laboratory studies of seed beetles (genus *Callosobruchus*), fruit flies (genus *Drosophila*), and nematodes (genus *Caenorhabditis*). In these studies, researchers used a range of experimental techniques to isolate the fitness effects of mt and N genes in different environments (Figure 9.9). One of the pioneering research programs to assess the potential for functional consequences of variation in mt genotypes within a population was undertaken by Göran Arnqvist

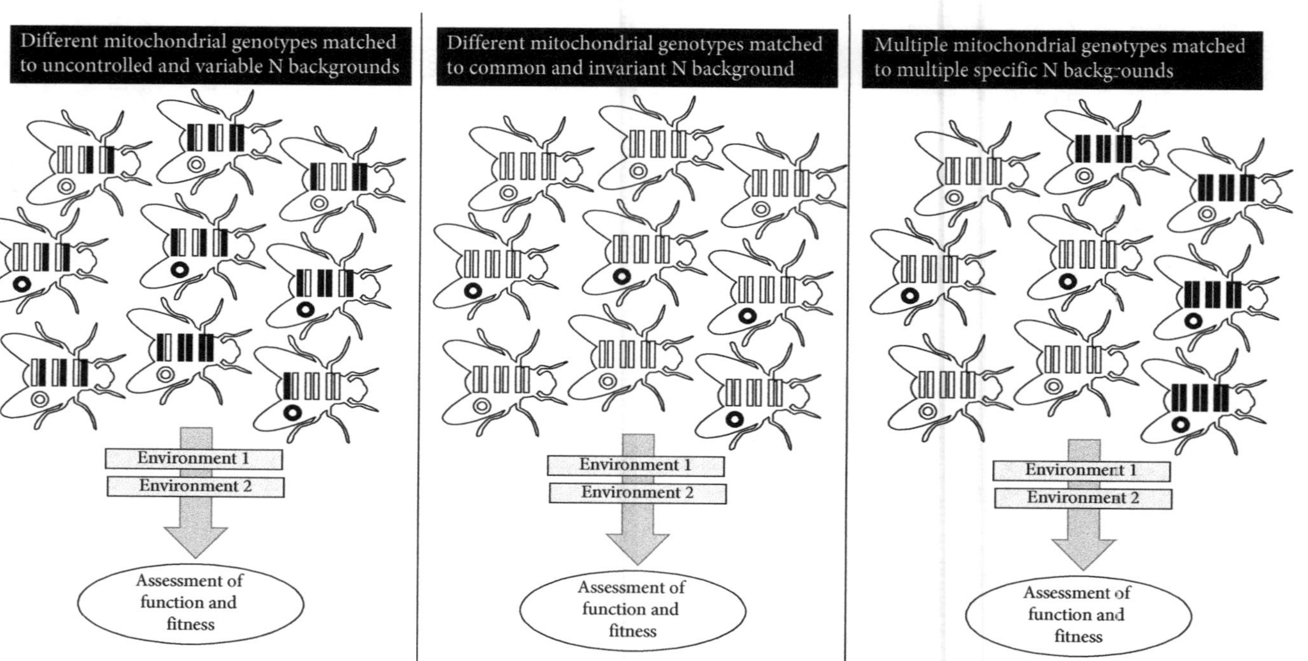

Figure 9.9 A pictorial overview of different experimental approaches used for the study of functional consequences of variation in mt genotype. The circles represent mitochondrial genomes and paired bars represent paired chromosomes. Shading represents different genotypes. Left panel: the function and/or fitness of mitochondrial genotypes is compared without controlling the N genotype. In this design, it is often assumed that the effects of variation in N genotype are randomized. (Middle panel) Different mt genotypes are expressed with a common N genotype eliminating variation from N genes. (Right panel) More than one mt genotype is paired to more than one N genotype. In the right panel, unlike the left panel, the N genotypes are known and limited to a few variants.

and colleagues focusing on laboratory colonies of seed beetles, *Callosobruchus chinensis*. There is standing variation in mt genotypes in wild populations of seed beetles that appears to be linked to adaptation to different habitats (Tuda et al., 2004). Building from these observations of wild seed beetles, Arnqvist and colleagues assessed the functional significance of standing variation in mt genotypes in lab populations of seed beetles by experimentally pairing different mt genotypes with different N genotypes and measuring the respiratory performance of individuals in different thermal environments. They found significant and substantial mt genotype by N-mt genotype by environment ($G_{mt} \times G_{N\text{-}mt} \times E$) effects on respiratory performance (Arnqvist et al., 2010) and development rate (Dowling et al., 2007a). Perhaps most importantly, variation in mt genotype had big effects on respiratory function—effects on a scale that would likely affect fitness in natural populations (Arnqvist et al., 2010).

In a follow-up study of seed beetles, Arnqvist's lab group investigated male reproductive performance in relation to mt genotype paired with different N genes and in different environments. Again, they found evidence for a significant effect of mt genotype on fitness (Immonen et al., 2016). Then, to experimentally test the role of selection in maintaining variation in populations, they conducted a selection experiment with the seed beetles (Đorđević et al., 2017). For 160 generations they selected for long life in one line of seed beetles and short life in another line of beetles. They found a strong response to experimental selection, with lifespan significantly altered in the experimental populations. Most of the effect was due to selection on N genes, but a substantial portion of the variation in longevity could be attributed specifically to the effects of mt genotype (Đorđević et al., 2017). Thus, standing variation in mt genotype in a population of beetles can affect lifespan.

Among the most extensive and multi-faceted investigations of adaptive evolution of standing variation in mt genotypes has focused on different species of fruit flies in the genus *Drosophila*. The lab group of Damian Dowling conducted some of the most detailed and well-controlled studies of the influence of environment on the fitness consequences of mitonuclear genotypes in *Drosophila*. First, they measured the fitness of different mt genotypes drawn from within laboratory populations of *Drosophila melanogaster* when those mt genotypes were expressed in different nuclear backgrounds and in different environments (the experimental design illustrated in the middle panel in Figure 9.9). They found substantial differences in fitness among different $G_{mt} \times G_{N\text{-}mt} \times E$ combinations, but there was no universally most fit or least fit genotype. Fitness was entirely contingent on the epistatic interactions of the mt and N-mt genes within a given environment (Dowling et al., 2007c). In a follow-up study, they studied the effects of variable environments on the fitness effects of an allelic variant of mt-encoded cytochrome *b* gene. This allele causes impaired male fertility when expressed in a standard laboratory environment (Clancy et al., 2011). Dowling and colleagues looked at the effects of nuclear background and thermal environment on the fitness effects of the allele (Wolff et al., 2016b). While the male-harming effects persisted across environments, the effects of the allele were nearly twice as great at 27°C compared with 18°C. The effects also varied substantially across nuclear backgrounds. Again this study is a clear demonstration of $G_{mt} \times G_{N\text{-}mt} \times E$ effects within *D. melanogaster.*

In another study of laboratory populations of *D. melanogaster* carrying different mt genotypes but with a shared nuclear background, populations were reared on different diets, with one approximating an optimal diet for wild *D. melanogaster*, while another shifted far from this optimum (Pichaud et al., 2013). Flies carrying different mt genotypes varied substantially in how well they fared on the different diets. Because the flies in this experiment shared a common N genotype, the differences in diet adaptation could be attributed solely to the effects of mt genes (Pichaud et al., 2013). David Rand and his lab group followed up on this "small scale" diet study with an ambitious investigation focused on assessing the fitness effects of $G_{mt} \times G_{N\text{-}mt} \times E$ (diet) interactions in laboratory populations of *Drosophila* (Zhu et al., 2014). (Refer to Figure 9.1 for an overview of possible interactions for each environmental variable.) To insure significant variation in mt genotype, they used mt genotypes from two sister species—*D. melanogaster* and *D. simulans*—and they expressed these two mt genotypes with one of three *D. melanogaster* N genotypes. The result was eighteen unique and discrete mitonuclear gene combinations. The goal of the study was to assess the effect of diet on longevity, and they tested the performance of each of the eighteen mitonuclear genotypes in five different dietary environments. In total, they assessed a staggering ninety distinct $G_{mt} \times G_{N\text{-}mt} \times E$ treatments. They found significant effects of mt genotype by diet, mt genotype by N genotype, and N genotype by diet interactions on longevity of individuals, once again demonstrating that standing variation in mt genotype can have significant effects on longevity. The recurring theme in all of these laboratory studies with seed beetles and fruit flies is that mt genes are not passive bystanders in the determination of individual fitness (Ballard and Pichaud, 2014; Wolff et al., 2016a). Mitochondrial genotype interacts with N genotype and the environment to shape the fitness of the organism.

Studies with laboratory populations of seed beetles and fruit flies clearly demonstrated the huge potential for mt genotype to play a role in adaptation, but these studies on laboratory lines of animals used artificial laboratory environments. Such studies allow for the tight control of variables, and they establish the plausibility of mt genes playing a role in adaptive evolution in natural environments. However, they leave open the questions: does natural selection really shape mt evolution in wild population in response to natural environments and can mt genotypes contribute to adaptation in the wild?

To begin to move investigations of the adaptive evolution of mt genotypes from the lab to the field, Dowling and colleagues collected *D. melangaster* from wild populations across eastern Australia from tropical regions near the equator in the north to chilly Melbourne in the south (Figure 9.10a) (Camus et al., 2017). These fruit fly populations descend from flies that were introduced to Australia in the nineteenth century. The founding fruit flies were of unknown genetic makeup, but one can guess that they were transported to Australia with fruit from Asia. The flies have subsequently had thousands of generations to adapt to local environments across the Australian continent. Dowling and colleagues first genotyped flies from across the collection transect, and they observed clinal variation in dominant mt genotypes with a preponderance of a genotype in the north that I will refer to as Red, and an over-representation of an

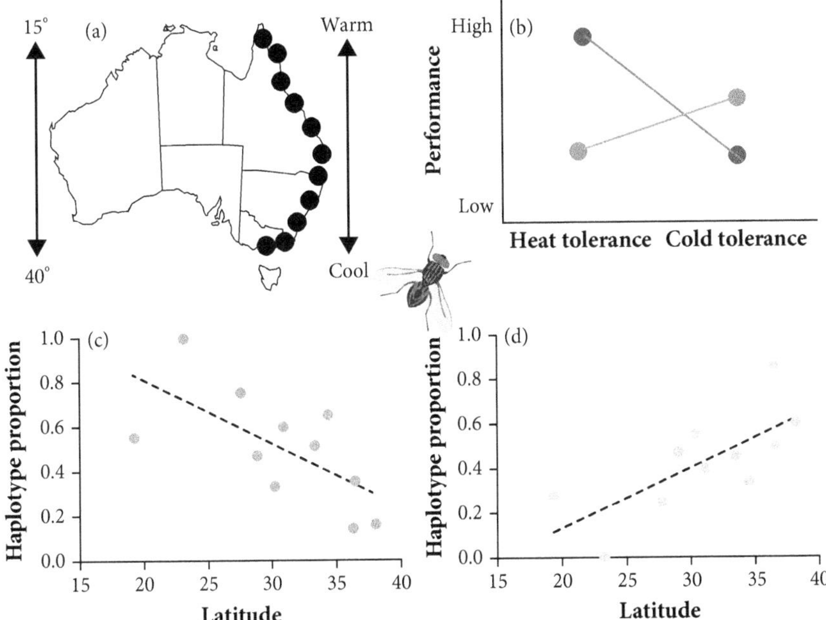

Figure 9.10 Standing variation in Australian *Drosophila melanogaster* populations is strongly associated with climate. (a) Collecting sites across Australia covering a thermal gradient from warm in the north to cool in the south. (b) The performance of the warm-adapted mt genotype versus cool-adapted mt genotype in relation to a heat challenge versus a cold challenge. Shown is the mean response of the two genotypes. Cold-adapted did best in the cold challenge and warm-adapted in the hot challenge. (c) The proportion of warm-adapted genotypes in populations from north to south. (d) The proportion of cool-adapted haplotypes in populations from north to south. Details in Camus et al. (2017), from which images were adapted.

alternative genotype in the south, which I will call Blue (Camus et al., 2017) (Figure 9.10c and d). (There is actually a bit of variation in mt genotype within the Red and Blue mt genotypes, but for this overview I'll only consider the major divide between Red and Blue genotypes.) As I mentioned in the previous discussion of variation in human mt genotypes, clinal variation in genotypes that is strongly associated with clinal variation in thermal environment suggests that natural selection has shaped the cline (Endler, 1977). And indeed, when Camus et al. (2017a) tested the cold and heat tolerance of Red and Blue mt genotypes, they found the Red performed better in heat challenges, and Blue performed better in cold challenges (Figure 9.10b). In these tests of thermal tolerance, the researchers created test strains by placing each of the two dominant mt genotypes against a standardized N background, so they could map the thermal tolerance effects directly to the mt DNA genotype. And finally, when Dowling and colleagues established populations with mixed mt genotypes in either warm or cool environments in the laboratory, they observed a significant increase in the frequency of the Red genotype in the warm environment and a significant increase in the Blue genotype in the cool environment (Lajbner et al., 2018).

The study of geographic variation in thermal adaptations suggests that standing variation in mt genotype within Australian *D. melanogaster* is maintained by natural selection for climate adaptation. This observation of mt genotypes linked to climate adaptation in Australian *D. melanogaster* echoed an earlier study of climate adaptation in *D. simulans*. Researchers discovered two wild populations carrying different mitochondrial types, and laboratory studies of these divergent mitochondrial types showed that they differed in thermal tolerance. Moreover, the differences in thermal tolerance affected muscle performance (Pichaud et al., 2010, 2011). These differences in thermal tolerance in *D. simulans* persisted even when the divergent mt genotypes were expressed against a common nuclear background, demonstrating that the climate adaptation was a product of mt and not N genes (Pichaud et al., 2012). Thus, in wild populations of two species of *Drosophila*, standing variation in mt genotype has been linked to thermal adaptation.

One of the puzzling features of the study of thermal adaptation in Australian *Drosophila* is that the genetic variation that differentiates warm-adapted mt genotypes from the cold-adapted mt genotypes is a change in fifteen nucleotides in protein-coding genes, each of which is a synonymous substitution (Camus et al., 2017). In other words, the changes to the nucleotide sequences that are associated with thermal adaptation are silent with respect to the amino acid sequence such that these variations can have no direct effect on protein function. And yet, functional change related to these fifteen synonymous substitutions has been experimentally demonstrated.

Observations from a detailed study of the mummichog (*Fundulus heteroclitis*), a small coastal fish, further underscore the challenges of connecting biochemical mechanisms of thermal adaptation to the associations of genes with thermal gradients. Mummichogs occur along the Altantic coast from warm waters in the south to cold waters in the north, and there are two mt genotypes in mummichogs that differ in five amino acid replacements in ETS subunits, with one genotype predominating in the north and the other predominating in the south (Whitehead, 2009). Populations of fish with the two genotypes overlap near New York City, and there is limited gene flow from one population into another. Baris et al. (2017) looked for nuclear alleles that were associated with the two mt genotypes. They focused on N-mt genes, and they found 349 single nucleotide polymorphisms that were significantly associated with the mt genotype. As was observed in the Australian fruit fly study, Baris et al. (2017) showed that mitonuclear gene combinations were associated with respriratory function, such that the changes appeared to have evolved for thermal adaptation. The clear prediction is that some of the 349 differences in N genotype between the populations would involve N-encoded subunits of ETS complexes and specifically subunits with close functional interaction with one of the five mt-encoded protein subunits of ETS complexes that differ between the populations. But frustratingly (if you like simple answers), none of the nucleotide substitutions to N genes that were geographically associated with mt genotype involved genes encoding subunits of the OXPHOS complexes (Baris et al., 2017). Several of the differentiated N genes are involved in the regulation of OXPHOS components, so the implication is that regulatory interactions mediate the epistatic effects that underlie thermal adaptation (Rand, 2017). Likewise, in the Camus et al. (2017b) study on fruit flies, it was suggested that the effects of the mt genotype came from transcriptional regulation of key protein-coding mt DNA genes. But the mechanisms by which a specific genotype bestows an adaptive advantage in a particular environment remain obscure in both the fruit fly and fish examples.

A study of natural variation among populations of the nematode *Caenorhabditis elegans* provided further evidence for climate adaptation linked to standing variation in mt genotype, and unlike in the mummichog and fruit fly examples above, the nematode study identified the likely biochemical mechanisms for adaptation. Researchers first established that there are fixed non-synonymous differences in COX1 genotype (a mt-encoded subunit of Complex IV of the ETS) in nematodes from the cool environment of England compared with nematodes from warm tropical Hawaii (Dingley et al., 2014). These two populations of nematodes share a recent common ancestor, and they have been in their current environments for only decades (Hawaii) to at most a couple of centuries (England). These researchers then showed experimentally that these differences in mt genotype give rise to functional variation. When the nematodes from Hawaii were maintained in a cool environment, they increased the enzyme activity of Complex IV and suffered increased production of free radicals and signs of increased oxidative stress compared with temperate nematodes in the same environment. Moreover, by controlling for nuclear background, researchers were able to show that thermal adaptation was largely due to mt genotype (Dingley et al., 2014). The authors concluded that the evolution of mt genotype may be at least in part shaped by selection for adaptation to local environment, and the mechanism for this adaptation was functional change in Complex IV.

The experimental evidence linking standing variation in mt genotypes to climate adaptation in populations of fruit flies and nematodes presents the most comprehensive studies on functional consequences of standing variation in mt genotypes in natural populations, but similar patterns of clinal variation in mt genotypes associated with a thermal gradient across the range of a species have now been documented for a marine mammal (killer whales, *Orcinus orca*; Foote et al., 2011); marine fish (European anchovy, *Engraulis encrasicolus*; Silva et al., 2014; Figure 9.11); Atlantic salmon, *Salmo salar* (Consuegra et al., 2015); Atlantic herring, *Clupea harengus* (Teacher et al., 2012), killifish, *Fundulus heteroclitus* (Baris et al., 2016); freshwater fish (rainbow trout, *Oncorhynchus mykiss*; Garvin et al. 2015); a freshwater gastropod (*Radix balthica* (Quintela et al., 2014); and a passerine bird (eastern yellow robin, *Eopsaltria australis*; Morales et al., 2017) (Box 9.3). Similar clinal or regional variation in mt genotypes related to altitude—and hence oxygen pressure—have been reported for an insect (migratory locust, *Locusta migratoria*; (Zhang et al., 2013b), a songbird (rufous-collared sparrow; *Zonotrichia capensis*; Cheviron and Brumfield, 2009), a gallinaceous bird (chicken; Zhao et al., 2015), and several placental mammals—white-toothed shrew, *Crocidura russula* (Fontanillas et al. 2005), domestic dog (Li et al., 2014), domestic horse (Xu et al., 2007; Ning et al., 2010), sheep (Hassanin et al., 2009), and humans (Simonson et al., 2010; Yi et al., 2010). Temperature and oxygen pressure

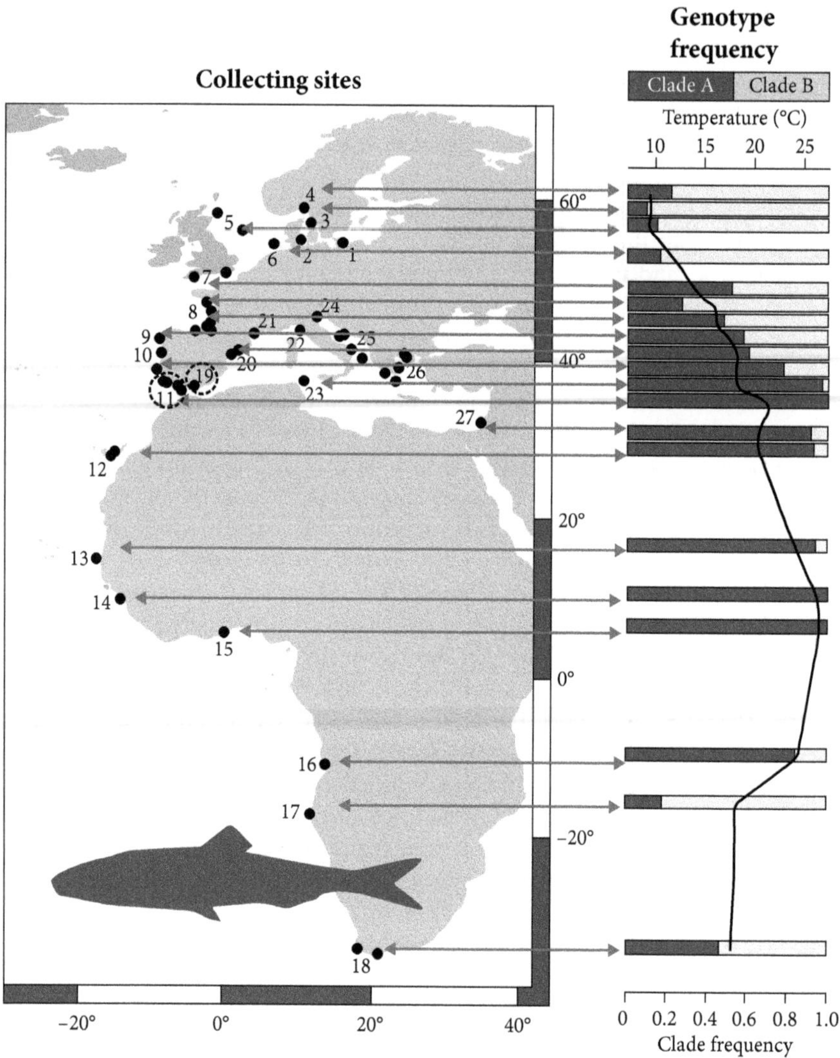

Figure 9.11 Change in mt genotype frequency of anchovies across a global thermal gradient from cold to warm to cold. A cline in gene frequency that correlates with an environmental gradient suggests adaptive variation but other explanations are also possible. Numbered dots on the map to the left show collecting sites with corresponding frequency of mt genotype at that location shown in the bar chart on the right. Arrows connect collecting site to bar chart. The line running through the histogram shows mean temperature at collecting sites. Reprinted with permission from Silva et al. (2014).

are often correlated—high elevations are colder than low elevations and cold water holds more oxygen than warm water—so adaptation to altitude could be adaptation to either hypoxia or temperature. Garvin et al. (2015) concluded that differences between populations of freshwater trout living at low and high elevations were a function of both temperature and oxygen pressure. However, they went on to report that changes in the N-mt genotypes between the high and low altitude populations of trout included protein changes that could affect assembly of supercomplexes. They argued that supercomplexes increase the efficiency of respiration and are particularly important under low-oxygen conditions.

One of the most compelling observations of directional natural selection on standing variation in mt genotype come from a study that tallied who lived and who died during mass die-off of the seabird common murre (*Uria aalge*). Murres and other seabirds that nest in huge colonies are well known for periodic episodes of mass mortality caused by a crashes in populations of schooling prey that is their primary food. Drovetski et al. (2012) fortuitously sampled individuals randomly from a common murre population immediately prior to a massive die-off that killed a significant fraction of the population. They then used their samples to assess the proportion of the population carrying the dominant mt genotype versus other mt genotypes that,

Box 9.3 Perched on a mitonuclear divide

In my discussions of mt DNA barcoding applied to birds in Chapter 6, I emphasized that the boundaries between species that emerged from DNA barcode analysis almost always followed established taxonomy. The "almost" qualifier in the above statement represents about 5 percent of Class Aves; about 5 percent of birds either have divergent phenotypes with no associated barcode gap or multiple mt genotypes within one phenotype (Kerr et al., 2007). The details of those "complicated" species can be fascinating. Case in point, a suite of thirteen Australian bird species (of the seventeen examined) was discovered to have two or more divergent mt genotypes with no associated difference in plumage coloration or song (at least no conspicuous difference that had been formerly recognized) (Lamb et al., 2018). In eight of these species, climate was a significant predictor of the occurrence of a particular mt genotype; in other words, beyond what could be predicted by geographic distance or simply geography (e.g. two sides of a mountain range), the mt genotypes tracked climate (Lamb et al., 2018).

Paul Sunnucks and his lab group followed up on the discovery of this fascinating pattern of diversity of mt genotypes by conducting a detailed study of the eastern yellow robin (*Eopsaltria australis*), a beautiful but easily overlooked songbird of eastern Australia (Morales et al., 2018). Eastern yellow robins occupy a 500-km-wide band around the coast of Australia from Queensland to southern Victoria (Figure 7.12; Figure Box 9.3). This range avoids the hot arid regions of the interior of the Australian continent, but it still holds a tremendous range of climates, with the coastal part of the robin's range much wetter and cooler than the more interior portion. Sunnucks and his team found a sharp divide in mt genotype where this climate transition occurred and, interestingly, the best climate predictor for where one robin mt genotype replaced another was related to the hottest days of summer, not average temperature (Pavlova et al., 2013) (Figure Box 9.3). In Australia, this pattern makes sense because the interior of the continent gets famously, baking hot in the summer and periodic north winds drive this heat out to coastal areas. As hot as it can feel on these days in Melbourne or Sydney on the coast, it is always significantly hotter on the other side of the coastal mountains in the interior portion of the range of the eastern yellow robin. As presented in Figure 7.12, there is a sharp coastal/interior divide not only in the mt genotype of eastern yellow robins but also in a linked set of N genes that includes many

(Continued)

Box 9.3 *Continued*

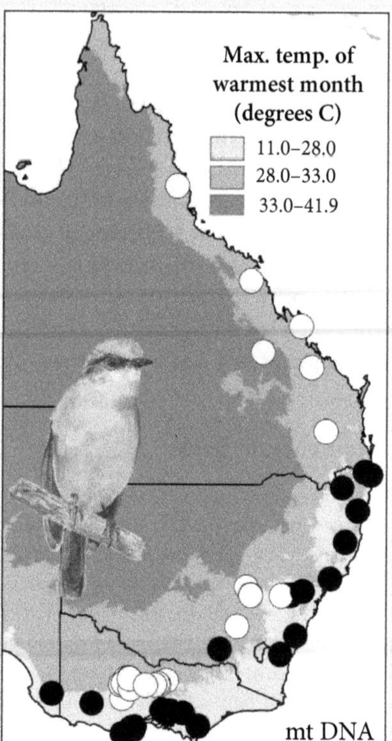

Figure Box 9.3 The distribution of mt genotypes in eastern yellow robin in relation to climate. Shaded regions on the map of Australia show annual maximum temperatures. Genotypes coded as white or black are divergent across more than 6 percent of the mt genome. Birds with different mt genotypes also differ from each other in a set of linked genes on chromosome 1A that contains many N-mt genes. Genotype/climate map adapted from Morales et al. (2017).

N-mt genes. There is nearly perfect congruence in the sorting of these mt and N-mt gene sets, such that they both show the same abrupt transition across temperature zones (Morales et al., 2017, 2018). This strong association between mitonuclear genotype and climate suggests that hot summer days have imposed natural selection on mt and N-mt genotypes creating the observed mitonuclear divide. The key remaining question is whether specific mitonuclear genotypes actually enable better climate adaptation.

combined, were carried by about 40 percent of the individuals in the population. They found that common murres carrying the dominant allele were significantly more likely to have survived the starvation event than murres with other genotypes, strongly implicating directional selection on mt genotype and a role of mt genotype in the adaptation of common murres (Figure 9.12). One can speculate that the

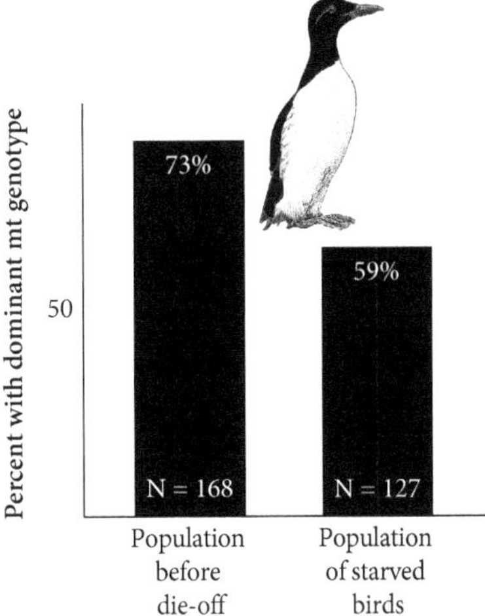

Figure 9.12 Evidence for functional differences in variants in mt genotype in a wild population of common murres (*Uria aalge*). Birds with the dominant genotype were less likely to starve to death when food became scarce. No N genes predicted survival. Drawn from data in Drovetski et al. (2012).

advantages of the alternative alleles under non-starvation conditions maintain them at high frequencies in the population, but no follow-up studies have yet been conducted.

The catalogue of publications documenting adaptive variation in mt genotype that is correlated to the external environment is now expansive, and it continues to grow rapidly. It is no longer a question of whether adaptive evolution can select for standing variation in mt genotypes within populations; the question now is whether adaptive evolution of mt genotypes is a ubiquitous feature of eukaryotes.

Adaptive divergence at species boundaries

As described in the last section, there is substantial and growing evidence that standing variation in mt genotypes within a population can be functional and enable local adaptation in different parts of a species' range. The divergence of mt genotypes within a population will always remain modest, however, because gene flow is the great homogenizer, dragging populations away from local adaptive peaks (Slatkin, 1985). When speciation occurs and gene flow is stopped, populations are free to move toward divergent adaptive peaks. Thus, adaptive divergence in mt genotype between closely related species can provide the best arena in which to assess the importance of mt variation to function and fitness.

The same environmental variables that exert selective pressure on aerobic respiration and that can shape the adaptive evolution of mt and N-mt genotypes are also the variables that shape adaptive divergence as populations split into distinct species. Whether or not adaptive divergence in mt and N-mt genes in response to environmental variation can actually drive speciation as a form of ecological speciation (see Chapter 7) is entirely speculative at this point. It is also entirely unnecessary to invoke adaptive divergence of populations as the genesis of species. Disruption of gene flow by separation of populations in space and time creates opportunities for adaptive divergence, because once gene flow between populations is disrupted natural selection can reshape their genotypes to rise toward local adaptive peaks (Slatkin, 1985). Here, I will focus on how species of eukaryotes can adapt to local environments through the evolution of mt and N-mt genes.

Garvin and colleagues published among the most detailed studies documenting positive selection on mt and N-mt genotypes relative to the thermal environments in their comparative studies of salmon species in the genus *Oncorhynchus* (Garvin et al., 2011) and trout in the genus *Salvelinus* (Garvin et al., 2015). They found evidence for positive selection in the mt genomes of fish in these lineages and they documented fixed differences between species in subunits of Complex I, IV, and V, as well as cytochrome *b* (Garvin et al., 2011). They focused particularly on the mt-encoded Complex I subunit ND5, because the structural changes to ND5 that they observed appeared to create functional changes to Complex I that could underlie adaptations to thermal environments (Garvin et al., 2014).

The majority of studies assessing functional changes in mt and N-mt genes in response to different environments experienced by closely related species concern adaption to variation in the partial pressure of oxygen (Luo et al., 2013). Comparisons are mostly made between closely related species living at high or low elevations, but comparisons have also been made between terrestrial and burrow-dwelling animals as well as between marine or aquatic organisms in water with different levels of dissolved oxygen. Perhaps the most detailed and well-known studies of adaptation to low partial pressure of oxygen involving adaptive changes to mt gene products were conducted by Graham Scott and colleagues on the bar-headed goose (*Anser indicus*), a species of Asian waterfowl that flies over the Himalayan Mountains as the population moves back and forth from a northern breeding range to a southern wintering range each year (Scott et al., 2015) (Figure 9.5). This over-mountain route is hundreds of kilometers shorter than routes around the vast mountain range, but flying above "the top of the world" requires flying through a very low-oxygen environment. Several species of geese that are closely related to the bar-headed goose take the much longer, low-elevation migration route, and Scott et al. considered whether a physiological adaptation may have enabled bar-headed geese to maintain aerobic respiration in the low-oxygen environment above the mountaintops (Scott et al., 2011, 2015). They found that compared with the closely related graylag goose (*Anser anser*), which never crosses high mountain ranges, the oxygen affinity of the Complex IV was much higher in bar-headed geese, enabling them to sustain aerobic respiration in the lower oxygen pressures encountered at high elevations (Scott et al., 2009). This Complex IV adaptation that enables respiratory function at high elevation was traced to a single amino acid substitution in the mt-encoded subunit COX3. Graylag and other low-elevation geese lack the COX3 structure of bar-headed geese.

Similar genomic and functional comparisons of the ETSs have been made for a number of animal taxa that have closely related species that are adapted to lowlands versus those adapted to the the most expansive high-elevation habitat on Earth, the Tibetan Plateau (Figure 9.13). Paired comparisons between high-elevation and

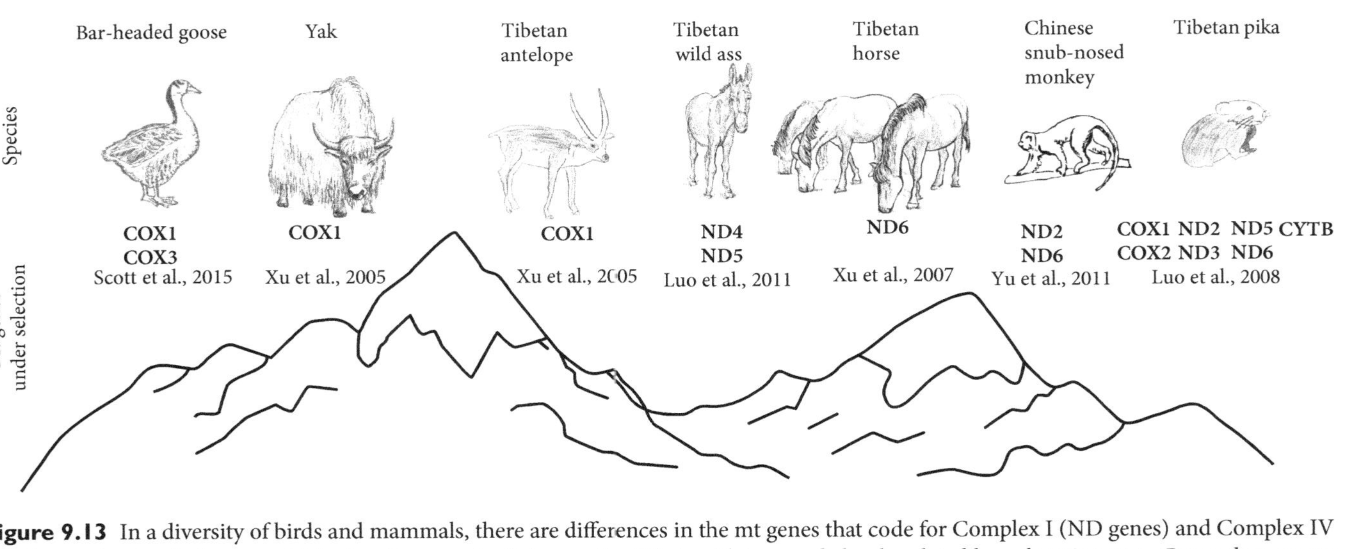

Figure 9.13 In a diversity of birds and mammals, there are differences in the mt genes that code for Complex I (ND genes) and Complex IV (COX genes) subunits between high-elevation species living on the Tibetan Plateau and closely related low-elevation taxa. Genes that appear to be the object of positive selection in high elevations are listed next to the taxon in which they were studied. These differences are hypothesized to be adaptations to the low oxygen pressure of high-elevation environments.

lowland taxa showed evidence for changes in the mt genotype that were implicated as adaptations to low oxygen pressure. These studies included an alpine pika (plateau pika, *Ochotona curzoniae*) compared with rabbits that were not high-altitude adapted; an alpine antelope (Tibetan antelope, *Pantholops hodgsonii*) compared with low-elevation domestic sheep (*Ovis aries*) and goats (*Capra hircus*) (Xu et al., 2005); the Himalayan cow known as the yak (*Bos grunniens*) compared with low-altitude cattle (*Bos taurus*) (Qiu et al., 2012); the high-altitude Tibetan wild ass (*Equus kiang*) compared with congeneric and low-elevation donkeys (*Equus asinus*) and horses (*Equus caballus*) (Luo et al., 2011); and the alpine snub-nosed monkey (*Rhinopithecus roxellana*) compared with low-elevation primate species (Yu et al., 2011). In all of these comparisons, the authors found functional changes to ETS subunits that could plausibly enable better function at high elevations. Most of the changes involve Complex IV subunits (COX genes), as in the bar-headed goose example, but several of the comparisons revealed changes to Complex I subunits (ND genes; Figure 9.13).

Oxygen pressure varies not only with altitude but also in aquatic and marine environments according to variables such as temperature, salinity, and surface exposure, as well as in burrow and cave environments. In comparisons of eight species of sculpins (family Cottidae, Tribe Actinopterygii), small fish that inhabit the shallow coastal water that varies substantially in oxygen availability, Lau et al. (2017) reported a strong association between hypoxia tolerance of a species and the affinity of that species' Complex IV enzyme for oxygen binding. The implication was that, among these fish taxa, Complex IV had evolved in response to the availability of oxygen in that species' specific coastal environment. Moreover, the variation in the stability of mt-encoded COX3, one of the core catalytic subunits of Complex IV, was implicated as the key mediator of oxygen binding affinity (Lau et al., 2017). A study comparing two species of sharks, one adapted to a very low-oxygen environment and another adapted to a higher-oxygen environment, found changes to mitochondrial function much like what was observed in the sculpin study: compared with sharks adapted to low oxygen, sharks adapted to high oxygen increased COX activity when held in a low-oxygen environment (Hickey et al., 2012). One would predict that the same functional changes to COX genes would be observed as different taxa adapt to variation in oxygen pressure (Lau et al., 2017). Other studies of adaptation to oxygen pressure, however, did not support this prediction. Just as in sculpins and sharks, closely related species of triplefin fish live in environments with high and low oxygen availability and yet triplefin fish species that were adapted to low-oxygen environments increased COX activity in oxygen-poor environments when compared with triplefin fish species from high-oxygen environments, exactly the opposite patterns observed for sculpin and sharks (Hilton et al., 2010). In the triplefin fish study, temperature was more the focus of experiments than oxygen pressure and that could account for the discrepancy, but clearly more studies focused on how changes to ETS complexes can facilitate evolutionary changes to available oxygen are needed.

Living in burrows or other closed chambers can also changes access to oxygen because compared with air at the surface, air in burrows has higher levels of carbon dioxide and lower levels of oxygen and hence a lower partial pressure of oxygen

(Culver and Pipan, 2009). Tomasco and Lessa (2014) assessed the evolution of functional changes to COX genes in two rodent species that independently adopted a burrow-dwelling life history compared with other, closely related terrestrial rodents. They found evidence for directional selection on mt-encoded COX2 and cytochrome *b* genes that appeared to be associated with functional changes that enabled better respiratory performance in low-oxygen burrow environments. Living in burrows with low oxygen is very common in eukaryotes, involving perhaps the majority of invertebrates as well as many mammals, so adaptation to such conditions could be major selective force among animals. Bird species that are otherwise adapted to high-oxygen sea-level air often nest in burrows or cavities, subjecting developing offspring to reduced oxygen (Birchard et al., 1984). Whether burrow-nesting birds have evolved mitochondrial adaptation to function better in the low-oxygen burrow environment remains unstudied.

Some of the cleanest examples of adaptive divergence between species in genes coding from aerobic respiration in mitochondria concern exposure to hydrogen sulfide, a substance that directly suppresses activity of Complex IV (Cooper and Brown, 2008). Hydrogen sulfide is rare in most aquatic environments, but at sulfur springs the concentration of hydrogen sulfide can be very high, rendering an otherwise suitable habitat uninhabitable for aquatic life. However, relatively simple structural changes to mt-encoded Complex IV subunits can block hydrogen sulfide binding thereby negating its poisonous effects. Pfenninger et al. (2014) studied three species of mollies (genus *Poecilia*) that occupy springs with high concentrations of hydrogen sulfide as well as sister species that occupy portions of the streams with essentially no hydrogen sulfide. (These molly populations might be considered subspecies by some taxonomists but they have limited gene flow (Plath et al., 2013) and I will discuss them as species). In two of the three species living in sulfur springs, changes to the same mt-encoded Complex IV subunits, COX1 and COX3, evolved independently, and the COX1 substitution was concluded to code for a conformational change that would block the binding of hydrogen sulfide (Pfenninger et al., 2014). Thus, this study presents very strong evidence for adaptive evolution of a mt-encoded protein subunit of Complex IV in response to the chemical environment. Interestingly, the third molly population living in a sulfur spring lacked the modifications to Complex IV that made the other sulfur-adapted molly populations capable of living in the hydrogen sulfide environment. This population of mollies seemed to have evolved the capacity to tolerate hydrogen sulfide rather than to block its binding (Pfenninger et al., 2015).

Adaptive changes to mitochondrial function may also enable eukaryotes to better cope with exposure to environmental salt (NaCl). Botanists and agricultural researchers have been intensively studying salinity tolerance in plants for over 100 years, and plants provide, by far, the most and the best examples of adaptive divergence for different salt environments. Still, the great majority of the literature is focused on acclimation to salt exposure within a species, and mitochondia are rarely mentioned. Among the few studies comparing the evolution of mitochondrial function for handling salt tolerance was a comparison among saltmarsh plant species in the genus *Spartina*. Compared with individuals from "high marsh" species, which avoided salt exposure, "low marsh" species with high salt exposure showed reduced aerobic respiration and a lower sensitivity to sulfide (Maricle et al., 2006). The adaptations of these plants were not linked specifically to genetic-based changes to mitochondria but such an effect seems likely given the central role of aerobic respiration.

Salinity gradients are most commonly associated with coastlines, as in the *Spartina* example above. Interestingly, the boundaries of numerous species lie at the transition from freshwater environments to saltwater environments (Attrill and Rundle, 2002). Adaptation to salt versus fresh water may have led to changes in mitochondrial function in fish (Whitehead, 2009) and mammals (Caballero et al., 2015). Among North American birds, there is a mt DNA barcode gap at the freshwater environment/saltwater environmental divide between the sister taxa clapper rail/king rail (*Rallus crepitans/R. elegans*), laughing gull/Franklin's gull (*Leucophaeus atricilla/L. pipixcan*), saltmarsh sparrow/Nelson's sparrow (*Ammodramus caudacutus/A. nelsoni*), western willet/eastern willet (*Tringa semipalmata semipalmata/T. s. inornata*), and mallard/mottled duck (*Anas platyrhynchos/A. fulvigula*). It is interesting to speculate that these transitions represented adaptive divergence in mt genotypes to the salinity gradient, although there is no proposed mechanism for mitochondrial function affecting saltwater adaptation in birds, except change in energy production to facility increased kidney function (Gutiérrez et al., 2011; Sabat et al., 2017).

Adaptation via mitochondrial introgression

Rather ironically, some of the best evidence for adaptive divergence between species that is a result of selection on mt genes comes from studies of introgession of mt genotypes from one species to another. Among the first and most detailed studies of adaptive introgression of mt genotypes was conducted by the lab group of Pierre Blier and focused on two closely related species of freshwater fish: the brook char (*Salvelinus fontinalis*) and the arctic char (*S. alpinus*). They found seven allopatric populations of brook char (defined as fish that had the typical N genotype of brook char) that carried the mt genome of arctic char. This pattern suggested several independent events of introgression of mt genomes from the more northerly and more cold-adapted species to the more southerly, more warm-adapted species (Glémet et al., 1998). In each of these mitochondrial-transfer events, the population of the southern species receiving the cold-adapted mt genotype was in a local environment with a very cold alpine climate. When this lab group assessed functional changes in the two char genotypes, they found nucleotide changes in the mt-encoded Complex I subunits ND2 and ND5 that coded for amino acid substitutions that models suggested could affect Complex I function and plausibly how well the ETS functioned in different temperatures (Doiron et al., 2002).

Interestingly, introgression of mt genotypes from red voles (*Myodes rutilus*) into bank voles (*M. glareolus*) in northern Europe seems also to have been driven by similar need for climate adaptation at the edge of a species' range. Red voles are an arctic-adapted rodent, while bank voles occupy more southern and warmer temperate regions. In numerous cold, northern populations of bank voles, red vole mt genomes have replaced bank vole mt genomes, with no introgression of N genes (Tegelström, 1987). Studies of the basal metabolic rates of different combinations of mt and N genes implicated the northern mt genotype in cold tolerance (Boratyński et al., 2011), and models of the different cytochrome *b* proteins encoded by the two mt genotypes indicated that there could be functional differences that would affect performance in different thermal environments (Boratyński et al., 2014). These authors concluded "These results suggest that the evolution of mtDNA in *Myodes* may have functional, ecological and adaptive significance" (Boratyński et al., 2014, p.277).

Climate adaptation may also explain introgression of mt genotypes among species of rabbits in the genus *Lepus*. Rabbits live in a wide range of thermal climates in Eurasia and North America, and there are multiple documented cases of introgression of mt genotypes from arctic-adapted species into northern populations of otherwise temperate-adapted species. Genomic analysis of species in the clade turned up evidence for positive selection, particularly in arctic-adapted species, on the Complex V subunit ATP8, on CYTB, and on several subunits in Complex I genes (Melo-Ferreira et al., 2014). There was also evolution of the control region of the mt DNA for rabbits in the arctic lineage. Models of the different OXPHOS complexes did not support the hypothesis that the arctic version of mt proteins would enable better adaptation to cold, although secondary effects of interaction with N-mt gene products could not be ruled out (Melo-Ferreira et al., 2014). It seems plausible that the changes to the control region affected replication and transcription of the mt genome to affect climate adaptation. Although these authors found it difficult to establish a tight link between the evolution of mt-encoded gene products and thermal adaptation, the evidence for strong selection on mt genes among different thermal environments was convincing.

Introgression of mt genomes from one population to another is predicted to occur only when the benefits of the new mt genotype outweigh the costs of incompatibilities between the resident N-mt genotype and the introduced mt genotype (Figure 3.12). Studies demonstrating patterns of mt introgression that are consistent with enhanced adaptation, such as a cold-adapted mt genotype introgressing into a population in a cold micro-climate, support this hypothesis of adaptive introgression. Better measures of the cost created by mitonuclear incompatiblities and the benefits of improved physiological function would enable better testing of hypotheses for both compensatory coevolution and adaptation related to mt genotype.

Signatures of adaptive evolution

The studies that I reviewed in the previous sections are the best examples of natural selection driving adaptive divergence in mt genotypes. But across all eukaryotes, how important is adaptation to the evolution of mt genotypes? In the early chapters of this book, I put a huge emphasis on mutational erosion and the prediction that most functional change in mt genotypes would be deleterious. So which is it? Are functional changes in mt genotypes the accumulation of slightly deleterious alleles or do such changes reflect local adaptation? A comparative genomic analysis by James et al. (2016) suggests that, not surprisingly, functional changes in mt genotypes emerge from a combination of the two forms of evolution. They assessed the mt genotype of more than 500 animal species representing essentially the full range of taxa in Kingdom Animalia. They looked at how the patterns of genomic evolution compared with predictions of neutral change, the accumulation of slightly deleterious alleles, or adaptive evolution. They found that, as predicted by theory, change in the mt genotype was dominated by the accumulation of slight deleterious alleles. However, when they controlled for the presence of deleterious mutations, they found that an estimated 26 percent of non-synonymous changes to the genotype could be attributed to adaptive evolution. In another comparative study of the evolution of amino acid sequences in forty-one placental mammalian species, da Fonseca et al. (2008) presented detailed assessments that supported a conclusion that there has been adaptive evolution in several key components of the ETS. Adaptive changes to protein components of Complex I were restricted to the loop regions that are embedded in the inner mitochondrial membrane and likely function as proton pumps. In Complex IV, changes that were associated with adaptive evolution concerned mt subunits with close association with N-encoded subunits. If the conclusions of these papers hold up, then adaptive evolution is a major force in the evolution of mt DNA.

If the adaptive evolution involving mt gene products is often directed to the same endpoint, such as the capacity to function in a colder or warmer environment, then it follows logically that the same functional changes to mt gene products should evolve repeatedly. Support for this basic prediction of adaptive evolution of mt genes was provided in a comparative study by Garvin et al. (2014). They identified 237 species of animals in genomic archives with detailed mitochondrial sequence information and then they tested for for positive selection for twelve mt-encoded subunits for Complex I, IV, and V as well as cytochrome *b* in these 237 taxa. They found evidence for positive selection among most of these mt-encoded products in various taxa, but they found especially consistent, strong, and potentially parallel evolution of ND5, a core subunit of Complex I, in many of the species studied. They speculated that similar selective pressures may have driven similar functional changes to ND5 across diverse groups of animals. The approach of Garvin and colleagues, combining phylogenetic reconstructions, detailed genomic analyses of pairs and groups of related species, structural modeling of the products of genotypes, and assessment of how changes might facilitate adaptation, provides a model for how to study the role of the mt genotype and co-functioning mt and N-mt genes in the adaptive evolution of eukaryotes (Garvin et al., 2016) (see also Finch et al., 2014).

Adaptive radiation via mt evolution

Adaptive radiation describes changes to morphology and physiology that enable organisms to occupy new niche spaces and thereby radiate through an ecosystem

(Glor, 2010; Losos, 2010). Many aspects of the adaptive radiation of organisms will have nothing to do with mitonuclear coevolution. In Darwin's finches, a small bird that radiated through the Galapogos Islands following a single colonization event, a set of N genes has now been identified that controls the size and shape of the beak that in turn determines adaptation to seed resources (Lamichhaney et al., 2015). The evolution of these growth and development N genes is not likely to involve specific coadaption with mt genes. Adaptive radiations in which mitochondrial function and mitonuclear coadaptation likely play a key role involve adaptation to the same external environments that shape standing variation in mt genotypes (Figure 9.2), but also adaptations that fundamentally change the life history of a lineage enabling it to utilize a previously unoccupied niche space. There is no clear line to distinguish simple adaptation to variable environments among species within a lineage from adaptive radiation: the latter is generally invoked when there is, in human judgement, a significant shift in type of environments that are occupied or in the life history of the organism.

Among the earliest and still most convincing investigations of the role of mitonuclear genes and OXPHOS function in the adaptive radiation of a lineage concerns the evolution of large brains in primates and especially humans. The brain is the most complex organ in the body of most vertebrates, requiring a huge energy investment during development (Koch and Hill, 2018). In humans the brain accounts for only about 2 percent of body mass, but it is estimated to account for about 20 percent of the energy used by the body at rest (Swaminathan, 2008). Thus, there is a close tie between mitochondrial function and brain function simply at the gross level of need for ATP. But mitochondria also play a key role in regulating synaptic transmission, brain function, and cognition (Picard and McEwen, 2014), so there is the potential for adaptive changes in mt genotype in the evolution larger or more complex brains that are much more targeted than simply an enhancement of ATP production.

Grossman et al. (2001) hypothesized that the evolution of an organ as developmentally challenging and energy-demanding as a brain would be associated with functional changes to aerobic respiration. To test this idea, they studied the evolution of OXPHOS genes in anthropoid primates (New World monkeys, Old World monkeys and apes including humans) relative to other mammals, and they reported evidence for significant functional changes to both mt and N-mt genes that encode subunits of Complex III and IV as well as the N-encoded electron carrier molecule cytochrome *c* in the primate lineage (Grossman et al., 2001, 2004) (Figure 9.14). The numbers of changes in these genes that were observed in the primate lineage far exceeded what was observed in other mammalian lineages. These changes to subunits of OXPHOS complexes are particularly significant because (1) they are almost certainly the product of selection and not drift (Wu et al., 1997l Grossman et al., 2004) and (2) the amino acid changes are clustered in regions of contact between mt and N-mt-encoded subunits (Wu et al., 2000; Schmidt et al., 2001; Uddin et al., 2008) (Figure 9.14). The presumption by Grossman and colleagues was that these changes were adaptions for better or different function of the OXPHOS system to support a large, energy-demanding brain and to enable adaptive radiation into the big-brain niche space.

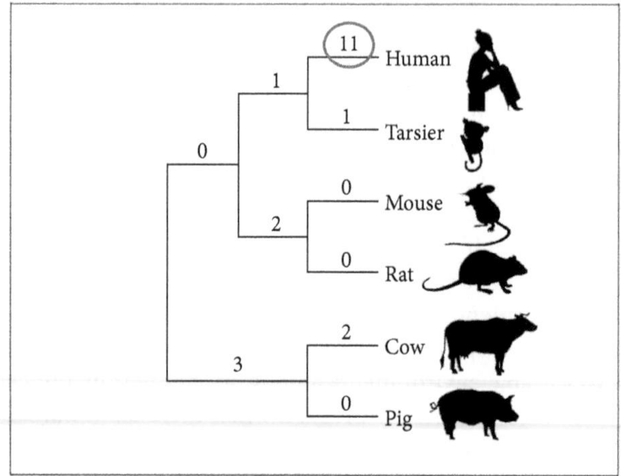

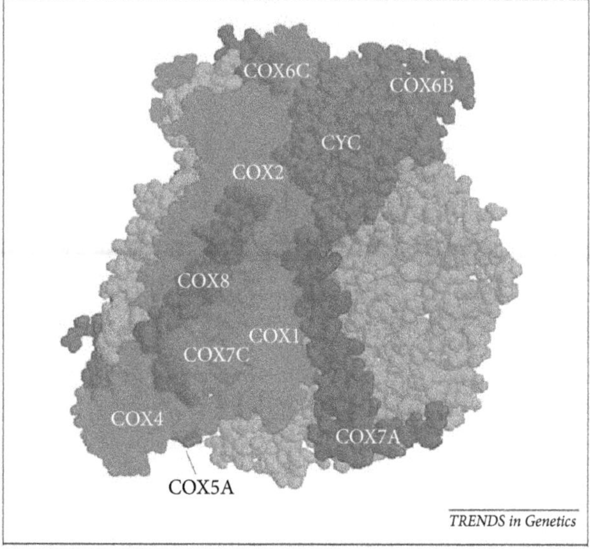

Figure 9.14 In the anthropoid primate lineage, which includes humans, there has been rapid, functional change in genes coding for subunits of Complex III and Complex IV as well as cytochrome *c*. (Top panel) A diagram showing the number of amino acid substitutions in the N-encoded Complex IV subunit, COX6B between three pairs of species with about the same time since divergence. The ten substitutions in the evolutionary lineage from a human/tarsier ancestor to modern humans (circled) is exceptional among mammal lineages. (Bottom panel) A crystal structure representation of the human Complex IV enzyme. The nine subunits (including the two mt-encoded COX1, COX2) that show accelerated functional change during primate evolution are darkly shaded. The N-encoded electron transporter, cytochrome *c*, with which Complex IV closely co-functions, has the darkest shading. The four subunits, all N-encoded, in which no acceleration has been detected, which includes mt-encoded COX3, are lightly shaded. From Grossman et al. (2004).

The idea that changes in the structure and function of the ETS reflect novel adaptations that enable expanded brain function is the brain–energy hypothesis (Grossman et al., 2004; Uddin et al., 2008). However, one could also imagine that the greater rate of functional change actually represents stronger selection for compensatory coevolution of mt and N-mt genes to maintain an ancestral OXPHOS system at full efficiency; in other words, instead of selection for novelty, there was selection against mutational erosion and loss of respiratory efficiency. Indeed, Osada and Akashi (2012) conducted a follow-up analysis looking in more detail at the functional changes to OXPHOS genes in the primate lineage. They concluded that most changes were likely the result of selection for compensatory coevolution, but they could not rule out the sort of adaptive innovations invoked by the brain–energy hypothesis. Whether the rapid evolution of primate OXPHOS genes arises to support innovative changes to the OXPHOS system or via selection for full performance of an existing system, there is strong evidence that mitonuclear evolution played a key role in the evolution of large brains in primates.

Another transition in mode of life that required increased capacity to generate significant power occurred during the evolution from terrestrial locomotion to powered flight (Pennycuick, 1975). A basic prediction regarding the evolution of flight is that in addition to the evolution of the wings, muscles, and skeletal structures that were needed for an organism to fly, there was also a need for new physiological mechanisms that could significantly increase the output of energy. Shen et al. (2010) tested this idea—that flight would be associated with changes to aerobic respiration—in bats, a large lineage of flying mammals in Order Chiroptera that evolved from terrestrial mammals around 50 million years ago. Although the evolution of gliding is relatively common in mammals, bats are unique among mammals in deploying powered flapping flight. Shen et al. (2010) looked at the rate of change in mt genes that code for OXPHOS complexes, N-mt genes that code for OXPHOS complexes, and N-mt genes that do not code for OXPHOS complexes. They found strong evidence for positive selection in both N-mt and mt genes that code for OXPHOS complexes, but the rate of functional change was greatest for mt genes (Figure 9.15). Almost a quarter (23 percent) of the mt genes for OXPHOS subunits had undergone functional change in the bat lineage. There was less change in the N-encoded OXPHOS subunits: only 5 percent of N-mt genes that code for OXPHOS subunits had also undergone function change. Most significantly, however, only 2 percent of N-mt genes that do not code for respiratory chain subunits showed functional change. This study implicates rapid change of OXPHOS genes in the evolution of flight, but Shen et al. (2010) did not provide direct evidence that the evolution of novel changes to OXPHOS genes promoted the evolution of flight.

As in the study of primate brain size, selection for compensatory coevolution could also have selected for the observed patterns of positive selection in the bat lineage. A subsequent analysis by Zhang et al. (2013a) turned up a number of N-encoded genes that evolved rapidly in the bat lineage, some of which were implicated in the evolution of flight, but this subsequent analysis is entirely complementary to and in no way a challenge to the findings of Shen et al. (2010) We expect most adaptive changes to be in N genes, but critical changes to core respiratory processes will necessarily involve mt and N-mt genes that enable aerobic respiration.

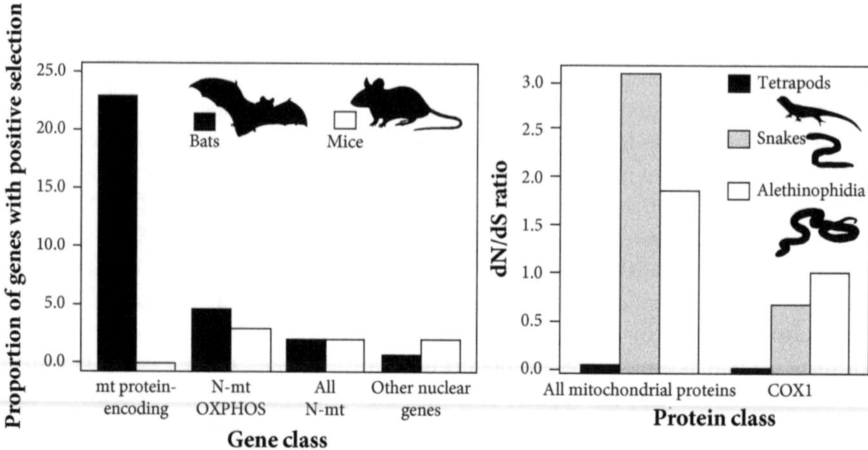

Figure 9.15 Changes in the genes that code for subunits of electron transport complexes have been implicated in the adaptive radiation of bats (left) and snakes (right). Flight requires high energy output, and in the transition from terrestrial locomotion to flight in bats, there was strong selection for changes to the electron transport system for increased ATP production to power flight muscles (Shen et al., 2010). In bats, 23 percent of the mt genes and about 5 percent of N-mt genes that code for OXPHOS proteins show evidence of positive selection, a proportion that is much higher than for non-OXPHOS N-mt genes or non-mt nuclear genes. The rates of positive selection on OXPHOS genes were also higher in bats than in mice. Similarly in the evolution of snakes, non-synonymous changes in mitochondrial OXPHOS genes, and particularly in Complex IV, suggest that a major shift in respiratory function accompanied the major shifts in body plan and prey consumption (Castoe et al., 2008). The alethinophidia lineage is the clade within all snakes with the greatest adaptive shifts. Deducing whether respiratory chain innovation leads or follows adaptive radiation will be a fascinating focus of future studies. Figure adapted from Hill (2015).

Perhaps the best study on the role of functional changes to mt and N-mt genes in the evolution of a major shift in life history change concerns the dietary adaptation of snakes. If you posed the question "What is the biggest change that occurred when snakes evolved from lizards?" Ninety-nine out of 100 biologists would surely respond, "loss of limbs." Surprisingly, however, loss of limbs takes a backseat to the digestive system innovations that characterized the snake lineage. Snakes not only slither around without any limbs, they also eat very large prey—the meal of a snake commonly weighs as much as it does. After such a meal, some snakes may not eat for a year. Snakes live in a boom-and-bust niche that would be death to nearly all other vertebrates. The boom-and-bust life history strategy of a snake is only feasible if snakes can elevate their digestive systems from inactive and atrophied to active at a level beyond the capacity of most other vertebrates very quickly when they, unpredictably, swallow a huge meal. Accordingly, snake oxidative metabolism can fluctuate by as much as 4000 percent, and the heart and gut of snakes can enlarge by 100 percent within 48 hours following the consumption of a big meal (Secor and Diamond, 1995, 1998). These changes emerge from a mitochondrial-produced energy and it makes sense that these changes in life history would require evolutionary changes to mitochondrial function.

These energy-driven processes in the life of a snake require not only massive output of energy following a meal but equally massive shutdown of aerobic respiration between meals. Castoe et al. (2008) hypothesized that such unique energy requirements would lead to selection for novel traits related to aerobic respiration. They compared the rates of evolution of respiratory chain proteins (1) across all tetrapods, (2) within all snakes, and (3) within just the Alethinophidia (sometimes referred to as "advanced snakes"). Only the Alethinophidia show the boom-and-bust life history that might lead to what Castoe et al. (2008) called "evolutionary redesign." They found that, compared with the mt-encoded proteins of other vertebrates examined, the mt-encoded proteins of Alethinophidia snakes experienced exceptional levels of amino acid substitition (Figure 9.15). They presented evidence for positive selection, coevolution, convergence, and reversion in ETS subunits. They conducted a detailed analysis of COX1 and found evidence for changes to COX1 structures that they interpreted as adaptation to the boom-and-bust lifestyle via altered flow of protons and changes to fundamental characteristics of aerobic metabolism.

Adaptive radiations into new life zones generally involve not only changes in the size, shape, or other N-encoded morphologies of the organism, but also major shifts in the degree and type of aerobic respiration needed to support the new life history. Large brains, the ability to fly, and capacity to digest a huge prey item opened entire new niche spaces to primates, bats, and snakes. Evidence suggests that these adaptive radiations would not have been possible without adaptive evolution of mitochondria with mt genes playing a central role.

Human mt genotypes and environment

A growing realization that mt genotype can be shaped by natural selection and that mt genotype can play an important role in adaptation to external environments has huge implications not only for an understanding of natural populations of organisms but also for better understanding of human biology (Hill et al., 2019). In the biomedical literature, genotypes related to mitochondrial function are generally characterized as being either "normal" or "pathological" (Ling et al., 2009; Konovalova and Tyynismaa, 2013; Sissler et al., 2017), with a pathological label pasted on any genotype that is associated with a dysfunction in a person living in the environments provided by western culture. As emphasized in this chapter, however, the outcome of mitonuclear interactions depends on the specific environmental backdrop. Environment, therefore, can play a major role in determining whether any given mt or N-mt allele is good or bad. For instance, in a cybrid line of *D. melanogaster/D. simulans* flies, in which reduced OXPHOS linked specifically to an allele that caused incompatibilities between mt-encoded tRNAs and N-encoded mtARS, the effects on growth and development were highly temperature dependent. At lab temperatures, the negative fitness effects of the allele were significant. At cooler temperatures, however, the flies carrying the "bad" allele showed essentially normal patterns of development (Hoekstra et al., 2013). The recasting of mt alleles from villains to neutral players to even saviors depending on the environment in which they are expressed is not unique to fruit flies. Indeed, it is a topic of growing interest in the biomedical community because the environment in which mitonuclear interactions play out may affect human health.

I will give one example to make the point, but different outcomes of mitonuclear interactions in different environments should be as prominent in humans as they are in other eukaryotes. Consider Leber's hereditary optic neuropathy (LHON), which is a maternally inherited condition that can cause blindness. It is associated with disease in some humans so it is almost always categorized as arising from an unambiguously "bad" and disease-causing genotype. But LHON has long been known to be a complex disease because its effects on a particular patient vary with sex and environmental factors (Man et al., 2002). It turns out that LHON is not an outcome of one specific genetic change to one mt-encoded product. Most LHON is caused by one of three different amino acid changes to Complex I subunits (ND1, ND4, and ND6) (Tońska et al., 2010). So, distinct changes to different mt-encoded subunits are classed as giving rise to the same pathogenic phenotype. Moreover, at least some of the genetic changes that can give rise to LHON do not affect people in all populations equally. At high elevations in Tibet and North Africa, the pathological phenotype LHON is virtually unknown, and yet paradoxically the frequency of what should be LHON-causing alleles for the amino acid sequence of ND1 is very high in these populations (Ji et al., 2012). This ND1 LHON allele is twenty-two times more common in Tibetan populations than in lowland Chinese populations. Not only do Tibetans tend to carry the LHON-causing allele with no ill effects, they also have a mt genotype that is distinct from the common genotype in the surrounding lowland. It seems that when it occurs in association with the high-elevation mt genotype, the potential LHON-causing ND1 genotype no longer causes blindness. There is some suggestion that, under hypoxic conditions, the Tibetan genotype with LHON ND1 allele may even have an adaptive advantage. The idea that standing variations in mt genotypes across human populations might have functional consequences that play out through gene × gene × environment interactions has been slow to gain acceptance in the medical community, but Douglas Wallace and other scientists have been emphatically making the case that such considerations would be a huge benefit if they were better integrated into modern medicine (Wallace, 2008; Wallace and Fan, 2010; Picard et al., 2016).

Summary

For most of the twentieth century when evolutionary biologists gave any consideration to mitochondrial evolution, they dismissed it as the product of drift. Conventional wisdom stated that purifying selection would anchor mt genotypes to a fixed point in an adaptive landscape leaving no potential for adaptive evolution. The past few decades of research has proven that this paradigm is not only incorrect, but that it obscured a fuller understanding of adaptive evolution that is only possible with consideration of the evolution of mt and interacting N-mt genes in different environments. New biochemical models and empirical observations implicate changes in mitochondrial function as a central component of adaptation related to temperature, oxygen pressure, and diet, among other external environmental factors. In addition to a role for mitonuclear interactions in microevolution, evidence is emerging that changes in mitochondrial function resulting from mitonuclear coevolution underlie key evolutionary innovations associated with major adaptive radiations including the transition from terrestrial locomotion to flight. Adaptive evolution of mt genotypes within and between populations has big implications for speciation and sexual selection. If changes in mt genotypes are a result not simply of drift but also of adaptation, then divergence in mt genotypes becomes a more plausible basis for speciation in eukaryotes. If there is standing variation in mt genotypes linked to adaptation to external environments in eukaryotic populations, then mate choice for functional mitonuclear genotypes within a population makes more sense. The concept of adaptation arising from mt genotype is already reshaping basic theory in population biology, but I think the revolution in mitonuclear thinking with regard to adaptive evolution has scarcely begun.

10

Epilogue

In the early twentieth century, the foundations of modern evolutionary ecology were laid with essentially no knowledge of the molecular basis of heredity, no understanding of the basic architecture of the N genome, and no idea that a mt genome was even a possibility. More fundamentally, until quite recently, biologists viewed the evolution of complex plants and animals as the endpoint of more-or-less continuous progression that began with the first spark of life and flowed as a steady ascension spanning 4 billion years (Gould, 1996). Although Lynn Margulis championed the endosymbiosis origin of mitochondria for decades (Margulis, 1970), it was only recently deduced—and still under substantial debate—that eukaryotes arose abruptly from a prokaryotic world that had persisted relatively unchanged for more than a billion years (Lane, 2005). According to this new view of the history of the biotic world, the ascension of life was idled for an eon for lack of a mitochondrion. Most evolutionary biologists have yet to come to grips with the enormous implications of a chimeric origin of eukaryotes and the partnership of the mt and N genomes. This book is an attempt to highlight the fundamental importance of mitonuclear considerations to basic concepts in evolutionary ecology and the defining characteristics of eukaryotes.

I propose that the necessity of mitonuclear coadaptation is the unifying concept of evolutionary ecology. To borrow from a famous declaration by Theodosius Dobzhansky: the evolution of complex life makes sense only in the light of mitonuclear coadaptation. The union of two formerly independent prokaryotes in the formation of a proto-eukaryote unleased a cataclysmic chain of events from which arose an entirely new type of organism with the novel potential to evolve complexity (Lane and Martin, 2010). The genomic interactions that were necessary to stabilize the two-genome design of the eukaryotic cell gave rise to the defining characteristics of complex life. Nick Lane (2005) brilliantly presented the evolutionary sequence by which the emergence of a cell with two genomes demanded the radical new architecture of eukaryotes including the genes-in-pieces arrangement of N DNA, the spliceosome, and the nuclear membrane, as well as the absolute necessity of a genome in the mitochondrion. He also proposed that defining features of eukaryotes, including sexual reproduction with recombination, a diploid N genome, senescence, a germ line, and even cancer, are inevitable consequences of an organism with two genomes (Lane, 2005, 2015a). Lane showed the relevance of mitonuclear coadaptation to processes that brought mitonuclear thinking right to the threshold of evolutionary ecology. In this book, I carry the idea forward by distilling the pioneering research of evolutionary biologists

Mitonuclear Ecology. Geoffrey E. Hill, Oxford University Press (2019). © Geoffrey E. Hill 2019.
DOI: 10.1093/oso/9780198818250.001.0001

over the past few decades that points to the conclusion that mitonuclear ecology is a foundational concept in evolutionary ecology.

With two genomes encoding the phenotype of a single organism, the need for mitonuclear coadaptation became an essential feature in the evolution of eukaryotes. But the maintenance of coadaptation across evolutionary time presents lineages with an unrelenting challenge because, as I present in Chapter 3, neither the N genome nor the mt genome is static. The mt genome, in particular, is subject to mutational erosion that can endanger the coordinated function of mitochondrial and nuclear components and degrade core respiration. This challenge to maintain close mitonuclear coadaptation is proposed to be the impetus for the evolution of two mating types with exclusively maternal transmission of mt genes in many eukaryotes (Chapter 5), for the evolution of a germ line in bilaterian animals (Chapter 6), and perhaps for the evolution of sexual reproduction (Chapter 5). Each of the hypotheses linking the necessity of mitonuclear coadaptation to the evolution of quintessential characteristics of eukaryotes is new, and these are far from settled explanations for sex, two sexes, and a sequestered germ line. But the new and enlightening hypotheses that have emerged from the injection of mitonuclear thinking into these old evolutionary debates reveal the potential for novel insights when the chimeric origin of complex eukaryotes is incorporated into evolutionary theory.

Speciation, sexual selection, and adaptation are the frontiers of mitonuclear ecology. These are emerging properties in the evolution of complex life, and the capacity of evolutionary biologists to explain these phenomena is a reflection on basic understanding of the process of natural selection. Frustratingly, for the study of speciation and sexual selection, we have lacked conceptual frameworks that can encompass the full suite of observations across eukaryotes and that are supportable by tested predictions. The current textbook explanations of both sexual selection and speciation are piecemeal and require corollary hypotheses or exceptions for a large set of observations. Basically, instead of general theories of sexual selection and speciation we have case-by-case explanations. Theories grounded in the necessity of mitonuclear coadaptation hold the potential to provide a cohesive and unified framework for understanding speciation and sexual selection and the relationships that bind these processes.

As I argue in detail in Chapter 7, a mitonuclear perspective can clarify understanding of the evolution of species boundaries among eukaryotes. From the awakening of abstract thought in humans, the division of plants, animals, and fungi into species has been an obvious and universally recognized feature of the natural world. The reality that living things naturally subdivide into species has been the starting point of essentially all biological investigations. It follows logically that such a universal and fundamental aspect of eukaryotic life would emerge from characteristics that are shared by all eukaryotes.

The pursuit of a universal concept of species and basic understanding of the process of speciation has been among the greatest challenges in all of the sciences. I propose that two approaches that were adopted early in the twentieth century significantly hampered and continue to impede the effort to understand speciation. First, evolutionary biologists have nearly always worked downward from the endpoint of extant species in an attempt to understand the underlying processes. Current explanations of the process of speciation are modifications of theories that were first proposed based on descriptions of morphology and biogeography made in the early twentieth century. I would argue that the fundamentally better approach to an investigation into the process of speciation is to work upward from first principles of biochemistry, cell biology, and genomics. As the entrepreneur Elon Musk has shown us, if one attempts to build a better automobile by tweaking models that already exist, one remains forever shackled to the mistakes and historical quirks that led to that particular design of car. On the other hand, if one works from first principles in restating the basic concept of an automobile, then one clearly perceives and avoids those previous errors and so reaches a much better endpoint.

Second, at the very inception of the speciation studies, evolutionary biologists rejected the hypothesis that elements in the cytoplasm play key roles in the process of speciation. I use the analogy of inadvertently diverting onto a side trail at the very beginning of a long hike. Instead of a relatively easy stroll down a clear path, one is left bushwhacking, and such a dogged struggle forward is unlikely to ever return a hiker to a correct path. The route to the destination is first back to the main trail and then forward. In the same way, I propose that investigations of speciation would benefit by completely recasting speciation theory based on first principles of genomics, cell biology, biochemistry, and the necessity of mitonuclear coadaptation.

I argue in Chapter 8 that a mitonuclear perspective is also the key to understanding sexual selection. Over the past 100 years, theories of sexual selection and the evolution of ornamental traits have drifted even farther than speciation theory from a coherent line of investigation. From its inception, the foundational principle of sexual selection has been that variance in reproductive success of males and females leads to non-random mating by females (Darwin, 1871; Bateman, 1948). But the key to understanding sexual selection lies not in deducing the predicted strength of choosiness or the sex that should be choosier; rather, it lies in understanding the object of choice. As I argued in Chapter 8, the fundamental function of non-random mating is correctly matching coadapted sets of mt and N genes each generation. All other aspects of mate assessment build on the fundamental necessity of ensuring mitonuclear compatibility and mitochondrial function in offspring. The cost and benefits of being choosy for males or females become descriptive details to be considered on a species-by-species basis, but such cost–benefit analyses are not central to understanding the fundamental need for mate choice.

I also contend that, as with studies of speciation, studies of sexual selection diverted from a path toward a clear understanding at the very inception of the field of study not only by focusing too intensely on variance in reproductive success, but also through a fixation on peacocks' tails and other extremely elaborate ornaments. While highly elaborate ornaments present a fascinating biological puzzle, they are observed in only a minute fraction of eukaryotes. Framing a theory of sexual selection around a peacock's tail is like framing a theory of the evolution of mating types around snails with a doubly uniparental inheritance of mitochondria. Both enormous tails and transmission of mitochondria via two lineages within the same species present extraordinary systems that are worthy of study, but both are sideshows to a core line of the investigation. We can only possibly understand the exceptional cases of doubly uniparental inheritance of mitochondria or tail feathers that are twice the length of a bird's body once we understand the basic principles of mitochondrial inheritance or mate choice. In my opinion, a new clarity of understanding of patterns of non-random mating will be gained by working from the first principle of a necessity of matching mt and N genes in typical eukaryotes—not by studying peafowl or other fantastically ornamented species.

In contrast to studies of the processes of speciation and sexual selection, which are inextricably bound to the maintenance of mitonuclear coadaptation—and hence were stifled by the lack of a mitonuclear perspective—adaptation can be effectively studied with no knowledge of a mt genome. The evolution of adaptations via natural selection can be and generally is a process involving primarily N genes. Nevertheless, I follow the lead of pioneering evolutionary biologists who contend, based on a growing body of experimental evidence, that an understanding of adaptation can only be complete with a consideration of the epistatic interactions of mt and N genes playing out in different environments (Chapter 9). Adaptation involving mt genes might be particularly important in understanding how organisms adapt to thermal environment, chemical environments, oxygen pressure associated with altitude and subterranean habitats, and diets. The evolution of mt genes might also play a fundamental role in adaptive radiations, such as the evolution of flight in birds and bats and the evolution of big brains in some primates. And finally, because human populations diverged under different selective pressures, an understanding of the role of mt genes in adaptation is one of the keys to understanding many inherited diseases in humans.

As I argue in the chapters of this book, mitonuclear ecology provides a theoretical foundation, framed on first principles of the genomic architecture of eukaryotes, that can move the study of adaptation, sexual selection, and speciation on a path of more productive investigation. In moving forward, I implore evolutionary biologists to do less searching for insights from 100-year-old tomes and instead to do more theorizing about the implications of the genomic organization of eukaryotes for whole-organism processes. The scientists working in the first three-quarters of the twentieth century made brilliant first cuts in an attempt to explain the natural world, but it is preposterous to think that evolutionary biologists working a century ago with such limited knowledge of molecular genetics and no concept of a mt genome could have correctly explained the processes of speciation, sexual selection, or adaptation. Restructuring the theoretical foundations of these fields from the first principle of the necessity of mitonuclear coadaptation holds potential to transform these disciplines into coherent and predictive fields of study.

References

Adams, K. L., and J. D. Palmer (2003). Evolution of mitochondrial gene content: Gene loss and transfer to the nucleus. Molecular Phylogenetics and Evolution 29:380–95.

Adrion, J. R., P. S. White, and K. L. Montooth (2016). The roles of compensatory evolution and constraint in aminoacyl tRNA synthetase evolution. Molecular Biology and Evolution 33:152–61.

Al-Attar, S., and S. de Vries (2013). Energy transduction by respiratory metallo-enzymes: From molecular mechanism to cell physiology. Coordination Chemistry Reviews 257:64–80.

Albon, S. D., and T. H. Clutton-Brock (1979). The roaring of red deer and the evolution of honest advertisement. Behaviour 69:145–70.

Alexander, H. J., J. M. L. Richardson, S. Edmands, and B. R. Anholt (2015). Sex without sex chromosomes: Genetic architecture of multiple loci independently segregating to determine sex ratios in the copepod *Tigriopus californicus*. Journal of Evolutionary Biology 28:2196–207.

Allen, J. F. (1993). Control of gene expression by redox potential and the requirement for chloroplast and mitochondrial genomes. Journal of Theoretical Biology 165:609–31.

Allen, J. F. (2003). The function of genomes in bioenergetic organelles. Philosophical Transactions of the Royal Society B: Biological Sciences 358:19–38.

Allen, J. F. (2015). Why chloroplasts and mitochondria retain their own genomes and genetic systems: Colocation for redox regulation of gene expression. Proceedings of the National Academy of Sciences 112:10231–8.

Allen, J. F. (2017). The CoRR hypothesis for genes in organelles. Journal of Theoretical Biology 434:50–7.

Allen, J., and J. Raven (1996). Free radical-induced mutation versus redox regulation: Costs and benefits of genes in organelles. Journal of Molecular Evolution 42:482–92.

Allio, R., S. Donega, N. Galtier, and B. Nabholz (2017). Large variation in the ratio of mitochondrial to nuclear mutation rate across animals: Implications for genetic diversity and the use of mitochondrial DNA as a molecular marker. Molecular Biology and Evolution 34:2762–72.

Alric, J., J. Lavergne, F. Rappaport, A. Verméglio, K. Matsuura, K. Shimada, and K. V. P. Nagashima (2006). Kinetic performance and energy profile in a roller coaster electron transfer chain: A study of modified tetraheme-reaction center constructs. Journal of the American Chemical Society 128:4136–45.

Alston, C. L., M. C. Rocha, N. Z. Lax, D. M. Turnbull, and R. W. Taylor (2017). The genetics and pathology of mitochondrial disease. Journal of Pathology 241:236–50.

Amarnath, D., I. Choi, A. R. Moawad, T. Wakayama, and K. H. S. Campbell (2011). Nuclear-cytoplasmic incompatibility and inefficient development of pig-mouse cytoplasmic hybrid embryos. Reproduction 142:295–307.

Amunts, A., A. Brown, J. Toots, S. H. W. Scheres, and V. Ramakrishnan (2015). The structure of the human mitochondrial ribosome. Science 348:95–8.

Andersson, M. (1994). Sexual Selection. Princeton University Press, Princeton, NJ.

Andersson, G. E., O. Karlberg, B. Canback, and C. G. Kurland (2003). On the origin of mitochondria: A genomics perspective. Philosophical Transactions of the Royal Society B: Biological Sciences 358:165–79.

Andersson, M., and L. W. Simmons (2006). Sexual selection and mate choice. Trends in Ecology and Evolution 21:296–302.

Andziak, B., and R. Buffenstein (2006). Disparate patterns of age-related changes in lipid peroxidation in long-lived naked mole-rats and shorter-lived mice. Aging Cell 5:525–32.

Andziak, B., T. P. O'Connor, W. Qi, E. M. Dewaal, A. Pierce, A. R. Chaudhuri, H. Van Remmen, and R. Buffenstein (2006). High oxidative damage levels in the longest-living rodent, the naked mole-rat. Aging Cell 5:463–71.

Angerer, H., K. Zwicker, Z. Wumaier, L. Sokolova, H. Heide, M. Steger, S. Kaiser, E. Nübel, B. Brutschy, M. Radermacher, U. Brandt, and V. Zickermann (2011). A scaffold of accessory subunits links the peripheral arm and the distal proton-pumping module of mitochondrial complex I. The Biochemical Journal 437:279–88.

Angilletta, M. J. Jr, and M. J. Angilletta (2009). Thermal Adaptation: A Theoretical and Empirical Synthesis. Oxford University Press, Oxford, UK.

Antonicka, H., F. Sasarman, T. Nishimura, V. Paupe, and E. A. Shoubridge (2013). The mitochondrial RNA-binding protein GRSF1 localizes to RNA granules and is required for post-transcriptional mitochondrial gene expression. Cell Metabolism 17:386–98.

Aon, M. A., S. Cortassa, and B. O'Rourke (2010). Redox-optimized ROS balance: A unifying hypothesis. Biochimica et Biophysica Acta—Bioenergetics 1797:865–77.

Arnqvist, G. (2006). Sensory exploitation and sexual conflict. Philosophical Transactions of the Royal Society B: Biological Sciences 361:375–86.

Arnqvist, G., D. K. Dowling, P. Eady, L. Gay, T. Tregenza, M. Tuda, and D. J. Hosken (2010). Genetic architecture of metabolic rate: Environment specific epistasis between mitochondrial and nuclear genes in an insect. Evolution 64:3354–63.

Asin-Cayuela, J., and C. M. Gustafsson (2007). Mitochondrial transcription and its regulation in mammalian cells. Trends in Biochemical Sciences 32:111–17.

Atkin, O. K., and D. Macherel (2009). The crucial role of plant mitochondria in orchestrating drought tolerance. Annals of Botany 103:581–97.

Attrill, M. J., and S. D. Rundle (2002). Ecotone or ecocline: Ecological boundaries in estuaries. Estuarine, Coastal and Shelf Science 55:929–36.

Austad, S. N., and J. M. Hoffman (2018). Is antagonistic pleiotropy ubiquitous in aging biology? Evolution, Medicine, and Public Health 2018:287–294.

Avadhani, N. G., M. C. Sangar, S. Bansal, and P. Bajpai (2011). Bimodal targeting of cytochrome P450s to endoplasmic reticulum and mitochondria: The concept of chimeric signals. FEBS Journal 278:4218–29.

Aw, W. C., N. A. Youngson, and B. J. W. O. (2016). Can we alter dietary macronutrient compositions and alleviate mitochondrial disease? Journal of Rare Diseases Research and Treatment 1:31–7.

Azevedo, L., J. Carneiro, B. van Asch, A. Moleirinho, F. Pereira, and A. Amorim (2009). Epistatic interactions modulate the evolution of mammalian mitochondrial respiratory complex components. BMC Genomics 10:266.

Azzu, V., and M. D. Brand (2010). The on-off switches of the mitochondrial uncoupling proteins. Trends in Biochemical Sciences 35:298–307.

Backström, N., and U. Väli (2011). Sex- and species-biased gene flow in a spotted eagle hybrid zone. BMC Evolutionary Biology 11:100.

Badyaev, A. V, and G. E. Hill (2003). Avian sexual dichromatism in relation to phylogeny and ecology. Annual Review of Ecology, Evolution, and Systematics 34:27–49.

Badyaev, A. V., A. L. Potticary, and E. S. Morrison (2017). Most colorful example of genetic assimilation? Exploring the evolutionary destiny of recurrent phenotypic accommodation. The American Naturalist 190:266–80.

Baker, A. J., E. S. Tavares, and R. F. Elbourne (2009). Countering criticisms of single mitochondrial DNA gene barcoding in birds. Molecular Ecology Resources 9:257–68.

Baklouti-Gargouri, S., M. Ghorbel, A. Ben Mahmoud, E. Mkaouar-Rebai, M. Cherif, N. Chakroun, A. Sellami, F. Fakhfakh, and L. Ammar-Keskes (2014). Identification of a novel m.9588G> A missense mutation in the mitochondrial COIII gene in asthenozoospermic Tunisian infertile men. Journal of Assisted Reproduction and Genetics 31:595–600.

Ballard, J. W. O., and M. Kreitman (1995). Is mitochondrial DNA a strictly neutral marker? Trends in Ecology and Evolution 10:485–8.

Ballard, J. W. O., and N. Pichaud (2014). Mitochondrial DNA: More than an evolutionary bystander. Functional Ecology 28:218–31.

Ballard, J. W. O., and M. C. Whitlock (2004). The incomplete natural history of mitochondria. Molecular Ecology 13:729–44.

Ballard, J. W., and N. A. Youngson (2015). Can diet influence the selective advantage of mitochondrial DNA haplotypes? Bioscience Reports 35:e00277.

Balloux, F., L.-J. L. Handley, T. Jombart, H. Liu, and A. Manica (2009). Climate shaped the worldwide distribution of human mitochondrial DNA sequence variation. Proceedings of the Royal Society B: Biological Sciences 276:3447–55.

Bandy, B., and A. J. Davison (1990). Mitochondrial mutations may increase oxidative stress: Implications for carcinogenesis and aging? Free Radical Biology and Medicine 8:523–39.

Bar Yaacov, D., K. Arbel-Thau, Y. Zilka, O. Ovadia, A. Bouskila, and D. Mishmar (2012). Mitochondrial DNA variation, but not nuclear DNA, sharply divides morphologically identical chameleons along an ancient geographic barrier. PloS One 7:e31372.

Barešić, A., and A. C. R. Martin (2011). Compensated pathogenic deviations. Biomolecular Concepts 2:281–92.

Barešić, A., L. E. M. Hopcroft, H. H. Rogers, J. M. Hurst, and A. C. R. Martin (2010). Compensated pathogenic deviations: Analysis of structural effects. Journal of Molecular Biology 396:19–30.

Baris, T. Z., P. U. Blier, N. Pichaud, D. L. Crawford, and M. F. Oleksiak (2016). Gene by environmental interactions affecting oxidative phosphorylation and thermal sensitivity. American Journal of Physiology—Regulatory, Integrative and Comparative Physiology 311:R157–R165.

Baris, T. Z., D. N. Wagner, D. I. Dayan, X. Du, P. U. Blier, N. Pichaud, M. F. Oleksiak, and D. L. Crawford (2017). Evolved genetic and phenotypic differences due to mitochondrial-nuclear interactions. PLoS Genetics 13:1–23.

Barja, G. (1998). Mitochondrial free radical production and aging in mammals and birds. Annals of the New York Academy of Sciences 854:224–38.

Barr, C. M., and L. Fishman (2010). The nuclear component of a cytonuclear hybrid incompatibility in mimulus maps to a cluster of pentatricopeptide repeat genes. Genetics 184:455–65.

Barr, C. M., M. Neiman, and D. R. Taylor (2005). Inheritance and recombination of mitochondrial genomes in plants, fungi and animals. New Phytologist 168:39–50.

Barreira, A. S., D. A. Lijtmaer, and P. L. Tubaro (2016). The multiple applications of DNA barcodes in avian evolutionary studies. Genome 59:899–911.

Barreto, F. S., and R. S. Burton (2013a). Evidence for compensatory evolution of ribosomal proteins in response to rapid divergence of mitochondrial rRNA. Molecular Biology and Evolution 30:310–14.

Barreto, F. S., and R. S. Burton (2013b). Elevated oxidative damage is correlated with reduced fitness in interpopulation hybrids of a marine copepod. Proceedings of the Royal Society B: Biological Sciences 280:20131521.

Barreto, F. S., E. T. Watson, G. T. Lima, C. S. Willett, S. Edmands, W. Li, and R. S. Burton (2018). Genomic signatures of mitonuclear coevolution across populations of *Tigriopus californicus*. Nature Ecology and Evolution 2:1250.

Barrientos, A., L. Kenyon, and C. T. Moraes (1998). Human xenomitochondrial cybrids. Journal of Biological Chemistry 273:14210.

Barrientos, A., S. Müller, R. Dey, J. Wienberg, and C. T. Moraes (2000). Cytochrome c oxidase assembly in primates is sensitive to small evolutionary variations in amino acid sequence. Molecular Biology and Evolution 17:1508–19.

Barton, N. H., and B. Charlesworth (1998). Why sex and recombination? Science 281:1986–90.

Bateman, A. J. (1948). Intra-sexual selection in *Drosophila*. Heredity 2:349.

Bateson, W. (1909). Heredity and variation in modern lights. In Darwin and Modern Science. Cambridge University Press, Cambridge, UK, pp. 85–101.

Bayona-Bafaluy, M. P., C. Muller, and C. T. Moraes (2005). Fast adaptive coevolution of nuclear and mitochondrial subunits of ATP synthetase in orangutan. Molecular Biology and Evolution 22:716–24.

Bazin, E., S. Glémin, and N. Galtier (2006). Population size does not influence mitochondrial genetic diversity in animals. Science 312:570–2.

Beale, G. H., and J. K. C. Knowles (1976). Interspecies transfer of mitochondria in *Paramecium aurelia*. MGG Molecular and General Genetics 143:197–201.

Beck, E. A., A. C. Thompson, J. Sharbrough, E. Brud, and A. Llopart (2015). Gene flow between *Drosophila yakuba* and *Drosophila santomea* in subunit V of cytochrome c oxidase: A potential case of cytonuclear cointrogression. Evolution 69:1973–86.

Beekman, M., D. K. Dowling, and D. K. Aanen (2014). The costs of being male: Are there sex-specific effects of uniparental mitochondrial inheritance? Philosophical Transactions of the Royal Society of London. Series B, Biological Sciences 369:20130440.

Bennett, N. C., and C. G. Faulkes (2000). African Mole-rats: Ecology and Eusociality. Cambridge University Press, Cambridge, UK.

Bensch, S., D. E. Irwin, J. H. Irwin, L. Kvist, and S. Åkesson (2006). Conflicting patterns of mitochondrial and nuclear DNA diversity in *Phylloscopus* warblers. Molecular Ecology 15:161–71.

Bergstrom, C. T., and J. Pritchard (1998). Germline bottlenecks and the evolutionary maintenance of mitochondrial genomes. Genetics 149:2135–46.

Berry, S. (2003). Endosymbiosis and the design of eukaryotic electron transport. Biochimica et Biophysica Acta—Bioenergetics 1606:57–72.

Beuning, P. J., and K. Musier-Forsyth (1999). Transfer RNA recognition by aminoacyl-tRNA synthetases. Biopolymers 52:1–28.

Beyhan, Z., A. E. Iager, and J. B. Cibelli (2007). Interspecies nuclear transfer: Implications for embryonic stem cell biology. Cell Stem Cell 1:502–12.

Billiard, S., M. López-Villavicencio, B. Devier, M. E. Hood, C. Fairhead, and T. Giraud (2011). Having sex, yes, but with whom? Inferences from fungi on the evolution of anisogamy and mating types. Biological Reviews 86:421–42.

Birchard, G. F., D. L. Kilgore Jr, and D. F. Boggs (1984). Respiratory gas concentrations and temperatures within the burrows of three species of burrow-nesting birds. The Wilson Bulletin 96:451–6.

Birky, C. W. (2001). The inheritance of genes in mitochondria and chloroplasts: Laws, mechanisms, and models. Annual Review of Genetics 35:125–48.

Björkholm, P., A. Harish, E. Hagström, A. M. Ernst, and S. G. E. Andersson (2015). Mitochondrial genomes are retained by selective constraints on protein targeting. Proceedings of the National Academy of Sciences 112:201421372.

Björkholm, P., A. M. Ernst, E. Hagström, and S. G. Andersson (2017). Why mitochondria need a genome revisited. FEBS Letters 59:65–75.

Blanchard, J. L., and M. Lynch (2000). Organellar genes: Why do they end up in the nucleus? Trends in Genetics 16:315–20.

Van Blerkom, J. (2011). Mitochondrial function in the human oocyte and embryo and their role in developmental competence. Mitochondrion 11:797–813.

Blondin, D. P., H. C. Tingelstad, O. L. Mantha, C. Gosselin, and F. Haman (2014). Maintaining thermogenesis in cold exposed humans: Relying on multiple metabolic pathways. Comprehensive Physiology 4:1383–402.

Blumberg, A., B. S. Sailaja, A. Kundaje, L. Levin, S. Dadon, S. Shmorak, E. Shaulian, E. Meshorer, and D. Mishmar (2014). Transcription factors bind negatively selected sites within human mtDNA genes. Genome Biology and Evolution 6:2634–46.

Bobay, L.-M., and H. Ochman (2017). Biological species are universal across life's domains. Genome Biology and Evolution 9:491–501.

Boczonadi, V., and R. Horvath (2014). Mitochondria: Impaired mitochondrial translation in human disease. International Journal of Biochemistry and Cell Biology 48:77–84.

Boesch, P., F. Weber-Lotfi, N. Ibrahim, V. Tarasenko, A. Cosset, F. Paulus, R. N. Lightowlers, and A. Dietrich (2011). DNA repair in organelles: Pathways, organization, regulation, relevance in disease and aging. Biochimica et Biophysica Acta—Molecular Cell Research 1813:186–200.

Bogenhagen, D. F., D. W. Martin, and A. Koller (2014). Initial steps in RNA processing and ribosome assembly occur at mitochondrial DNA nucleoids. Cell Metabolism 19:618–29.

Bolender, N., A. Sickmann, R. Wagner, C. Meisinger, and N. Pfanner (2008). Multiple pathways for sorting mitochondrial precursor proteins. EMBO Reports 9:42–9.

Bolnick, D. I., and T. J. Near (2005). Tempo of hybrid inviability in centrarchid fishes (Teleostei: Centrarchidae). Evolution 59:1754–67.

Bolnick, D. I., M. Turelli, H. Lopez-Fernández, P. C. Wainwright, and T. J. Near (2008). Accelerated mitochondrial evolution and "Darwin's corollary": Asymmetric viability of reciprocal F1 hybrids in centrarchid fishes. Genetics 178:1037–48.

Bonawitz, N. D., and G. S. Shadel (2007). Rethinking the mitochondrial theory of aging: The role of mitochondrial gene expression in lifespan determination. Cell Cycle 6:1574–8.

Bonawitz, N. D., D. A. Clayton, and G. S. Shadel (2006). Initiation and beyond: Multiple functions of the human mitochondrial transcription machinery. Molecular Cell 24:813–25.

Bonnet, T., R. Leblois, F. Rousset, and P.-A. Crochet (2017). A reassessment of explanations for discordant introgressions of mitochondrial and nuclear genomes. Evolution 71:2140–58.

De Bont, R., and N. van Larebeke (2004). Endogenous DNA damage in humans: A review of quantitative data. Mutagenesis 19:169–85.

Boore, J. L. (1999). Animal mitochondrial genomes. Nucleic Acids Research 27:1767–80.

Boratyński, Z., P. Alves, S. Berto, E. Koskela, T. Mappes, and J. Melo-Ferreira (2011). Introgression of mitochondrial DNA among *Myodes* voles: Consequences for energetics? BMC Evolutionary Biology 11:355.

Boratyński, Z., J. Melo-Ferreira, P. C. Alves, S. Berto, E. Koskela, O. T. Pentikäinen, P. Tarroso, M. Ylilauri, and T. Mappes (2014). Molecular and ecological signs of mitochondrial adaptation: Consequences for introgression? Heredity 113:277–86.

Boratyński, Z., T. Ketola, E. Koskela, and T. Mappes (2016). The sex specific genetic variation of energetics in bank voles, consequences of introgression? Evolutionary Biology 43:37–47.

Borst, P., and L. A. Grivell (1973). Mitochondrial nucleic acids. Biochimie 55:801–4.

Bowles, E. J., R. T. Tecirlioglu, A. J. French, M. K. Holland, and J. C. St. John (2008). Mitochondrial DNA transmission and transcription after somatic cell fusion to one or more cytoplasts. Stem Cells 26:775–82.

Bradbury, J. W., M. B. Andersson, and L. Heisler (1987). Sexual Selection: Testing the Alternatives. Dahlem Workshop Reports. John Wiley & Sons, Hoboken, NJ.

Brand, M. D. (2000). Uncoupling to survive? The role of mitochondrial inefficiency in ageing. Experimental Gerontology 35:811–20.

Brand, M. D., and D. G. Nicholls (2011). Assessing mitochondrial dysfunction in cells. Biochemical Journal 435:297–312.

Brand, M. D., N. Turner, A. Ocloo, P. L. Else, and A. J. Hulbert (2003). Proton conductance and fatty acyl composition of liver mitochondria correlates with body mass in birds. Biochemical Journal 376:741–8.

Brandvain, Y., and M. J. Wade (2009). The functional transfer of genes from the mitochondria to the nucleus: The effects of selection, mutation, population size and rate of self-fertilization. Genetics 182:1129–39.

Bratic, A., and N. G. Larsson (2013). The role of mitochondria in aging. The Journal of Clinical Investigation, 123:951–7.

Breen, M. S., C. Kemena, P. K. Vlasov, C. Notredame, and F. A. Kondrashov (2012). Epistasis as the primary factor in molecular evolution. Nature 490:535–8.

Breeuwer, J. A. J., and J. H. Werren (1995). Hybrid breakdown between two haplodiploid species: The role of nuclear and cytoplasmic genes. Evolution 49:705–17.

Breton, S., L. Milani, F. Ghiselli, D. Guerra, D. T. Stewart, and M. Passamonti (2014). A resourceful genome: Updating the functional repertoire and evolutionary role of animal mitochondrial DNAs. Trends in Genetics 30:555–64.

Brookes, P. S. (2005). Mitochondrial H+ leak and ROS generation: An odd couple. Free Radical Biology and Medicine 38:12–23.

Brown, M. D., E. Starikovskaya, O. Derbeneva, S. Hosseini, J. C. Allen, I. E. Mikhailovskaya, R. I. Sukernik, and D. C. Wallace (2002). The role of mtDNA background in disease expression: A new primary LHON mutation associated with Western Eurasian haplogroup J. Human Genetics 110:130–8.

Brown, W. M., M. George, and A. C. Wilson (1979). Rapid evolution of animal mitochondrial DNA. Proceedings of the National Academy of Sciences 76:1967–71.

Brustovetsky, N., and M. Klingenberg (1994). The reconstituted ADP/ATP carrier can mediate H+ transport by free fatty acids, which is further stimulated by mersalyl. Journal of Biological Chemistry 269:27329–36.

Bucklin, A., D. Steinke, and L. Blanco-Bercial (2011). DNA barcoding of marine metazoa. Annual Review of Marine Science 3:471–508.

Burger, G., M. W. Gray, and B. Franz Lang (2003). Mitochondrial genomes: Anything goes. Trends in Genetics 19:709–16.

Burley, N. (1986). Sexual selection for aesthetic traits in species with biparental care. American Naturalist 127:415–45.

Burns, J. A., H. Zhang, E. Hill, E. Kim, and R. Kerney (2017). Transcriptome analysis illuminates the nature of the intracellular interaction in a vertebrate-algal symbiosis. eLife 6:e22054.

Burt, A., and R. L. Trivers (2008). Genes in Conflict: The Biology of Selfish Genetic Elements. Belnap Press, New York.

Burton, R. S., and F. S. Barreto (2012). A disproportionate role for mtDNA in Dobzhansky–Muller incompatibilities? Molecular Ecology 21:4942–57.

Burton, R. S., R. J. Byrne, and P. D. Rawson (2007). Three divergent mitochondrial genomes from California populations of the copepod *Tigriopus californicus*. Gene 403:53–9.

Burton, R. S., C. K. Ellison, and J. S. Harrison (2006). The sorry state of F-2 hybrids: Consequences of rapid mitochondrial DNA evolution in allopatric populations. American Naturalist 168:S14–S24.

Burton, R. S., R. J. Pereira, and F. S. Barreto (2013). Cytonuclear genomic interactions and hybrid breakdown. Annual Review of Ecology, Evolution, and Systematics 44:281–302.

Busiello, R. A., S. Savarese, and A. Lombardi (2015). Mitochondrial uncoupling proteins and energy metabolism. Frontiers in Physiology 6:36.

Butlin, R., and C. M. Smadja (2017). Coupling, reinforcement, and speciation. The American Naturalist 191:155–72.

Butlin, R., A. Debelle, C. Kerth, R. R. Snook, L. W. Beukeboom, R. F. Castillo Cajas, W. Diao, M. E. Maan, S. Paolucci, F. J. Weissing, L. van de Zande, et al. (2012). What do we need to know about speciation? Trends in Ecology and Evolution 27:27–39.

Caballero, S., S. Duchêne, M. F. Garavito, B. Slikas, and C. S. Baker (2015). Initial evidence for adaptive selection on the NADH subunit two of freshwater dolphins by analyses of mitochondrial genomes. PLoS ONE 10:1–17.

Cagnone, G. L. M., T. S. Tsai, Y. Makanji, P. Matthews, J. Gould, M. S. Bonkowski, K. D. Elgass, A. S. A. Wong, L. E. Wu, M. McKenzie, D. A. Sinclair, and J. C. S. John (2016). Restoration of normal embryogenesis by mitochondrial supplementation in pig oocytes exhibiting mitochondrial DNA deficiency. Scientific Reports 6:27–31.

Calvo, S. E., and V. K. Mootha (2010). The mitochondrial proteome and human disease. Annual Review of Genomics and Human Genetics 11:25–44.

Calvo, S. E., K. R. Clauser, and V. K. Mootha (2016). MitoCarta2.0: An updated inventory of mammalian mitochondrial proteins. Nucleic Acids Research 44:D1251–D1257.

Cameron, S. L., K. Yoshizawa, A. Mizukoshi, M. F. Whiting, and K. P. Johnson (2011). Mitochondrial genome deletions and minicircles are common in lice (Insecta: Phthiraptera). BMC Genomics 12:394.

Camus, M. F., D. J. Clancy, and D. K. Dowling (2012). Mitochondria, maternal inheritance, and male aging. Current Biology 22:1717–21.

Camus, M. F., J. N. Wolff, C. M. Sgrò, and D. K. Dowling (2017). Experimental support that natural selection has shaped the latitudinal distribution of mitochondrial haplotypes in Australian *Drosophila melanogaster*. Molecular Biology and Evolution 34:2600–12.

Cannon, M. V., D. A. Dunn, M. H. Irwin, A. I. Brooks, F. F. Bartol, I. A. Trounce, and C. A. Pinkert (2011). Xenomitochondrial mice: Investigation into mitochondrial compensatory mechanisms. Mitochondrion 11:33–9.

Cantarero, A., and C. Alonso-Alvarez (2017). Mitochondria-targeted molecules determine the redness of the zebra finch bill. Biology Letters 13:20170455.

Carling, M. D., and R. T. Brumfield (2008). Haldane's rule in an avian system: Using cline theory and divergence population genetics to test for differential introgression of mitochondrial, autosomal, and sex-linked loci across the *Passerina* bunting hybrid zone. Evolution 62:2600–15.

Carneiro, M., and D. L. Hartl (2010). Adaptive landscapes and protein evolution. Proceedings of the National Academy of Sciences 107:1747–51.

Carrier, D. R., A. K. Kapoor, T. Kimura, M. K. Nickels, E. C. Scott, J. K. So, and E. Trinkaus (1984). The energetic paradox of human running and hominid evolution [and comments and reply]. Current Anthropology 25:483–95.

Caruso, C. M., A. L. Case, and M. F. Bailey (2012). The evolutionary ecology of cytonuclear interactions in angiosperms. Trends in Plant Science 17:638–43.

Castellana, S., S. Vicario, and C. Saccone (2011). Evolutionary patterns of the mitochondrial genome in Metazoa: Exploring the role of mutation and selection in mitochondrial protein-coding genes. Genome Biology and Evolution 3:1067–79.

Castoe, T. A., Z. J. Jiang, W. Gu, Z. O. Wang, and D. D. Pollock (2008). Adaptive evolution and functional redesign of core metabolic proteins in snakes. PLoS One 3:e2201.

Chang, C.-C., J. Rodriguez, and J. Ross (2015). Mitochondrial-nuclear epistasis impacts fitness and mitochondrial physiology of interpopulation *Caenorhabditis briggsae* hybrids. G3 6:209–19.

Charlat, S., A. Duplouy, E. A. Hornett, E. A. Dyson, N. Davies, G. K. Roderick, N. Wedell, and G. D. D. Hurst (2009). The joint evolutionary histories of *Wolbachia* and mitochondria in *Hypolimnas bolina*. BMC Evolutionary Biology 9:64.

Charlesworth, B. (2008). Mutation-selection balance and the evolutionary advantage of sex and recombination. Genetics Research 89:451–73.

Charlesworth, B., and D. Charlesworth (2000). The degeneration of Y chromosomes. Philosophical Transactions of the Royal Society B: Biological Sciences 355:1563–72.

Charlesworth, B., J. A. Coyne, and N. H. Barton (1987). The relative rates of evolution of sex chromosomes and autosomes. The American Naturalist 130:113–46.

Charlesworth, D., and B. Charlesworth (2010). Evolutionary biology: The origins of two sexes. Current Biology 20:R519–R521.

Charlesworth, D., B. Charlesworth, and G. Marais (2005). Steps in the evolution of heteromorphic sex chromosomes. Heredity 95:118–28.

Che-Othman, M. H., A. H. Millar, and N. L. Taylor (2017). Connecting salt stress signalling pathways with salinity-induced changes in mitochondrial metabolic processes in C3 plants. Plant, Cell and Environment, 40:2875–905.

Cheviron, Z. A., and R. T. Brumfield (2009). Migration-selection balance and local adaptation of mitochondrial haplotypes in rufous-collared sparrows (*Zonotrichia capensis*) along an elevational gradient. Evolution 63:1593–605.

Chinnery, P. F. (2014). Mitochondrial disorders overview. In GeneReviews® [Internet] (M. P. Adam, H. H. Ardinger, R. A. Pagon, S. E. Wallace, L. J. H. Bean, K. Stephens, A. Amemiya, editors). University of Washington, Seattle, WA, pp. 1–22.

Chinnery, P. F. (2015). Mitochondrial disease in adults: What's old and what's new ? EMBO Molecular Medicine 7:1503–12.

Chinnery, P. F., D. R. Thorburn, D. C. Samuels, S. L. White, H. H. M. Dahl, D. M. Turnbull, R. N. Lightowlers, and N. Howell (2000). The inheritance of mitochondrial DNA heteroplasmy: Random drift, selection or both? Trends in Genetics 16:500–5.

Chittenden, R. J. (1927). Cytoplasmic inheritance in flax. Journal of Heredity 18:337–43.

Choi, S. S., W. Li, and B. T. Lahn (2005). Robust signals of coevolution of interacting residues in mammalian proteomes identified by phylogeny-aided structural analysis. Nature Genetics 37:1367–71.

Chou, J. Y., and J. Y. Leu (2010). Speciation through cytonuclear incompatibility: Insights from yeast and implications for higher eukaryotes. BioEssays 32:401–11.

Chrétien, D., P. Bénit, H. H. Ha, S. Keipert, R. El-Khoury, Y. T. Chang, M. Jastroch, H. T. Jacobs, P. Rustin, and M. Rak (2018). Mitochondria are physiologically maintained at close to 50°C. PLoS Biology 16:e2003992.

Christie, J. R., and M. Beekman (2016). Uniparental inheritance promotes adaptive evolution in cytoplasmic genomes. Molecular Biology and Evolution 34:msw266.

Christie, J. R., T. M. Schaerf, and M. Beekman (2015). Selection against heteroplasmy explains the evolution of uniparental inheritance of mitochondria. PLoS Genetics 11:1–30.

Christy, S. F., R. I. Wernick, M. J. Lue, G. Velasco, D. K. Howe, D. R. Denver, and S. Estes (2017). Adaptive evolution under extreme genetic drift in oxidatively stressed *Caenorhabditis elegans*. Genome Biology and Evolution 9:3008–22.

Chun, S., and J. C. Fay (2011). Evidence for hitchhiking of deleterious mutations within the human genome. PLoS Genetics 7:e1002240.

Clancy, D. J., G. R. Hime, and A. D. Shirras (2011). Cytoplasmic male sterility in *Drosophila melanogaster* associated with a mitochondrial CYTB variant. Heredity 107:374–6.

Claros, M. G., J. Perea, Y. Shu, F. A. Samatey, J. L. Popot, and C. Jacq (1995). Limitations to in vivo import of hydrophobic proteins into yeast mitochondria. The case of a cytoplasmically synthesized apocytochrome b. European Journal of Biochemistry 228:762–71.

Clayton, D. A. (1982). Replication of animal mitochondrial DNA. Cell 28:693–705.

Clayton, D. A. (2000a). Transcription and replication of mitochondrial DNA. Human Reproduction 15:11–17.

Clayton, D. A. (2000b). Vertebrate mitochondrial DNA—A circle of surprises. Experimental Cell Research 255:4–9.

Cline, S. D. (2012). Mitochondrial DNA damage and its consequences for mitochondrial gene expression. Biochimica et Biophysica Acta (BBA)—Gene Regulatory Mechanisms 1819:979–91.

Cock, H. R., S. J. Tabrizi, J. M. Cooper, and A. H. V Schapira (1998). The influence of nuclear background on the biochemical expression of 3460 Leber's hereditary optic neuropathy. Annals of Neurology 44:187–93.

Comeron, J. M., A. Williford, and R. M. Kliman (2008). The Hill–Robertson effect: Evolutionary consequences of weak selection and linkage in finite populations. Heredity 100:19–31.

Confer, J. L., and K. Knapp (2016). Golden-winged warblers and blue-winged warblers: The relative success of a habitat specialist and a generalist. The Auk 98:108–14.

Connallon, T., M. F. Camus, E. H. Morrow, and D. K. Dowling (2018). Coadaptation of mitochondrial and nuclear genes, and the cost of mother's curse. Proceedings of the Royal Society B: Biological Sciences 285:1–7.

Consuegra, S., E. John, E. Verspoor, and C. G. De Leaniz (2015). Patterns of natural selection acting on the mitochondrial genome of a locally adapted fish species. Genetics Selection Evolution 47:1–10.

Cooper, B. S., C. R. Burrus, C. Ji, M. W. Hahn, and K. L. Montooth (2015). Similar efficacies of selection shape mitochondrial and nuclear genes in both *Drosophila melanogaster* and *Homo sapiens*. G3 5:1–45.

Cooper, C. E., and G. C. Brown (2008). The inhibition of mitochondrial cytochrome oxidase by the gases carbon monoxide, nitric oxide, hydrogen cyanide and hydrogen sulfide: Chemical mechanism and physiological significance. Journal of Bioenergetics and Biomembranes 40:533.

Cosmides, L. M., and J. Tooby (1981). Cytoplasmic inheritance and intergenomic conflict. Journal of Theoretical Biology 89:83–129.

Couvillion, M. T., I. C. Soto, G. Shipkovenska, and L. S. Churchman (2016). Synchronized mitochondrial and cytosolic translation programs. Nature 533:499–503.

Coyne, J. A. (1974). The evolutionary origin of hybrid inviability. Evolution 28:505–6.

Coyne, J. A. (1985). The genetic basis of Haldane's rule. Nature 314:736–8.

Coyne, J. A., and H. A. Orr (2004). Speciation. Sunderland, MA: Sinauer Associates.

Cracraft, J. (1983). Species concepts and speciation analysis. Current Ornithology 1:159–87.

Cree, L. M., D. C. Samuels, S. C. de Sousa Lopes, H. K. Rajasimha, P. Wonnapinij, J. R. Mann, H. H. Dahl, and P. F. Chinnery (2008). A reduction of mitochondrial DNA molecules during embryogenesis explains the rapid segregation of genotypes. Nature Genetics 40:249–54.

Criscuolo, F., M. del M. Gonzalez-Barroso, F. Bouillaud, D. Ricquier, B. Miroux, and G. Sorci (2005). Mitochondrial uncoupling proteins: New perspectives for evolutionary ecologists. The American Naturalist 166:686–99.

Crochet, P.-A., J. Z. Chen, J.-M. Pons, J. D. Lebreton, P. D. N. Hebert, and F. Bonhomme (2003). Genetic differentiation at nuclear and mitochondrial loci among large white-headed gulls: Sex-biased interspecific gene flow? Evolution 57:2865–78.

Crochet, P.-A., J.-D. Lebreton, and F. Bonhomme (2018). Systematics of large white-headed gulls: Patterns of mitochondrial DNA variation in western European taxa. The Auk 119:603–20.

Cronin, H. (1991). The Ant and the Peacock. Cambridge University Press, Cambridge, UK.

Crow, J. F. (1997). The high spontaneous mutation rate: Is it a health risk? Proceedings of the National Academy of Sciences 94:8380–6.

Cui, H., Y. Kong, and H. Zhang (2012). Oxidative stress, mitochondrial dysfunction, and aging. Journal of Signal Transduction 2012:1–13.

Culver, D. C., and T. Pipan (2009). The Biology of Caves and Other Subterranean Habitats. Oxford University Press, Oxford, UK.

Cutter, A. D. (2012). The polymorphic prelude to Bateson-Dobzhansky-Muller incompatibilities. Trends in Ecology and Evolution 27:209–18.

D'Aurelio, M., C. D. Gajewski, G. Lenaz, and G. Manfredi (2006). Respiratory chain supercomplexes set the threshold for respiration defects in human mtDNA mutant cybrids. Human Molecular Genetics 15:2157–69.

D'Souza, A. R., and M. Minczuk (2018). Mitochondrial transcription and translation: Overview. Essays in Biochemistry 62:309–20.

Dale, J. (2006). Intraspecific variation in bird colors. In Bird Coloration, Volume 2. Function and Evolution (G. E. Hill and K. J. McGraw, Editors). Harvard University Press, Cambridge, MA.

Darwin, C. (1859). On the Origin of the Species. John Murray, London.

Darwin, C. (1871). Sexual Selection and the Descent of Man. D. Appleton and Company, New York.

Davies, R., K. E. Mathers, A. D. Hume, K. Bremer, Y. Wang, and C. D. Moyes (2012). Hybridization in sunfish influences the muscle metabolic phenotype. Physiological and Biochemical Zoology 85:321–31.

Dawkins, R. (1976). The Selfish Gene. Oxford University Press, Oxford, UK.

Dawson, W. R., and J. W. Hudson (1970). Birds. In Comparative Physiology of Thermoregulation, vol. 1. (G. C. Whittow, Editor). Academic Press, New York, pp. 223–310.

Dean, R., F. Zimmer, and J. E. Mank (2014). The potential role of sexual conflict and sexual selection in shaping the genomic distribution of mito-nuclear genes. Genome Biology and Evolution 6:1096–104.

DeLong, J. P., J. G. Okie, M. E. Moses, R. M. Sibly, and J. H. Brown (2010). Shifts in metabolic scaling, production, and efficiency across major evolutionary transitions of life. Proceedings of the National Academy of Sciences 107:12941–5.

Derelle, R., G. Torruella, V. Klimeš, H. Brinkmann, E. Kim, Č. Vlček, B. F. Lang, and M. Eliáš (2015). Bacterial proteins pinpoint a single eukaryotic root. Proceedings of the National Academy of Sciences 112:E693–E699.

Desmond, E., C. Brochier-Armanet, P. Forterre, and S. Gribaldo (2011). On the last common ancestor and early evolution of eukaryotes: Reconstructing the history of mitochondrial ribosomes. Research in Microbiology 162:53–70.

Detto, T., P. R. Backwell, J. M. Hemmi, and J. Zeil (2006). Visually mediated species and neighbour recognition in fiddler crabs (*Uca mjoebergi* and *Uca capricornis*). Proceedings of the Royal Society of London B: Biological Sciences 273:1661–6.

Devault, D. (1980). Quantum mechanical tunnelling in biological systems. Quarterly Review of Biophysics 13:387–564.

Dhami, K. K., L. Joseph, D. A. Roshier, and J. L. Peters (2016). Recent speciation and elevated Z-chromosome differentiation between sexually monochromatic and dichromatic species of Australian teals. Journal of Avian Biology 47:92–102.

Dill, K. A., K. Ghosh, and J. D. Schmit (2011). Physical limits of cells and proteomes. Proceedings of the National Academy of Sciences 108:17876–82.

Dimroth, P., G. Kaim, and U. Matthey (2000). Crucial role of the membrane potential for ATP synthesis by F(1)F(o) ATP synthases. The Journal of Experimental Biology 203:51–9.

Dingley, S. D., E. Polyak, J. Ostrovsky, S. Srinivasan, I. Lee, A. B. Rosenfeld, M. Tsukikawa, R. Xiao, M. A. Selak, J. J. Coon, A. S. Hebert, et al. (2014). Mitochondrial DNA variant in COX1 subunit significantly alters energy metabolism of geographically divergent wild isolates in *Caenorhabditis elegans*. Journal of Molecular Biology 426:2199–216.

Dion-Côté, A. M., and D. A. Barbash (2017). Beyond speciation genes: An overview of genome stability in evolution and speciation. Current Opinion in Genetics and Development 47:17–23.

Divakaruni, A. S., and M. D. Brand (2011). The regulation and physiology of mitochondrial proton leak. Physiology 26:192–205.

Dobzhansky, T. (1936). Studies on hybrid sterility. II. Localization of sterility factors in *Drosophila pseudoobscura* hybrids. Genetics 21:113.

Dobzhansky, T. (1937a). Genetic nature of species differences. The American Naturalist 71:404–20.

Dobzhansky, T. (1937b). Genetics and the Origin of Species. Columbia University Press, New York.

Dobzhansky, T. (1948). Genetics of natural populations; experiments on chromosomes of *Drosophila pseudoobscura* from different geographic regions. Genetics 33:588–602.

Dobzhansky, T. (1973). Nothing in biology makes sense except in light of evolution. The American Biology Teacher 35:125–9.

Doiron, S., L. Bernatchez, and P. U. Blier (2002). A comparative mitogenomic analysis of the potential adaptive value of arctic charr mtDNA introgression in brook charr populations (*Salvelinus fontinalis mitchill*). Molecular Biology and Evolution 19:1902–9.

Dong, Y., T. Yoshitomi, J.-F. Hu, and J. Cui (2017). Long noncoding RNAs coordinate functions between mitochondria and the nucleus. Epigenetics and Chromatin 10:41.

Donoghue, M. J. (1985). A critique of the biological species concept and recommendations for a phylogenetic alternative. The Bryologist 88:172.

Đorđević, M., B. Stojković, U. Savković, E. Immonen, N. Tucić, J. Lazarević, and G. Arnqvist (2017). Sex-specific mitonuclear epistasis and the evolution of mitochondrial bioenergetics, ageing, and life history in seed beetles. Evolution 71:274–88.

Dowling, D. K. (2014a). Evolutionary perspectives on the links between mitochondrial genotype and disease phenotype. Biochimica et Biophysica Acta—General Subjects 1840:1393–403.

Dowling, D. K., K. C. Abiega, and G. Arnqvist (2007a). Temperature-specific outcomes of cytoplasmic-nuclear interactions on egg-to-adult development time in seed beetles. Evolution 61:194–201.

Dowling, D. K., U. Friberg, and G. Arnqvist (2007b). A comparison of nuclear and cytoplasmic genetic effects on sperm competitiveness and female remating in a seed beetle. Journal of Evolutionary Biology 20:2113–25.

Dowling, D. K., U. Friberg, F. Hailer, and G. Arnqvist (2007c). Intergenomic epistasis for fitness: Within-population interactions between cytoplasmic and nuclear genes in *Drosophila melanogaster*. Genetics 175:235–44.

Dowling, D. K., A. L. Nowostawski, and G. Arnqvist (2007d). Effects of cytoplasmic genes on sperm viability and sperm morphology in a seed beetle: Implications for sperm competition theory? Journal of Evolutionary Biology 20:358–68.

Dowling, D. K., U. Friberg, and J. Lindell (2008). Evolutionary implications of non-neutral mitochondrial genetic variation. Trends in Ecology and Evolution 23:546–54.

Drovetski, S. V., A. S. Kitaysky, N. A. Mode, R. M. Zink, U. Iqbal, and C. Barger (2012). MtDNA haplotypes differ in their probability of being eliminated by a mass die-off in an abundant seabird. Heredity 109:29–33.

Drovetski, S. V, G. Semenov, Y. A. Red'Kin, V. N. Sotnikov, I. V Fadeev, and E. A. Koblik (2015). Effects of asymmetric nuclear introgression, introgressive mitochondrial sweep, and purifying selection on phylogenetic reconstruction and divergence estimates in the pacific clade of *Locustella* warblers. PLoS One 10:1–14.

Drown, D. M., K. M. Preuss, and M. J. Wade (2012). Evidence of a paucity of genes that interact with the mitochondrion on the X in mammals. Genome Biology and Evolution 4:875–80.

Du, S. N. N., F. Khajali, N. J. Dawson, and G. R. Scott (2017). Hybridization increases mitochondrial production of reactive oxygen species in sunfish. Evolution 71:1643–52.

Dumollard, R., M. Duchen, and J. Carroll (2007). The role of mitochondrial function in the oocyte and embryo. Current Topics in Developmental Biology 77:21–49.

Dunthorn, M., and L. A. Katz (2010). Secretive ciliates and putative asexuality in microbial eukaryotes. Trends in Microbiology 18:183–8.

Durham, S. E., E. Bonilla, D. C. Samuels, S. DiMauro, and P. F. Chinnery (2005). Mitochondrial DNA copy number threshold in mtDNA depletion myopathy. Neurology 65:453–5.

Dzeja, P. P., and A. Terzic (2003). Phosphotransfer networks and cellular energetics. The Journal of Experimental Biology 206:2039–47.

Ebert, K. M., H. Liem, and N. B. Hecht (1988). Mitochondrial DNA in the mouse preimplantation embryo. Journal of Reproduction and Fertility 82:145–9.

Edmands, S. (2001). Phylogeography of the intertidal copepod *Tigriopus californicus* reveals substantially reduced population differentiation at northern latitudes. Molecular Ecology 10:1743–50.

Edmands, S. (2008). Recombination in interpopulation hybrids of the copepod *Tigriopus californicus*: Release of beneficial variation despite hybrid breakdown. Journal of Heredity 99:316–18.

Edmands, S., and R. S. Burton (1999). Cytochrome C oxidase activity in interpopulation hybrids of a marine copepod: A test for nuclear-nuclear or nuclear-cytoplasmic coadaptation. Evolution 53:1972–8.

Einsiedel, E. F. (2000). Cloning and its discontents—a Canadian perspective. Nature Biotechnology 18:943–4.

Eldredge, N., and J. Cracraft (1980). Phylogenetic Patterns and the Evolutionary Process. Method and Theory in Comparative Biology. Columbia University Press, New York.

Ellegren, H. (2009a). The different levels of genetic diversity in sex chromosomes and autosomes. Trends in Genetics 25:278–84.

Ellegren, H. (2009b). Genomic evidence for a large-Z effect. Proceedings of the Royal Society B: Biological Sciences 276:361–6.

Ellison, C. K., and R. S. Burton (2006). Disruption of mitochondrial function in interpopulation hybrids of *Tigriopus californicus*. Evolution 60:1382–91.

Ellison, C. K., and R. S. Burton (2008a). Genotype-dependent variation of mitochondrial transcriptional profiles in interpopulation hybrids. Proceedings of the National Academy of Sciences 105:15831–6.

Ellison, C. K., and R. S. Burton (2008b). Interpopulation hybrid breakdown maps to the mitochondrial genome. Evolution 62:631–8.

Ellison, C. K., and R. S. Burton (2010). Cytonuclear conflict in interpopulation hybrids: The role of RNA polymerase in mtDNA transcription and replication. Journal of Evolutionary Biology 23:528–38.

Ellison, C. K., O. Niehuis, and J. Gadau (2008). Hybrid breakdown and mitochondrial dysfunction in hybrids of *Nasonia* parasitoid wasps. Journal of Evolutionary Biology 21:1844–51.

Elson, J. L., D. M. Turnbull, and N. Howell (2004). Comparative genomics and the evolution of human mitochondrial DNA: Assessing the effects of selection. American Journal of Human Genetics 74:229–38.

Elurbe, D. M. and M. A. Huynen (2016). The origin of the supernumerary subunits and assembly factors of complex I: A treasure trove of pathway evolution. Biochimica et Biophysica Acta (BBA)—Bioenergetics 1857(7):971–9.

Endler, J. A. (1977). Geographic Variation, Speciation, and Clines. Princeton University Press, Princeton, NJ.

Enoki, S., A. Shimizu, C. Hayashi, H. Imanishi, O. Hashizume, K. Mekada, H. Suzuki, T. Hashimoto, K. Nakada, and J.-I. Hayashi (2014). Selection of rodent species appropriate for mtDNA transfer to generate transmitochondrial mito-mice expressing mitochondrial respiration defects. Experimental Animals 63:21–30.

Esterházy, D., M. S. King, G. Yakovlev, and J. Hirst (2008). Production of reactive oxygen species by complex I (NADH:ubiquinone oxidoreductase) from *Escherichia coli* and comparison to the enzyme from mitochondria. Biochemistry 47:3964–71.

Extavour, C. G. M. (2007). Evolution of the bilaterian germ line: Lineage origin and modulation of specification mechanisms. Integrative and Comparative Biology 47:770–85.

Falkenberg, M., N.-G. Larsson, and C. M. Gustafsson (2007). DNA replication and transcription in mammalian mitochondria. Annual Review of Biochemistry 76:679–99.

Fan, W., K. G. Waymire, N. Narula, P. Li, C. Rocher, P. E. Coskun, M. A. Vannan, J. Narula, G. R. MacGregor, and D. C. Wallace (2008). A mouse model of mitochondrial disease reveals germline selection against severe mtDNA mutations. Science 319:958–62.

Fay, J. C., and C. I. Wu (2000). Hitchhiking under positive Darwinian selection. Genetics 155:1405–13.

Felsenstein, J. (1974). The evolution advantage of recombination. Genetics 78:737–56.

Fenster, C. B., L. F. Galloway, and L. Chao (1997). Epistasis and its consequences for the evolution of natural populations. Trends in Ecology and Evolution 12:282–6.

Fernández-Vizarra, E., V. Tiranti, and M. Zeviani (2009). Assembly of the oxidative phosphorylation system in humans: What we have learned by studying its defects. Biochimica et Biophysica Acta (BBA)—Molecular Cell Research 1793:200–11.

Fernández, A., and R. Scott (2003). Dehydron: A structurally encoded signal for protein interaction. Biophysical Journal 85:1914–28.

Fiedorczuk, K., J. A. Letts, G. Degliesposti, K. Kaszuba, M. Skehel, and L. A. Sazanov (2016). Atomic structure of the entire mammalian mitochondrial complex I. Nature 538:406–10.

Finch, T. M., N. Zhao, D. Korkin, K. H. Frederick, and L. S. Eggert (2014). Evidence of positive selection in mitochondrial complexes I and V of the African elephant. PLoS ONE 9:e92587.

Fisher, R. A. (1930). The Genetical Theory of Natural Selection. Clarendon Press, Oxford, UK.

Fishman, L., and J. H. Willis (2006). A cytonuclear incompatibility causes anther sterility in Mimulus hybrids. Evolution 60:1372–81.

Foley, B. R., C. G. Rose, D. E. Rundle, W. Leong, and S. Edmands (2013). Postzygotic isolation involves strong mitochondrial and sex-specific effects in *Tigriopus californicus*, a species lacking heteromorphic sex chromosomes. Heredity 111:391–401.

da Fonseca, R. R., W. E. Johnson, S. J. O'Brien, M. J. Ramos, and A. Antunes (2008). The adaptive evolution of the mammalian mitochondrial genome. BMC Genomics 9:119.

Fontanillas, P., A. Depraz, M. S. Giorgi, and N. Perrin (2005). Nonshivering thermogenesis capacity associated to mitochondrial DNA haplotypes and gender in the greater white-toothed shrew, *Crocidura russula*. Molecular Ecology 14:661–70.

Foote, A. D., P. A. Morin, J. W. Durban, R. L. Pitman, P. Wade, E. Willerslev, M. T. P. Gilbert, and R. R. da Fonseca (2011). Positive selection on the killer whale mitogenome. Biology Letters 7:116–18.

Formosa, L. E., M. G. Dibley, D. A. Stroud, and M. T. Ryan (2017). Building a complex complex: Assembly of mitochondrial respiratory chain complex I. Seminars in Cell and Developmental Biology 76:154–62.

Frank, S. A., and L. D. Hurst (1996). Mitochondria and male disease. Nature 383:224.

Frerman, F. E., and S. I. Goodman (2001). Defects of electron transfer flavoprotein and electron transfer flavoprotein-ubiquinone oxidoreductase: Glutaric acidemia type II. In The Metabolic and Molecular Bases of Inherited Disease (C. R. Scriver, W. S. Sly, B. Childs, A. L. Beaudet and D. Valle, Editors). McGraw-Hill, New York, pp. 2357–65.

Friedberg, E. C. (2008). A brief history of the DNA repair field. Cell Research 18:3–7.

Froman, D. P., and J. D. Kirby (2005). Sperm mobility: Phenotype in roosters (*Gallus domesticus*) determined by mitochondrial function 1. Biology of Reproduction 72:562–7.

Fuhrmann, D. C., and B. Brüne (2017). Mitochondrial composition and function under the control of hypoxia. Redox Biology 12:208–15.

Fujii, S., C. S. Bond, and I. D. Small (2011). Selection patterns on restorer-like genes reveal a conflict between nuclear and mitochondrial genomes throughout angiosperm evolution. Proceedings of the National Academy of Sciences 108:1723–8.

Fujita, M. K., and C. Moritz (2010). Origin and evolution of parthenogenetic genomes in lizards: Current state and future directions. Cytogenetic and Genome Research 127:261–72.

Funk, D. J., and K. E. Omland (2003). Species-level paraphyly and polyphyly: Frequency, causes, and consequences, with insights from animal mitochondrial DNA. Annual Review of Ecology, Evolution, and Systematics 34:397–423.

Gabaldón, T., and M. A. Huynen (2004). Shaping the mitochondrial proteome. Biochimica et Biophysica Acta—Bioenergetics 1659:212–20.

Galtier, N., R. W. Jobson, B. Nabholz, S. Glemin, and P. U. Blier (2009a). Mitochondrial whims: Metabolic rate, longevity and the rate of molecular evolution. Biology Letters 5:413–16.

Galtier, N., B. Nabholz, S. Glémin, and G. D. D. Hurst (2009b). Mitochondrial DNA as a marker of molecular diversity: A reappraisal. Molecular Ecology 18:4541–50.

Garg, S. G., and W. F. Martin (2016). Mitochondria, the cell cycle, and the origin of sex via a syncytial eukaryote common ancestor. Genome Biology and Evolution 8:1950–70.

Garner, A., B. Goulet, M. Farnitano, Y. Molina-Henao, and R. Hopkins (2018). Genomic signatures of reinforcement. Genes 9:191.

Garvin, M. R., J. P. Bielawski, and A. J. Gharrett (2011). Positive Darwinian selection in the piston that powers proton pumps in Complex I of the mitochondria of Pacific salmon. PLoS One 6:e24127.

Garvin, M. R., J. P. Bielawski, L. A. Sazanov, and A. J. Gharrett (2014). Review and meta-analysis of natural selection in mitochondrial complex I in metazoans. Journal of Zoological Systematics and Evolutionary Research 53:1–17.

Garvin, M. R., G. H. Thorgaard, and S. R. Narum (2015). Differential expression of genes that control respiration contribute to thermal adaptation in redband trout (*Oncorhynchus mykiss gairdneri*). Genome Biology and Evolution 7:1404–14.

Garvin, M. R., W. D. Templin, A. J. Gharrett, N. Decovich, C. M. Kondzela, J. R. Guyon, and M. V. Mcphee (2016). Potentially adaptive mitochondrial haplotypes as a tool to identify divergent nuclear loci. Methods in Ecology and Evolution 8:821–34.

Gaspari, M., M. Falkenberg, N.-G. Larsson, and C. M. Gustafsson (2004). The mitochondrial RNA polymerase contributes critically to promoter specificity in mammalian cells. The EMBO Journal 23:4606–14.

Gavrilets, S. (2004). Fitness Landscapes and the Origin of Species. Princeton University Press, Princeton, NJ.

Ge, Z., J. D. Johnson, P. A. Cobine, K. J. McGraw, Garcia Rosana, and G. E. Hill (2015). High concentrations of ketocarotenoids in hepatic mitochondria of *Haemorhous mexicanus*. Physiological and Biochemical Zoology 88:444–50.

Gemmell, N., and J. N. Wolff (2015). Mitochondrial replacement therapy: Cautiously replace the master manipulator. BioEssays 37:584–5.

Gemmell, N. J., V. J. Metcalf, and F. W. Allendorf (2004). Mother's curse: The effect of mtDNA on individual fitness and population viability. Trends in Ecology and Evolution 19:238–44.

Gems, D., and R. Doonan (2009). Antioxidant defense and aging in *C. elegans*: Is the oxidative damage theory of aging wrong? Cell Cycle 8:1681–7.

Genova, M. L., and G. Lenaz (2013). A critical appraisal of the role of respiratory supercomplexes in mitochondria. Biological Chemistry 394:631–9.

Gershoni, M., L. Levin, O. Ovadia, Y. Toiw, N. Shani, S. Dadon, N. Barzilai, A. Bergman, G. Atzmon, J. Wainstein, A. Tsur, et al. (2014). Disrupting mitochondrial-nuclear coevolution affects OXPHOS complex i integrity and impacts human health. Genome Biology and Evolution 6:2665–80.

Gershoni, M., A. R. Templeton, and D. Mishmar (2009). Mitochondrial bioenergetics as a major motive force of speciation. Bioessays 31:642–50.

Ghezzi, D., and M. Zeviani (2018). Human diseases associated with defects in assembly of OXPHOS complexes. Essays in Biochemistry 62:271–86.

Gilbert, W. (1978). Why genes in pieces? Nature 271:501–1.

Gill, F. B. (1997). Local cytonuclear extinction of the golden-winged warbler. Evolution 51:519–25.

Gill, F. B. (2004). Blue-winged warblers (*Vermivora pinus*) versus golden-winged warblers (*V. chrysoptera*). The Auk 121:1014–18.

Gill, R. E., T. L. Tibbitts, D. C. Douglas, C. M. Handel, D. M. Mulcahy, J. C. Gottschalck, N. Warnock, B. J. McCaffery, P. F. Battley, and T. Piersma (2009). Extreme endurance flights by landbirds crossing the Pacific Ocean: Ecological corridor rather than barrier? Proceedings of the Royal Society B: Biological Sciences 276:447–57.

Glémet, H., P. Blier, and L. Bernatchez (1998). Geographical extent of arctic char (*Salvelinus alpinus*) mtDNA introgression in brook char populations (*S. fontinalis*) from eastern Quebec, Canada. Molecular Ecology 7:1655–62.

Glor, R. E. (2010). Phylogenetic insights on adaptive radiation. Annual Review of Ecology, Evolution, and Systematics 41:251–70.

Gómez, M. C., J. A. Jenkins, A. Giraldo, R. F. Harris, A. King, B. L. Dresser, and C. E. Pope (2003). Nuclear transfer of synchronized African wild cat somatic cells into enucleated domestic cat oocytes. Biology of Reproduction 69:1032–41.

Goodenough, U., and J. Heitman (2014). Origins of eukaryotic sexual reproduction. Cold Spring Harbor Perspectives in Biology 6:a016154.

Goodsell, D. S., S. Dutta, C. Zardecki, M. Voigt, H.M. Berman, and S.K. Burley (2015). The RCSB PDB "Molecule of the Month": Inspiring a molecular view of biology. PLoS Biology 13:e1002140.

Goodwin, T. W. (1984). The Biochemistry of Carotenoids 2, Animals. Chapman and Hall, New York.

Gorman, G. S., A. M. Schaefer, Y. Ng, N. Gomez, E. L. Blakely, C. L. Alston, C. Feeney, R. Horvath, P. Yu-Wai-Man, P. F. Chinnery, R. W. Taylor, et al. (2015). Prevalence of nuclear and mitochondrial DNA mutations related to adult mitochondrial disease. Annals of Neurology 77:753–9.

Gould, S. J. (1996). Full House: The Spread of Excellence from Plato to Darwin. Harmony Books, New York.

Gowen, F. C., J. M. Maley, C. Cicero, A. T. Peterson, B. C. Faircloth, T. C. Warr, and J. E. McCormack (2014). Speciation in western scrub-jays, Haldane's rule, and genetic clines in secondary contact. BMC Evolutionary Biology 14:1.

Grant, P. R. (2006). Evolution of character displacement in Darwin's finches. Science 313:224–6.

Grant, P. R., and B. R. Grant (2008). How and Why Species Multiply: The Radiation of Darwin's Finches. Princeton University Press, Princeton, NJ.

Gray, M. W. (2012). Mitochondrial evolution. Cold Spring Harbor Perspectives in Biology 4:a011403.

Gray, M. W. (2015). Mosaic nature of the mitochondrial proteome: Implications for the origin and evolution of mitochondria. Proceedings of the National Academy of Sciences 112:10133–8.

Gray, M. W., R. Cedergren, Y. Abel, D. Sankoff, R. Cedergrent, Y. Abelt, and D. Sankofft (1989). On the evolutionary origin of the plant mitochondrion and its genome. Proceedings of the National Academy of Sciences 86:2267–71.

Gray, M. W., B. F. Lang, R. Cedergren, G. B. Golding, C. Lemieux, D. Sankoff, M. Turmel, N. Brossard, E. Delage, T. G. Littlejohn, I. Plante, et al. (1998). Genome structure and gene content in protist mitochondrial DNAs. Nucleic Acids Research 26:865–78.

Greiner, S., and R. Bock (2013). Tuning a menage a trois: Co-evolution and co-adaptation of nuclear and organellar genomes in plants. BioEssays 35:354–65.

Greiner, S., U. Rauwolf, J. Meurer, and R. G. Herrmann (2011). The role of plastids in plant speciation. Molecular Ecology 20:671–91.

Greiner, S., J. Sobanski, and R. Bock (2015). Why are most organelle genomes transmitted maternally? BioEssays 37:80–94.

Grossman, L. I., T. R. Schmidt, D. E. Wildman, and M. Goodman (2001). Molecular evolution of aerobic energy metabolism in primates. Molecular Phylogenetics and Evolution 18:26–36.

Grossman, L. I., D. E. Wildman, T. R. Schmidt, and M. Goodman (2004). Accelerated evolution of the electron transport chain in anthropoid primates. Trends in Genetics 20:578–85.

Guaras, A. M., and J. A. Enríquez (2017). Building a beautiful beast: Mammalian respiratory Complex I. Cell Metabolism 25:4–5.

Gupta, S., A. Agarwal, J. Banerjee, and J. G. Alvarez (2007). The role of oxidative stress in spontaneous abortion and recurrent pregnancy loss: A systematic review. Obstetrical and Gynecological Survey 62:335–47.

Gustafsson, C. M., M. Falkenberg, and N.-G. Larsson (2016). Maintenance and expression of mammalian mitochondrial DNA. Annual Review of Biochemistry 85:133–60.

Gutiérrez, J. S., J. A. Masero, J. M. Abad-Gómez, A. Villegas, and J. M. Sánchez-Guzmán (2011). Understanding the energetic costs of living in saline environments: Effects of salinity on basal metabolic rate, body mass and daily energy consumption of a long-distance migratory shorebird. The Journal of Experimental Biology 214:829–35.

Hadjivasiliou, Z., A. Pomiankowski, R. M. Seymour, and N. Lane (2012). Selection for mitonuclear co-adaptation could favour the evolution of two sexes. Proceedings of the Royal Society B: Biological Sciences 279:1865–72.

Hadjivasiliou, Z., N. Lane, R. M. Seymour, and A. Pomiankowski (2013). Dynamics of mitochondrial inheritance in the evolution of binary mating types and two sexes. Proceedings of the Royal Society B: Biological Sciences 280:20131920.

Hagman, M., and T. J. Ord (2016). Many paths to a common destination: Morphological differentiation of a functionally convergent visual signal. The American Naturalist 188:306–18.

Hagström, E., C. Freyer, B. J. Battersby, J. B. Stewart, and N. G. Larsson (2014). No recombination of mtDNA after heteroplasmy for 50 generations in the mouse maternal germline. Nucleic Acids Research 42:1111–16.

Haig, D. (2016). Intracellular evolution of mitochondrial DNA (mtDNA) and the tragedy of the cytoplasmic commons. BioEssays 38:549–55.

Hajibabaei, M., G. A. C. Singer, E. L. Clare, and P. D. N. Hebert (2007). Design and applicability of DNA arrays and DNA barcodes in biodiversity monitoring. BMC Biology 5:24.

Haldane, J. B. S. (1922). Sex ratio and unisexual sterility in hybrid animals. Journal of Genetics 12:101–9.

Halestrap, A. P., C. Brenner, and C. Brennerb (2003). The adenine nucleotide translocase: A central component of the mitochondrial permeability transition pore and key player in cell death. Current Medicinal Chemistry 10:1507–25.

Hall, V. J., P. Stojkovic, and M. Stojkovic (2006). Using therapeutic cloning to fight human disease: A conundrum or reality? Stem Cells 24:1628–37.

Hamanaka, R. B., S. E. Weinberg, C. R. Reczek, and N. S. Chandel (2016). The mitochondrial respiratory chain is required for organismal adaptation to hypoxia. Cell Reports 15:451–9.

Hamel, S., J. L. Gala, S. Dufour, P. Vannuffell, N. Zammatteo, and J. Remacle (2001). Consensus PCR and microarray for diagnosis of the genus *Staphylococcus*, species, and methicillin resistance. BioTechniques 31:1364–72.

Hamilton, E. W. (2001). Mitochondrial adaptations to NaCl. Complex I is protected by antioxidants and small heat shock proteins, whereas complex II is protected by proline and betaine. Plant Physiology 126:1266–74.

Hamilton, W. D., R. Axelrod, and R. Tanese (1990). Sexual reproduction as an adaptation to resist parasites (a review). Proceedings of the National Academy of Sciences 87:3566–73.

Hancock, A. M., V. J. Clark, Y. Qian, and A. Di Rienzo (2011). Population genetic analysis of the uncoupling proteins supports a role for UCP3 in human cold resistance. Molecular Biology and Evolution 28:601–14.

Hao, H., and C. T. Moraes (1997). A disease-associated G5703A mutation in human mitochondrial DNA causes a conformational change and a marked decrease in steady-state levels of mitochondrial tRNA(Asn). Molecular and Cellular Biology 17:6831–7.

Harada, A. E., and R. S. Burton (2017). The mitochondrial basis of thermal tolerance in the intertidal copepod *Tigriopus californicus*. Society for Integrative and Comparative Biology Abstract 2017:169.

Harman, D. (1956). Aging: A theory based on free radical and radiation chemistry. Journal of Gerontology 11:298–300.

Harman, D. (1994). Free radical theory of aging: An update. Annals of the New York Academy of Sciences 1067:10–21.

Hartfield, M., and S. P. Otto (2011). Recombination and hitchhiking of deleterious alleles. Evolution 65:2421–34.

Hartman, P. S., N. Ishii, E. B. Kayser, P. G. Morgan, and M. M. Sedensky (2001). Mitochondrial mutations differentially affect aging, mutability and anesthetic sensitivity in *Caenorhabditis elegans*. Mechanisms of Ageing and Development 122:1187–201.

Hashimoto, Y., T. Niikura, H. Tajima, T. Yasukawa, H. Sudo, Y. Ito, Y. Kita, M. Kawasumi, K. Kouyama, M. Doyu, G. Sobue, et al. (2001). A rescue factor abolishing neuronal cell death by a wide spectrum of familial Alzheimer's disease genes and Abeta. Proceedings of the National Academy of Sciences 98:6336–41.

Hassanin, A., A. Ropiquet, A. Couloux, and C. Cruaud (2009). Evolution of the mitochondrial genome in mammals living at high altitude: New insights from a study of the tribe *Caprini* (Bovidae, Antilopinae). Journal of Molecular Evolution 68:293–310.

Hastings, I. M. (1994). Manifestations of sexual selection may depend on the genetic basis of sex determination. Proceedings of the Royal Society B: Biological Sciences 258:83–7.

Havird, J. C., and D. B. Sloan (2016). The roles of mutation, selection, and expression in determining relative rates of evolution in mitochondrial versus nuclear genomes. Molecular Biology and Evolution 33:3042–53.

Havird, J. C., M. D. Hall, and D. K. Dowling (2015a). The evolution of sex: A new hypothesis based on mitochondrial mutational erosion. BioEssays 37:951–8.

Havird, J. C., N. S. Whitehill, C. D. Snow, and D. B. Sloan (2015b). Conservative and compensatory evolution in oxidative phosphorylation complexes of angiosperms with highly divergent rates of mitochondrial genome evolution. Evolution 69:3069–81.

Hazkani-Covo, E., R. M. Zeller, and W. Martin (2010). Molecular poltergeists: Mitochondrial DNA copies (numts) in sequenced nuclear genomes. PLoS Genetics 6.

Hebert, P. D. N., and T. R. Gregory (2005). The promise of DNA barcoding for taxonomy. Systematic Biology 54:852–9.

Hebert, P. D. N., A. Cywinska, S. L. Ball, and J. R. deWaard (2003a). Biological identifications through DNA barcodes. Proceedings of the Royal Society B: Biological Sciences 270:313–21.

Hebert, P. D. N., S. Ratnasingham, and J. Waard (2003b). Barcoding animal life: Cytochrome c oxidase subunit 1 divergences among closely related species. Proceedings of the Royal Society B: Biological Sciences 270:S96–S99.

Hebert, P. D. N., M. Y. Stoeckle, T. S. Zemlak, and C. M. Francis (2004). Identification of birds through DNA barcodes. PLoS Biology 2:e312.

Hénault, M., and C. R. Landry (2017). When nuclear-encoded proteins and mitochondrial RNAs do not get along, species split apart. EMBO Reports 18:8–10.

Hickerson, M. J., C. P. Meyer, and C. Moritz (2006). DNA barcoding will often fail to discover new animal species over broad parameter space. Systematic Biology 55:729–39.

Hickey, A. J. R., G. M. C. Renshaw, B. Speers-Roesch, J. G. Richards, Y. Wang, A. P. Farrell, and C. J. Brauner (2012). A radical approach to beating hypoxia: Depressed free radical release from heart fibres of the hypoxia-tolerant epaulette shark (*Hemiscyllum ocellatum*). Journal of Comparative Physiology B: Biochemical, Systemic, and Environmental Physiology 182:91–100.

Hill, G. E. (1993). Male mate choice and the evolution of female plumage coloration in the house finch. Evolution 47:1515–25.

Hill, G. E. (1994a). Trait elaboration via adaptive mate choice: Sexual conflict in the evolution of signals of male quality. Ethology Ecology and Evolution 6:351–70.

Hill, G. E. (1994b). Geographic variation in male ornamentation and female mate preference in the house finch—A comparative test of models of sexual selection. Behavioral Ecology 5:64–73.

Hill, G. E. (2002). A Red Bird in a Brown Bag: The Function and Evolution of Colorful Plumage in the House Finch. Oxford University Press, Oxford, UK.

Hill, G. E. (2006a). Environmental regulation of ornamental coloration. In Bird Coloration: Vol. I. Mechanisms and Measurements (G. E. Hill and K. J. McGraw, Editors). Harvard University Press, Cambridge MA, pp. 507–60.

Hill, G. E. (2006b). Female mate choice for ornamental coloration. In Bird Coloration: Vol. II. Function and Evolution (G. E. Hill and K. J. McGraw, Editors). Harvard University Press, Cambridge MA, pp. 137–200.

Hill, G. E. (2014a). Sex linkage of nuclear-encoded mitochondrial genes. Heredity 112:469–70.

Hill, G. E. (2014b). Cellular respiration: The nexus of stress, condition, and ornamentation. Integrative and Comparative Biology 54:645–57.

Hill, G. E. (2014c). The evolution of armaments and ornaments. In How and Why Animals Do the Things They Do. Volume 2: Function and Evolution (K. Yasakawa, Editor). Praeger/ABC-CLIO, New York, pp. 145–172.

Hill, G. (2015a). Selection for reinforcement versus selection for signals of quality and attractiveness. Ideas in Ecology and Evolution 8:67–9.

Hill, G. E. (2015b). Mitonuclear ecology. Molecular Biology and Evolution 32:1917–27.

Hill, G. E. (2015c). Sexiness, individual condition, and species identity: The information signaled by ornaments and assessed by choosing females. Evolutionary Biology 42:251–9.

Hill, G. E. (2016). Mitonuclear coevolution as the genesis of speciation and the mitochondrial DNA barcode gap. Ecology and Evolution 6:5831–42.

Hill, G. E. (2017). The mitonuclear compatibility species concept. The Auk 134:393–409.

Hill, G. E. (2018). Mitonuclear mate choice: A missing component of sexual selection theory? BioEssays 40: 1700191.

Hill, G. E., and J. D. Johnson (2012). The vitamin A-redox hypothesis: A biochemical basis for honest signaling via carotenoid pigmentation. The American Naturalist 180:E127–E150.

Hill, G. E., and J. D. Johnson (2013). The mitonuclear compatibility hypothesis of sexual selection. Proceedings of the Royal Society B: Biological Sciences 280:20131314.

Hill, G. E., and K. J. Mcgraw (2004). Correlated changes in male plumage coloration and female mate choice in cardueline finches. Animal Behaviour 67:27–35.

Hill, G. E., and R. M. Zink (2018). Hybrid speciation in birds, with special reference to Darwin's finches. Journal of Avian Biology 49: e01879.

Hill, G. E., J. C. Havird, D. B. Sloan, R. S. Burton, C. Greening, and D. K. Dowling (2019). Assessing the fitness consequences of mitonuclear interactions in natural populations. Biological Reviews.

Hill, W. G., and A. Robertson (1966). The effect of linkage on limits to artificial selection. Genetical Research 8:269.

Hillis, D. M., and M. T. Dixon (1991). Ribosomal DNA : Molecular evolution and phylogenetic inference. Quarterly Review of Biology 66:411–53.

Hilton, Z., K. D. Clements, and A. J. R. Hickey (2010). Temperature sensitivity of cardiac mitochondria in intertidal and subtidal triplefin fishes. Journal of Comparative Physiology B: Biochemical, Systemic, and Environmental Physiology 180:979–90.

Hirst, J. (2011). Why does mitochondrial complex I have so many subunits? Biochemical Journal 437:e1–e3.

Hirst, J. (2013). Mitochondrial complex I. Annual Review of Biochemistry 82:551–75.

Hochachka, P. W., and G. N. Somero (2002). Biochemical Adaptation: Mechanism and Process in Physiological Evolution. Oxford University Press, New York.

Hoekstra, L. A., M. A. Siddiq, and K. L. Montooth (2013). Pleiotropic effects of a mitochondrial-nuclear incompatibility depend upon the accelerating effect of temperature in Drosophila. Genetics 195:1129–39.

Hoekstra, R. F. (1987). The evolution of sexes. In The Evolution of Sex and Its Consequences (S. C. Stearns, Editor). Springer, New York, pp. 59–91.

Hohtola, E., and H. Visser (1998). Development of locomotion and endothermy in altricial and precocial birds. Oxford Ornithology Series 8:157–73.

Horn, R., K. J. Gupta, and N. Colombo (2014). Mitochondrion role in molecular basis of cytoplasmic male sterility. Mitochondrion 19:198–205.

Houtman, A. M., and J. B. Falls (1994). Negative assortative mating in the white-throated sparrow, *Zonotrichia albicollis*: The role of mate choice and intra-sexual competition. Animal Behaviour 48:377–83.

Howard, R. S., and C. M. Lively (1994). Parasitism, mutation accumulation and the maintenance of sex. Nature 367:554–7.

Huang, S., and A. H. Millar (2013). Succinate dehydrogenase: The complex roles of a simple enzyme. Current Opinion in Plant Biology 16:344–9.

Hubert, N., and R. Hanner (2015). DNA barcoding, species delineation and taxonomy: A historical perspective. DNA Barcodes 3:44–58.

Hudson, E. J., and T. D. Price (2014). Pervasive reinforcement and the role of sexual selection in biological speciation. Journal of Heredity 105:821–33.

Hudson, G., S. Keers, P. Yu-Wai-Man, P. Griffiths, K. Huoponen, M.-L. Savontaus, E. Nikoskelainen, M. Zeviani, F. Carrara, R. Horvath, V. Karcagi, et al. (2005). Identification of an X-chromosomal locus and haplotype modulating the phenotype of a mitochondrial DNA disorder. American Journal of Human Genetics 77:1086–91.

Hudson, R. R., and J. A. Coyne (2002). Mathematical consequences of the genealogical species concept. Evolution 56:1557–65.

Hudson, R. R., and M. Turelli (2003). Stochasticity overrules the "three-times rule": Genetic drift, genetic draft, and coalescence times for nuclear loci versus mitochondrial DNA. Evolution 57:182–90.

Hui, J. H. L. (2018). Copepod incompatibilities. Nature Ecology and Evolution 2:1203–4.

Hulbert, A. J. (2008). The links between membrane composition, metabolic rate and lifespan. Comparative Biochemistry and Physiology—A Molecular and Integrative Physiology 150:196–203.

Hulbert, A. J., and P. L. Else (1999). Membranes as possible pacemakers of metabolism. Journal of Theoretical Biology 199:257–74.

Hulsey, C. D., K. L. Bell, F. J. García-de-León, C. C. Nice, and A. Meyer (2016). Do relaxed selection and habitat temperature facilitate biased mitogenomic introgression in a narrowly endemic fish? Ecology and Evolution 6:3684–98.

Hung, C.-M., and R. M. Zink (2014). Distinguishing the effects of selection from demographic history in the genetic variation of two sister passerines based on mitochondrial-nuclear comparison. Heredity 113:1–10.

Hung, C.-M., S. V. Drovetski, and R. M. Zink (2016). Matching loci surveyed to questions asked in phylogeography. Proceedings of the Royal Society B: Biological Sciences 283:20152340.

Hurst, G. D. D., and C. L. Frost (2015). Reproductive parasitism: Maternally inherited symbionts in a biparental world. Cold Spring Harbor Perspectives in Biology 7:1–21.

Hurst, L. D., and W. D. Hamilton (1992). Cytoplasmic fusion and the nature of sexes. Proceedings of the Royal Society B: Biological Sciences 247:189–94.

Hutson, V., and R. Law (1993). Four steps to two sexes. Proceedings of the Royal Society B: Biological Sciences 253:43–51.

Huynen, M. A., T. Dandekar, and P. Bork (1999). Variation and evolution of the citric-acid cycle: A genomic perspective. Trends in Microbiology 7:281–91.

Hyatt, H. W., Y. Zhang, W. R. Hood, and A. N. Kavazis (2017). Lactation has persistent effects on a mother's metabolism and mitochondrial function. Scientific Reports 7:17118.

Hyvärinen, A. K., J. L. O. Pohjoismäki, A. Reyes, S. Wanrooij, T. Yasukawa, P. J. Karhunen, J. N. Spelbrink, I. J. Holt, and H. T. Jacobs. (2007). The mitochondrial transcription termination factor mTERF modulates replication pausing in human mitochondrial DNA. Nucleic Acids Research 35(19): 6458–74.

Immonen, E., J. Rönn, C. Watson, D. Berger, and G. Arnqvist (2016). Complex mitonuclear interactions and metabolic costs of mating in male seed beetles. Journal of Evolutionary Biology 29:360–70.

Irwin, D. E. (2018). Sex chromosomes and speciation in birds and other ZW systems. Molecular Ecology 27:3831–3851.

Irwin, D. E., A. Brelsford, D. P. L. Toews, C. MacDonald, and M. Phinney (2009). Extensive hybridization in a contact zone between MacGillivray's warblers *Oporornis tolmiei* and mourning warblers *O. philadelphia* detected using molecular and morphological analyses. Journal of Avian Biology 40:539–52.

Ishizaki, K., T. R. T. Larson, N. Schauer, A. R. Fernie, I. A. Graham, and C. J. Leaver (2005). The critical role of Arabidopsis electron-transfer flavoprotein: Ubiquinone oxidoreductase during dark-induced starvation. The Plant Cell 17:2587–600.

Itoh, T., W. Martin, and M. Nei (2002). Acceleration of genomic evolution caused by enhanced mutation rate in endocellular symbionts. Proceedings of the National Academy of Sciences 99:12944–8.

Itsara, L. S., S. R. Kennedy, E. J. Fox, S. Yu, J. J. Hewitt, M. Sanchez-Contreras, F. Cardozo-Pelaez, and L. J. Pallanck (2014). Oxidative stress is not a major contributor to somatic mitochondrial DNA mutations. PLoS Genetics 10:e1003974.

Iwata, S. (1998). Complete structure of the 11-subunit bovine mitochondrial cytochrome bc1 complex. Science 281:64–71.

Jacoby, R. P., N. L. Taylor, and A. H. Millar (2011). The role of mitochondrial respiration in salinity tolerance. Trends in Plant Science 16:614–23.

James, J. E., G. Piganeau, and A. Eyre-Walker (2016). The rate of adaptive evolution in animal mitochondria. Molecular Ecology 25:67–78.

Jastroch, M., A. S. Divakaruni, S. Mookerjee, J. R. Treberg, and M. D. Brand (2010). Mitochondrial proton and electron leaks. Essays in Biochemistry 47:53–67.

Jhuang, H., H. Lee, and J. Leu (2017). Mitochondrial–nuclear co-evolution leads to hybrid incompatibility through pentatricopeptide repeat proteins. EMBO Reports 18:87–101.

Ji, F., M. S. Sharpley, O. Derbeneva, L. S. Alves, P. Qian, Y. Wang, D. Chalkia, M. Lvova, J. Xu, W. Yao, M. Simon, et al. (2012). Mitochondrial DNA variant associated with Leber hereditary optic neuropathy and high-altitude Tibetans. Proceedings of the National Academy of Sciences 109:7391–6.

Johannsen, W. (1923). Some remarks about units in heredity. Hereditas 4:133–41.

Johnson, J. D., and G. E. Hill (2013). Is carotenoid ornamentation linked to the inner mitochondria membrane potential? A hypothesis for the maintenance of signal honesty. Biochimie 95:436–44.

Johnson, N. A. (2010). Hybrid incompatibility genes: Remnants of a genomic battlefield? Trends in Genetics 26:317–25.

Johnstone, R. A., J. D. Reynolds, and J. C. Deutsch (1996). Mutual mate choice and sex differences in choosiness. Evolution 50:1382.

Jones, A. G., and N. L. Ratterman (2009). Mate choice and sexual selection: What have we learned since Darwin? Proceedings of the National Academy of Sciences 106:10001–8.

Jones, A. G., D. Walker, and J. C. Avise (2001). Genetic evidence for extreme polyandry and extraordinary sex-role reversal in a pipefish. Proceedings of the Royal Society B: Biological Sciences 268:2531–5.

Jones, O. R., A. Scheuerlein, R. Salguero-Gómez, C. G. Camarda, R. Schaible, B. B. Casper, J. P. Dahlgren, J. Ehrlén, M. B. García, E. S. Menges, P. F. Quintana-Ascencio, et al. (2014). Diversity of ageing across the tree of life. Nature 505:169–73.

de Juan, D., F. Pazos, and A. Valencia (2013). Emerging methods in protein co-evolution. Nature Reviews Genetics 14:249–61.

Kadenbach, B., M. Hüttemann, S. Arnold, I. Lee, and E. Bender (2000). Mitochondrial energy metabolism is regulated via nuclear-coded subunits of cytochrome c oxidase. Free Radical Biology and Medicine 29:211–21.

Kao, S. H., H. T. Chao, and Y. H. Wei (1995). Mitochondrial deoxyribonucleic-acid 4977-bp deletion is associated with diminished fertility and motility of human sperm. Biology of Reproduction 52:729–36.

Kazancioğlu, E., and G. Arnqvist (2014). The maintenance of mitochondrial genetic variation by negative frequency-dependent selection. Ecology Letters 17:22–7.

Keene, J. D. (2007). RNA regulons: Coordination of post-transcriptional events. Nature Reviews Genetics 8:533–43.

Kenyon, L., and C. T. Moraes (1997). Expanding the functional human mitochondrial DNA database by the establishment of primate xenomitochondrial cybrids. Proceedings of the National Academy of Sciences 94:9131–5.

Kern, A. D., and M. W. Hahn (2018). The neutral theory in light of natural selection. Molecular Biology and Evolution 35:1366–1371.

Kern, A. D., and F. A. Kondrashov (2004). Mechanisms and convergence of compensatory evolution in mammalian mitochondrial tRNAs. Nature Genetics 36:1207–12.

Kerr, K. C. R. (2011). Searching for evidence of selection in avian DNA barcodes. Molecular Ecology Resources 11:1045–55.

Kerr, K. C. R., M. Y. Stoeckle, C. J. Dove, L. A. Weigt, C. M. Francis, and P. D. N. Hebert (2007). Comprehensive DNA barcode coverage of North American birds. Molecular Ecology Notes 7:535–43.

Kim, J. W., I. Tchernyshyov, G. L. Semenza, and C. V. Dang (2006). HIF-1-mediated expression of pyruvate dehydrogenase kinase: A metabolic switch required for cellular adaptation to hypoxia. Cell Metabolism 3:177–85.

Kimura, M. (1977). Preponderance of synonymous changes as evidence for the neutral theory of molecular evolution. Nature 267:275–6.

Kimura, M. (1983). The Neutral Theory of Molecular Evolution. Cambridge University Press, Cambridge, UK.

Kingston, S. E., R. W. Jernigan, W. F. Fagan, D. Braun, and M. J. Braun (2012). Genomic variation in cline shape across a hybrid zone. Ecology and Evolution 2:2737–48.

Kirkpatrick, M. (1987). Sexual selection by female choice in polygynous animals. Annual Review of Ecology and Systematics 18:43–70.

Kirkpatrick, M., and D. W. Hall (2004). Sexual selection and sex linkage. Evolution 58:683.

Kivisild, T., P. Shen, D. P. Wall, B. Do, R. Sung, K. Davis, G. Passarino, P. A. Underhill, C. Scharfe, A. Torroni, R. Scozzari, et al. (2006). The role of selection in the evolution of human mitochondrial genomes. Genetics 172:373–87.

Klein, R. R., J. J. Burke, and R. F. Wilson (1986). Effect of osmotic stress on ion transport processes and phospholipid composition of wheat (*Triticum aestivum* L.) mitochondria. Plant Physiology 82:936–41.

Kleine, T., U. G. Maier, and D. Leister (2009). DNA transfer from organelles to the nucleus: The idiosyncratic genetics of endosymbiosis. Annual Review of Plant Biology 60:115–38.

Klopfstein, S., C. Kropf, and H. Baur (2016). *Wolbachia* endosymbionts distort DNA barcoding in the parasitoid wasp genus *Diplazon* (Hymenoptera: Ichneumonidae). Zoological Journal of the Linnean Society 177:541–57.

Knapton, R. W., and J. B. Falls (1983). Differences in parental contribution among pair types in the polymorphic white-throated sparrow. Canadian Journal of Zoology 61:1288–92.

Koch, R. E., and G. E. Hill (2018). Behavioural mating displays depend on mitochondrial function: A potential mechanism for linking behaviour to individual condition. Biological Reviews 93:1387–98.

Koch, R. E., C. Josefsson, and G. E. Hill (2017). Mitochondrial function, ornamentation, and immunocompetence. Biological Reviews 92:1459–74.

Kodric-Brown, A., and J. H. Brown (1984). Truth in advertising: The kinds of traits favored by sexual selection. The American Naturalist 124:309–23.

Koenig, W. (1982). Ecological and social factors affecting hatchability of eggs. The Auk 99:526–36.

Kokko, H., R. Brooks, M. D. Jennions, and J. Morley (2003). The evolution of mate choice and mating biases. Proceedings of the Royal Society B: Biological Sciences 270:653–64.

Kondrashov, A. S. (1988). Deleterious mutations and the evolution of sexual reproduction. Nature 336:435–40.

Kondrashov, A. S., S. Sunyaev, and F. A. Kondrashov (2002). Dobzhansky-Muller incompatibilities in protein evolution. Proceedings of the National Academy of Science 99:14878–83.

Kondrashov, F. A., and E. V. Koonin (2004). A common framework for understanding the origin of genetic dominance and evolutionary fates of gene duplications. Trends in Genetics 20:287–91.

Konovalova, S., and H. Tyynismaa (2013). Mitochondrial aminoacyl-tRNA synthetases in human disease. Molecular Genetics and Metabolism 108:206–11.

Koonin, E. V. (2009). Evolution of genome architecture. The International Journal of Biochemistry and Cell Biology 41:298–306.

Koonin, E. V. (2010). The origin and early evolution of eukaryotes in the light of phylogenomics. Genome Biology 11:209.

Koonin, E. V., and Y. I. Wolf (2008). Genomics of bacteria and archaea: The emerging dynamic view of the prokaryotic world. Nucleic Acids Research 36:6688–719.

Kovalchuk, I. (2016). Genome stability: An evolutionary perspective. In Genome Stability (I. Kovalchuk and O. Kovalchuk, Editors). Elsevier, Amsterdam, pp. 1–18.

Krakauer, D. C., and A. Mira (1999). Mitochondria and germ-cell death. Nature 400:125–6.

Kühlbrandt, W. (2014). The resolution revolution. Science 343:1443–4.

Kuijper, B., I. Pen, and F. J. Weissing (2012). A guide to sexual selection theory. Annual Review of Ecology, Evolution, and Systematics 43:287–311.

Kujoth, G. C., A. Hiona, T. D. Pugh, S. Someya, K. Panzer, S. E. Wohlgemuth, T. Hofer, A. Y. Seo, R. Sullivan, W. A. Jobling, J. D. Morrow, et al. (2005). Mitochondrial DNA mutations, oxidative stress, and apoptosis in mammalian aging. Science 309:481–4.

Kulathinal, R. J., B. R. Bettencourt, and D. L. Hartl (2004). Compensated deleterious mutations in insect genomes. Science 306:1553–4.

Kvist, L., and S. Rytkonen (2006). Characterization of a secondary contact zone of the great tit *Parus major* and the Japanese tit *P. minor* (Aves: Passeriformes) in Far Eastern Siberia with DNA markers. Zootaxa 73:55–73.

Kwong, S., A. Srivathsan, G. Vaidya, and R. Meier (2012). Is the COI barcoding gene involved in speciation through intergenomic conflict? Molecular Phylogenetics and Evolution 62:1009–12.

Lachaise, D., M.-L. Cariou, J. R. David, F. Lemeunier, L. Tsacas, and M. Ashburner (1988). Historical biogeography of the *Drosophila melanogaster* species subgroup. Evolutionary Biology 22:159–225.

Lagouge, M., and N. G. Larsson (2013). The role of mitochondrial DNA mutations and free radicals in disease and ageing. Journal of Internal Medicine 273:529–43.

Lai, L. B., A. Vioque, L. A. Kirsebom, and V. Gopalan (2010). Unexpected diversity of RNase P, an ancient tRNA processing enzyme: Challenges and prospects. FEBS Letters 584:287–96.

Lajbner, Z., R. Pnini, M. Camus, J. Miller, and D. Dowling (2018). Experimental evidence that thermal selection shapes mitochondrial genome evolution. Scientific Reports 8:9500.

Lamb, A. M., H. M. Gan, C. Greening, L. Joseph, Y. P. Lee, A. Morán-Ordóñez, P. Sunnucks, and A. Pavlova (2018). Climate-driven mitochondrial selection: A test in Australian songbirds. Molecular Ecology 27:898–918.

Lambert, J. D., and N. A. Moran (1998). Deleterious mutations destabilize ribosomal RNA in endosymbiotic bacteria. Proceedings of the National Academy of Sciences 95:4458–62.

Lamichhaney, S., J. Berglund, M. S. Almén, K. Maqbool, M. Grabherr, A. Martinez-Barrio, M. Promerová, C. J. Rubin, C. Wang, N. Zamani, B. R. Grant, et al. (2015). Evolution of Darwin's finches and their beaks revealed by genome sequencing. Nature 518:371–5.

Lamichhaney, S., F. Han, M. T. Webster, L. Andersson, B. R. Grant, and P. R. Grant (2018). Rapid hybrid speciation in Darwin's finches. Science 359:224–8.

Lande, R. (1981). Models of speciation by sexual selection on polygenic traits. Proceedings of the National Academy of Sciences 78:3721–5.

Lande, R. (1994). Risk of population extinction from fixation of new deleterious mutations. Evolution 48:1460–9.

Lane, N. (2002). Oxygen: The Molecule That Made the World. Oxford University Press, Oxford, UK.

Lane, N. (2005). Power, Sex, Suicide: Mitochondria and the Meaning of Life. Oxford University Press, New York.

Lane, N. (2006). Cell biology: Power games. Nature 443:901–3.

Lane, N. (2009). Biodiversity: On the origin of bar codes. Nature 462:272–4.

Lane, N. (2011a). Energetics and genetics across the prokaryote-eukaryote divide. Biology Direct 6:35.

Lane, N. (2011b). The costs of breathing. Science 334:184–5.

Lane, N. (2011c). Mitonuclear match: Optimizing fitness and fertility over generations drives ageing within generations. Bioessays 33:860–9.

Lane, N. (2014). Bioenergetic constraints on the evolution of complex life. Cold Spring Harbor Perspectives in Biology 6:a015982.

Lane, N. (2015a). The Vital Question: Energy, Evolution, and the Origins of Complex Life. WW Norton & Company, New York.

Lane, N. (2015b). Life force: Why energy shapes evolution. The Biochemist 37:6–11.

Lane, N. (2018). Hot mitochondria? PLoS Biology 16:e2005113.

Lane, N., and W. Martin (2010). The energetics of genome complexity. Nature 467:929–34.

Lang, A. S., O. Zhaxybayeva, and J. T. Beatty (2012). Gene transfer agents: Phage-like elements of genetic exchange. Nature Reviews Microbiology 10:472–82.

Lang, B. F., M. W. Gray, and G. Burger (1999). Mitochondrial genome evolution and the origin of eukaryotes. Annual Review of Genetics 33:351–97.

Lannig, G., L. G. Eckerle, I. Serendero, F. J. Sartoris, T. Fischer, R. Knust, T. Johansen, and H. O. Pörtner (2003). Temperature adaptation in eurythermal cod (*Gadus morhua*): A comparison of mitochondrial enzyme capacities in boreal and Arctic populations. Marine Biology 142:589–99.

Lannig, G., D. Storch, and H. O. Pörtner (2005). Aerobic mitochondrial capacities in Antarctic and temperate eelpout (Zoarcidae) subjected to warm versus cold acclimation. Polar Biology 28:575–84.

Lanza, R. P., B. L. Dresser, and P. Damiani (2000). Cloning Noah's ark. Scientific American 283:84–9.

Lapointe, J., and S. Hekimi (2010). When a theory of aging ages badly. Cellular and Molecular Life Sciences 67:1–8.

Larsson, N.-G. (2010). Somatic mitochondrial DNA mutations in mammalian aging. Annual Review of Biochemistry 79:683–706.

Laser, K. D., and N. R. Lersten (1972). Anatomy and cytology of microsporogenesis in cytoplasmic male sterile angiosperms. The Botanical Review 38:425–54.

Latorre-Pellicer, A., R. Moreno-Loshuertos, A. V. Lechuga-Vieco, F. Sánchez-Cabo, C. Torroja, R. Acín-Pérez, E. Calvo, E. Aix, A. González-Guerra, A. Logan, M. L. Bernad-Miana, et al. (2016). Mitochondrial and nuclear DNA matching shapes metabolism and healthy ageing. Nature 535:561–5.

Lau, G. Y., M. Mandic, and J. G. Richards (2017). Evolution of cytochrome c oxidase in hypoxia tolerant sculpins (Cottidae, Actinopterygii). Molecular Biology and Evolution 34:2153–62.

Lavretsky, P., J. M. Dacosta, B. E. Hernández-Baños, A. Engilis, M. D. Sorenson, and J. L. Peters (2015). Speciation genomics and a role for the Z chromosome in the early stages of divergence between Mexican ducks and mallards. Molecular Ecology 24:5364–78.

Lazarou, M., D. R. Thorburn, M. T. Ryan, and M. McKenzie (2009). Assembly of mitochondrial complex I and defects in disease. Biochimica et Biophysica Acta—Molecular Cell Research 1793:78–88.

Lee, C., K. Yen, and P. Cohen (2013). Humanin: A harbinger of mitochondrial-derived peptides? Trends in Endocrinology and Metabolism 24:222–8.

Lee, H. Y., J. Y. Chou, L. Cheong, N. H. Chang, S. Y. Yang, and J. Y. Leu (2008). Incompatibility of nuclear and mitochondrial genomes causes hybrid sterility between two yeast species. Cell 135:1065–73.

Letts, J. A., and L. A. Sazanov (2017). Clarifying the supercomplex: The higher-order organization of the mitochondrial electron transport chain. Nature Structural and Molecular Biology 24:800–8.

Levin, D. A. (2003). The cytoplasmic factor in plant speciation. Systematic Botany 28:5–11.

Lewontin, R. (1975). The genetic basis of evolutionary change. The American Journal of Human Genetics 27:249–51.

Li, W. H. (1997). Molecular Evolution. Sinauer, Sunderland, MA.

Li, Y., D. D. Wu, A. R. Boyko, G. D. Wang, S. F. Wu, D. M. Irwin, and Y. P. Zhang (2014). Population variation revealed high-altitude adaptation of Tibetan mastiffs. Molecular Biology and Evolution 31:1200–5.

Liebenberg, L. (2008). The relevance of persistence hunting to human evolution. Journal of Human Evolution 55:1156–9.

Lightowlers, R. N., R. W. Taylor, and D. M. Turnbull (2015). Mutations causing mitochondrial disease: What is new and what challenges remain? Science 349:1494–9.

Ling, J., N. Reynolds, and M. Ibba (2009). Aminoacyl-tRNA synthesis and translational quality control. Annual Review of Microbiology 63:61–78.

Litonin, D., M. Sologub, Y. Shi, M. Savkina, M. Anikin, M. Falkenberg, C. M. Gustafsson, and D. Temiakov (2010). Human mitochondrial transcription revisited: Only TFAM and TFB2M are required for transcription of the mitochondrial genes in vitro. Journal of Biological Chemistry 285:18129–33.

Little, A. G., G. Lau, K. E. Mathers, S. C. Leary, and C. D. Moyes (2017). Comparative biochemistry of cytochrome c oxidase in animals. Comparative Biochemistry and Physiology Part—B: Biochemistry and Molecular Biology 224:170–84.

Liu, P., and B. Demple (2010). DNA repair in mammalian mitochondria: Much more than we thought? Environmental and Molecular Mutagenesis 51:417–26.

Liu, Y., and X. J. Chen (2013). Adenine nucleotide translocase, mitochondrial stress, and degenerative cell death. Oxidative Medicine and Cellular Longevity 2013:146860.

Lively, C. M., and L. T. Morran (2014). The ecology of sexual reproduction. Journal of Evolutionary Biology 27:1292–303.

Lleonart, M. E., R. Grodzicki, D. M. Graifer, and A. Lyakhovich (2017). Mitochondrial dysfunction and potential anticancer therapy. Medicinal Research Reviews 37:1275–98.

Loewe, L. (2006). Quantifying the genomic decay paradox due to Muller's ratchet in human mitochondrial DNA. Genetical Research 87:133–59. doi:10.1017/S0016672306008123

Long, Y. C., T. M. Tan, I. Takao, and B. L. Tang (2014). The biochemistry and cell biology of aging: Metabolic regulation through mitochondrial signaling. American Journal of Physiology—Endocrinology and Metabolism 306:E581–E591.

Lopes, R. J., J. D. Johnson, M. B. Toomey, M. S. Ferreira, P. M. Araujo, J. Melo-Ferreira, L. Andersson, G. E. Hill, J. C. Corbo, and M. Carneiro (2016). Genetic basis for red coloration in birds. Current Biology 26:1427–34.

López-Fernández, H., and D. I. Bolnick (2007). What causes partial F1 hybrid viability? Incomplete penetrance versus genetic variation. PLoS ONE 2:1–8.

Losos, J. B. (2010). Adaptive radiation, ecological opportunity, and evolutionary determinism. The American Naturalist 175:623–39.

Lott, M.T., J. N. Leipzig, O. Derbeneva, H. M. Xie, D. Chalkia, M. Sarmady, V. Procaccio, and D. C. Wallace (2013). mtDNA variation and analysis using MITOMAP and MITOMASTER. Current Protocols in Bioinformatics 123:1–26.

Lotz, C., A. J. Lin, C. M. Black, J. Zhang, E. Lau, N. Deng, Y. Wang, N. C. Zong, J. H. Choi, T. Xu, D. A. Liem, et al. (2014). Characterization, design, and function of the mitochondrial proteome: From organs to organisms. Journal of Proteome Research 13:433–46.

Luo, Y., Y. Chen, F. Liu, C. Jiang, and Y. Gao (2011). Mitochondrial genome sequence of the Tibetan wild ass (*Equus kiang*). Mitochondrial DNA 22:6–8.

Luo, Y., X. Yang, and Y. Gao (2013). Mitochondrial DNA response to high altitude: A new perspective on high-altitude adaptation. Mitochondrial DNA 24:313–19.

Lv, J., S. Wu, Y. Zhang, Y. Chen, C. Feng, X. Yuan, G. Jia, J. Deng, C. Wang, Q. Wang, L. Mei, and X. Lin (2014). Assessment of four DNA fragments (COI, 16S rDNA, ITS2, 12S rDNA) for species identification of the Ixodida (Acari: Ixodida). Parasites and Vectors 7:7–93.

Ly, J. D., D. R. Grubb, and A. Lawen (2003). The mitochondrial membrane potential in apoptosis; an update. Apoptosis 8:115–28.

Lymbery, R. A., W. J. Kennington, and J. P. Evans (2017). Egg chemoattractants moderate intraspecific sperm competition. Evolution Letters 1:317–27.

Lynch, M. (1997). Mutation accumulation in nuclear, organelle, and prokaryotic transfer RNA genes. Molecular Biology and Evolution 14:914–25.

Lynch, M., and J. L. Blanchard (1998). Deleterious mutation accumulation in organelle genomes. Genetica 102:29–39.

Lynch, M., and K. Hagner (2015). Evolutionary meandering of intermolecular interactions along the drift barrier. Proceedings of the National Academy of Sciences 112:E30–E38.

Lynch, M., B. Koskella, and S. Schaack (2006). Mutation pressure and the evolution of organelle genomic architecture. Science 311:1727–30.

Lynch, M., A. Seyfert, B. Eads, and E. Williams (2008). Localization of the genetic determinants of meiosis suppression in *Daphnia pulex*. Genetics 180:317–27.

Ma, H., N. Marti Gutierrez, R. Morey, C. Van Dyken, E. Kang, T. Hayama, Y. Lee, Y. Li, R. Tippner-Hedges, D. P. Wolf, L. C. Laurent, and S. Mitalipov (2016). Incompatibility between nuclear and mitochondrial genomes contributes to an interspecies reproductive barrier. Cell Metabolism 24:283–94.

Magnúsdóttir, E., and M. A. Surani (2014). How to make a primordial germ cell. Development 141:245–52.

Maheshwari, S., and D. A. Barbash (2011). The genetics of hybrid incompatibilities. Annual Review of Genetics 45:331–55.

Mailloux, R. J., and M.-E. Harper (2011). Uncoupling proteins and the control of mitochondrial reactive oxygen species production. Free Radical Biology and Medicine 51:1106–15.

Man, P. Y., D. M. Turnbull, and P. F. Chinnery (2002). Leber hereditary optic neuropathy. Journal of Medical Genetics 39:162–9.

Mank, J. E., E. Axelsson, and H. Ellegren (2007). Fast-X on the Z: Rapid evolution of sex-linked genes in birds. Genome Research 17:618–24.

Mank, J. E., D. J. Hosken, and N. Wedell (2014). Conflict on the sex chromosomes: Cause, effect, and complexity. Cold Spring Harbor Perspectives in Biology 6:a017715.

Margulis, L. (1970). Origin of Eukaryotic Cells. Yale University Press, New Haven, CT.

Maricle, B. R., J. J. Crosier, B. C. Bussiere, and R. W. Lee (2006). Respiratory enzyme activities correlate with anoxia tolerance in salt marsh grasses. Journal of Experimental Marine Biology and Ecology 337:30–7.

Martikainen, M. H., J. P. Grady, Y. S. Ng, C. L. Alston, G. S. Gorman, R. W. Taylor, R. McFarland, and D. M. Turnbull (2017). Decreased male reproductive success in association with mitochondrial dysfunction. European Journal of Human Genetics 25:1163–5.

Martin, A. P. (1995). Metabolic rate and directional nucleotide substitution in animal mitochondrial DNA. Molecular Biology and Evolution 12:1124–31.

Martin, P. R. (2015). The paradox of the birds-of-paradise: Persistent hybridization as a signature of historical reinforcement. Ideas in Ecology and Evolution 8:1. 58–66.

Martin, W., and E. V. Koonin (2006). Introns and the origin of nucleus–cytosol compartmentalization. Nature 440:41–5.

Martin, W., and M. Müller (1998). The hydrogen hypothesis for the first eukaryote. Nature 392:37–41.

Martin, W., S. Garg, and V. Zimorski (2015). Endosymbiotic theories for eukaryote origin. Philosophical Transactions of the Royal Society B 370:20140330.

Mata, J. (2016). Translation factors specify cellular metabolic state. Cell Reports 16:1787–8.

Matysioková, B., N. Friedman, L. Turčoková, and V. Remeš (2017). The evolution of feather coloration and song in Old World orioles (genus *Oriolus*). Journal of Avian Biology 48:1015–24.

Matyushov, D. V. (2013). Protein electron transfer: Dynamics and statistics. Journal of Chemical Physics 139:07B606_1.

Maynard Smith, J. (1978). The Evolution of Sex. Cambridge University Press, Cambridge, UK.

Maynard Smith, J. (1991). Theories of sexual selection. Trends in Ecology and Evolution 6:146–51.

Maynard Smith, J., and J. Haigh (1974). The hitch-hiking effect of a favorable gene. Genetical Research 23:23–35.

Maynard Smith, J., and E. Szathmary (1998). The Major Transitions of Evolution. Oxford University Press, Oxford, UK.

Mayr, E. (1942). Systematics and the Origin of Species. Columbia University Press, New York.

Mayr, E. (1963). Animal Species and Evolution. Harvard University Press, Cambridge, MA.

Mayr, E. (2000). The biological species concept. In Species Concepts and Phylogenetic Theory: A Debate (Q. D. Wheeler and R. Meier, Editors). Columbia University Press, New York, pp. 17–29.

McCutcheon, J. P., and N. A. Moran (2012). Extreme genome reduction in symbiotic bacteria. Nature Reviews Microbiology 10:13–26.

McFarland, R., R. W. Taylor, and D. M. Turnbull (2007). Mitochondrial disease—its impact, etiology, and pathology. Current Topics in Developmental Biology 77:113–55.

Mcgraw, K. J. (2006). Mechanics of carotenoid coloration. In Bird Coloration, Vol. 1: Measurements and Mechanisms (G. E. Hill and K. J. McGraw, Editors). Harvard University Press, Cambridge MA, pp. 177–242.

McKay, B. D., and R. M. Zink (2010). The causes of mitochondrial DNA gene tree paraphyly in birds. Molecular Phylogenetics and Evolution 54:647–50.

McKay, B. D., and R. M. Zink (2015). Sisyphean evolution in Darwin's finches. Biological Reviews 90:689–98.

McKenzie, M., and I. Trounce (2000). Expression of *Rattus norvegicus* mtDNA in *Mus musculus* cells results in multiple respiratory chain defects. Journal of Biological Chemistry 275:31514–19.

McKenzie, M., M. Chiotis, C. A. Pinkert, and I. A. Trounce (2003). Functional respiratory chain analyses in murid xenomitochondrial cybrids expose coevolutionary constraints of cytochrome b and nuclear subunits of complex III. Molecular Biology and Evolution 20:1117–24.

McNab, B. K., and P. Morrison (1963). Body temperature and metabolism in subspecies of Peromyscus from arid and mesic environments. Ecological Monographs 33:63–82.

McVean, G. A. T., and B. Charlesworth (2000). The effects of Hill-Robertson interference between weakly selected mutations on patterns of molecular evolution and variation. Genetics 155:929–44.

Medawar, P. B. (1946). Old age and natural death. Modern Quarterly 2:30–49.

Meer, M. V., A. S. Kondrashov, Y. Artzy-Randrup, and F. A. Kondrashov (2010). Compensatory evolution in mitochondrial tRNAs navigates valleys of low fitness. Nature 464:279–82.

Meiklejohn, C. D., K. L. Montooth, and D. M. Rand (2007). Positive and negative selection on the mitochondrial genome. Trends in Genetics 23:259–63.

Meiklejohn, C. D., M. A. Holmbeck, M. A. Siddiq, D. N. Abt, D. M. Rand, and K. L. Montooth (2013). An incompatibility between a mitochondrial tRNA and its nuclear-encoded tRNA synthetase compromises development and fitness in *Drosophila*. PLoS Genetics 9:e1003238.

Meisinger, C., A. Sickmann, and N. Pfanner (2008). The mitochondrial proteome: From inventory to function. Cell 134:22–4.

Melnikov, S., A. Ben-Shem, N. Garreau de Loubresse, L. Jenner, G. Yusupova, and M. Yusupov (2012). One core, two shells: Bacterial and eukaryotic ribosomes. Nature Structural and Molecular Biology 19:560–7.

Melo-Ferreira, J., J. Vilela, M. M. Fonseca, R. R. Da Fonseca, P. Boursot, and P. C. Alves (2014). The elusive nature of adaptive mitochondrial DNA evolution of an arctic lineage prone to frequent introgression. Genome Biology and Evolution 6:886–96.

Melvin, R. G., and J. W. O. Ballard (2017). Cellular and population level processes influence the rate, accumulation and observed frequency of inherited and somatic mtDNA mutations. Mutagenesis 32:323–34.

el Meziane, A., J. C. Callen, and J. C. Mounolou (1989). Mitochondrial gene expression during Xenopus laevis development: A molecular study. The EMBO Journal 8:1649–55.

Michod, R. E. (1996). Cooperation and conflict in the evolution of individuality. II. Conflict mediation. Proceedings of the Royal Society B: Biological Sciences 263:813–22.

Michod, R. E., and D. Roze (2001). Cooperation and conflict in the evolution of multicellularity. Heredity 86:1–7.

Mick, D. U., T. D. Fox, and P. Rehling (2011). Inventory control: Cytochrome c oxidase assembly regulates mitochondrial translation. Nature Reviews Molecular Cell Biology 12:14–20.

Milenkovic, D., J. N. Blaza, N. G. Larsson, and J. Hirst (2017). The enigma of the respiratory chain supercomplex. Cell Metabolism 25:765–76.

Mimaki, M., X. Wang, M. McKenzie, D. R. Thorburn, and M. T. Ryan (2012). Understanding mitochondrial complex I assembly in health and disease. Biochimica et Biophysica Acta (BBA)—Bioenergetics 1817:851–62.

Mishmar, D., E. Ruiz-Pesini, P. Golik, V. Macaulay, A. G. Clark, S. Hosseini, M. Brandon, K. Easley, E. Chen, M. D. Brown, R. I. Sukernik, et al. (2003). Natural selection shaped regional mtDNA variation in humans. Proceedings of the National Academy of Sciences 100:171–6.

Mishmar, D., E. Ruiz-Pesini, M. Mondragon-Palomino, V. Procaccio, B. Gaut, and D. C. Wallace (2006). Adaptive selection of mitochondrial complex I subunits during primate radiation. Gene 378:11–18.

Mishra, P., and D. C. Chan (2014). Mitochondrial dynamics and inheritance during cell division, development and disease. Nature Reviews Molecular Cell Biology 15:634–46.

Mitchell, P. (1961). Coupling of phosphorylation to electron and hydrogen transfer by a chemiosmotic type of mechanism. Nature 192:963–4.

Moison, M., F. Roux, M. Quadrado, R. Duval, M. Ekovich, D. H. Lê, M. Verzaux, and F. Budar (2010). Cytoplasmic phylogeny and evidence of cyto-nuclear co-adaptation in *Arabidopsis thaliana*. Plant Journal 63:728–38.

Moore, J. C., and J. R. Pannell (2011). Sexual selection in plants. Current Biology 21:R176–R182.

Moraes, C. R., and S. Meyers (2018). The sperm mitochondrion: Organelle of many functions. Animal Reproduction Science 194:71–80.

Morales, H. E., A. Pavlova, L. Joseph, and P. Sunnucks (2015). Positive and purifying selection in mitochondrial genomes of a bird with mitonuclear discordance. Molecular Ecology 24:2820–37

Morales, H. E., P. Sunnucks, L. Joseph, and A. Pavlova (2017). Perpendicular axes of differentiation generated by mitochondrial introgression. Molecular Ecology, 26:3241–55.

Morales, H. E., A. Pavlova, N. Amos, R. Major, A. Kilian, C. Greening, and P. Sunnucks (2018). Concordant divergence of mitogenomes and a mitonuclear gene cluster in bird lineages inhabiting different climates. Nature Ecology and Evolution 2:1258–1267.

Moran, N. A. (2003). Tracing the evolution of gene loss in obligate bacterial symbionts. Current Opinion in Microbiology 6:512–18.

Moran, N. A., and P. Baumann (2000). Bacterial endosymbionts in animals. Current Opinion in Microbiology 3:270–5.

Moran, N. A., and J. J. Wernegreen (2000). Lifestyle evolution in symbiotic bacteria: Insights from genomics. Trends in Ecology and Evolution 15:321–6.

Moran, N. A., J. P. McCutcheon, and A. Nakabachi (2008). Genomics and evolution of heritable bacterial symbionts. Annual Review of Genetics 42:165–90.

Moreno-Loshuertos, R., G. Ferrín, R. Acín-Pérez, M. E. Gallardo, C. Viscomi, A. Pérez-Martos, M. Zeviani, P. Fernández-Silva, and J. A. Enríquez (2011). Evolution meets disease: Penetrance and functional epistasis of mitochondrial tRNA mutations. PLoS Genetics 7:e1001379.

Morgan, T. H. (1926). Genetics and the physiology of development. The American Naturalist 60:489–515.

Morrison, D. A. (2018). Review of species concepts in biology: Historical development, theoretical foundations and practical relevance. Systematic Biology 67:177–80.

Morrow, E. H., K. Reinhardt, J. N. Wolff, and D. K. Dowling (2015). Risks inherent to mitochondrial replacement. EMBO Reports 16:541–5.

Moya, A., J. Peretó, R. Gil, and A. Latorre (2008). Learning how to live together: Genomic insights into prokaryote–animal symbioses. Nature Reviews Genetics 9:218–29.

Muir, R., A. Diot, and J. Poulton (2016). Mitochondrial content is central to nuclear gene expression: Profound implications for human health. BioEssays 38:150–6.

Muller, H. J. (1942). Isolating mechanisms, evolution and temperature. In Biological Symposia: A Series of Volumes Devoted to Current Symposia in the Field of Biology (T. Dobzhansky, Editor). Jaques Cattell Press, Lancaster, PA, pp. 71–125.

Muller, H. J. (1964). The relation of recombination to mutational advance. Mutation Research 1:2–9.

Mundy, N. I. I., J. Stapley, C. Bennison, R. Tucker, H. Twyman, K. W. Kim, T. Burke, T. R. R. Birkhead, S. Andersson, and J. Slate (2016). Red carotenoid coloration in the zebra finch is controlled by a cytochrome P450 gene cluster. Current Biology 26:1435–40.

Murphy, H. A., H. A. Kuehne, C. A. Francis, and P. D. Sniegowski (2006). Mate choice assays and mating propensity differences in natural yeast populations. Biology Letters 2:553–6.

Murphy, M. P. (1989). Slip and leak in mitochondrial oxidative phosphorylation. BBA—Bioenergetics 977:123–41.

Myhill, S., N. E. Booth, and J. McLaren-Howard (2009). Chronic fatigue syndrome and mitochondrial dysfunction. International Journal of Clinical and Experimental Medicine 2: 1–16.

Nabholz, B., H. Ellegren, and J. B. W. Wolf (2013). High levels of gene expression explain the strong evolutionary constraint of mitochondrial protein-coding genes. Molecular Biology and Evolution 30:272–84.

Nabholz, B., S. Glémin, and N. Galtier (2008). Strong variations of mitochondrial mutation rate across mammals—The longevity hypothesis. Molecular Biology and Evolution 25:120–30.

Nabholz, B., S. Glémin, and N. Galtier (2009). The erratic mitochondrial clock: Variations of mutation rate, not population size, affect mtDNA diversity across birds and mammals. BMC Evolutionary Biology 9:54.

Nabholz, B., R. Lanfear, and J. Fuchs (2016). Body mass-corrected molecular rate for bird mitochondrial DNA. Molecular Ecology 25:4438–49.

Nagao, Y., Y. Totsuka, Y. Atomi, H. Kaneda, K. F. Lindahl, H. Imai, and H. Yonekawa (1998). Decreased physical performance of congenic mice with mismatch between the nuclear and the mitochondrial genome. Genes and Genetic Systems 73:21–7.

Nakada, K., A. Sato, K. Yoshida, T. Morita, H. Tanaka, S. Inoue, H. Yonekawa, and J. Hayashi (2006). Mitochondria-related male infertility. Proceedings of the National Academy of Science 103:15148–53.

Narra, H. P., and H. Ochman (2006). Of what use is sex to bacteria? Current Biology 16:R705–R710.

Nass, M. M., and S. Nass (1963). Intramitochondrial fibers with DNA characteristics: I. Fixation and electron staining reactions. The Journal of Cell Biology 19:593–611.

Nei, M., and M. Nozawa (2011). Roles of mutation and selection in speciation: From Hugo de Vries to the modern genomic era. Genome Biology and Evolution 3:812–29.

Nei, M., Y. Suzuki, and M. Nozawa (2010). The neutral theory of molecular evolution in the genomic era. Annual Review of Genomics and Human Genetics 11:265–89.

Neiman, M., and T. A. Linksvayer (2006). The conversion of variance and the evolutionary potential of restricted recombination. Heredity 96:111–21.

Neiman, M., and D. R. Taylor (2009). The causes of mutation accumulation in mitochondrial genomes. Proceedings of the Royal Society B: Biological Sciences 276:1201–9.

Neiman, M., S. Meirmans, and P. G. Meirmans (2009). What can asexual lineage age tell us about the maintenance of sex? Annals of the New York Academy of Sciences 1168:185–200.

Nelson, G., and N. I. Platnick (1981). Systematics and Biogeography: Cladistics and Vicariance. Columbia University Press, New York.

Neubauer, G., P. Nowicki, and M. Zagalska-Neubauer (2014). Haldane's rule revisited: Do hybrid females have a shorter lifespan? Survival of hybrids in a recent contact zone between two large gull species. Journal of Evolutionary Biology 27:1248–55.

Ng, L. F., J. Gruber, I. K. Cheah, C. K. Goo, W. F. Cheong, G. Shui, K. P. Sit, M. R. Wenk, and B. Halliwell (2014). The mitochondria-targeted antioxidant MitoQ extends lifespan and improves healthspan of a transgenic *Caenorhabditis elegans* model of Alzheimer disease. Free Radical Biology and Medicine 71:390–401.

Nicolas, V., B. Schaeffer, A. D. Missoup, J. Kennis, M. Colyn, C. Denys, C. Tatard, C. Cruaud, and C. Laredo (2012). Assessment of three mitochondrial genes (16S, Cytb, CO1) for identifying species in the Praomyini tribe (Rodentia: Muridae). PLoS One 7:e36586.

Ning, T., H. Xiao, J. Li, S. Hua, and Y. P. Zhang (2010). Adaptive evolution of the mitochondrial ND6 gene in the domestic horse. Genetics and Molecular Research 9:144–50.

Nishimura, T., T. Katsumura, M. Motoi, H. Oota, and S. Watanuki (2017). Experimental evidence reveals the UCP1 genotype changes the oxygen consumption attributed to non-shivering thermogenesis in humans. Scientific Reports 7:1–7.

Nosek, J., L. Tomáška, H. Fukuhara, Y. Suyama, and L. Kováč (1998). Linear mitochondrial genomes: 30 years down the line. Trends in Genetics 14:184–8.

Nosil, P., and D. Schluter (2011). The genes underlying the process of speciation. Trends in Ecology and Evolution 26:160–7.

Nouws, J., L. G. J. Nijtmans, J. A. Smeitink, and R. O. Vogel (2012). Assembly factors as a new class of disease genes for mitochondrial complex i deficiency: Cause, pathology and treatment options. Brain 135:12–22.

O'Brien, K. M., and I. A. Mueller (2010). The unique mitochondrial form and function of Antarctic channichthyid icefishes. Integrative and Comparative Biology 50:993–1008.

Ochman, H., J. G. Lawrence, and E. A. Groisman (2000). Lateral gene transfer and the nature of bacterial innovation. Nature 405:299–304.

Olson-Manning, C. F., M. R. Wagner, and T. Mitchell-Olds (2012). Adaptive evolution: Evaluating empirical support for theoretical predictions. Nature Reviews Genetics 13:867–77.

Omland, K. E., C. L. Tarr, W. I. Boarman, J. M. Marzluff, and R. C. Fleischer (2000). Cryptic genetic variation and paraphyly in ravens. Proceedings of the Royal Society B: Biological Sciences 267:2475–82.

Orr, H. A. (1995). The population genetics of speciation: The evolution of hybrid incompatibilities. Genetics 139:1805–13.

Orr, H. A. (1996). Dobzhansky, Bateson, and the genetics of speciation. Genetics 144:1331–5.

Orr, H. A., and M. Turelli (2001). The evolution of postzygotic isolation: Accumulating Dobzhansky-Muller incompatibilities. Evolution 55:1085–94.

Ortiz-Barrientos, D., B. A. Counterman, and M. A. F. Noor (2004). The genetics of speciation by reinforcement. PLoS Biology 2:e416.

Osada, N., and H. Akashi (2012). Mitochondrial–nuclear interactions and accelerated compensatory evolution: Evidence from the primate cytochrome c oxidase complex. Molecular Biology and Evolution 29:337–46.

Oswald, J. A., M. G. Harvey, R. C. Remsen, D. U. Foxworth, S. W. Cardiff, D. L. Dittmann, L. C. Megna, M. D. Carling, and R. T. Brumfield (2016). Willet be one species or two? A genomic view of the evolutionary history of *Tringa semipalmata*. The Auk 133:593–614.

Otto, S. P. (2009). The evolutionary enigma of sex. The American Naturalist 174:S1–S14.

Otto, S. P., and T. Lenormand (2002). Resolving the paradox of sex and recombination. Nature Reviews Genetics 3:252–61.

Pace, N. R. (1997). A molecular view of microbial diversity and the biosphere. Science 276:734–40.

Packer, L., J. Gibbs, C. Sheffield, and R. Hanner (2009). DNA barcoding and the mediocrity of morphology. Molecular Ecology Resources 9:42–50.

Pagliarini, D. J., S. E. Calvo, B. Chang, S. A. Sheth, S. B. Vafai, S. E. Ong, G. A. Walford, C. Sugiana, A. Boneh, W. K. Chen, D. E. Hill, et al. (2008). A mitochondrial protein compendium elucidates complex I disease biology. Cell 134:112–23.

Palikaras, K., E. Lionaki, and N. Tavernarakis (2015). Balancing mitochondrial biogenesis and mitophagy to maintain energy metabolism homeostasis. Cell Death and Differentiation 22:1399–401.

Parker, G. A. (1978). Evolution of competitive mate searching. Annual Review of Entomology 23:173–96.

Parker, G. A., R. R. Baker, and V. G. F. Smith (1972). The origin and evolution of gamete dimorphism and the male-female phenomenon. Journal of Theoretical Biology 36:529–53.

Partridge, L., and L. D. Hurst (1998). Sex and conflict. Science 281:2003–8.

Patel, M. R., G. K. Miriyala, A. J. Littleton, H. Yang, K. Trinh, J. M. Young, S. R. Kennedy, Y. M. Yamashita, L. J. Pallanck, and H. S. Malik (2016). A mitochondrial DNA hypomorph of cytochrome oxidase specifically impairs male fertility in *Drosophila melanogaster*. eLife 5:1–27.

De Paula, W. B. M., A. N. A. Agip, F. Missirlis, R. Ashworth, G. Vizcay-Barrena, C. H. Lucas, and J. F. Allen (2013). Female and male gamete mitochondria are distinct and complementary in transcription, structure, and genome function. Genome Biology and Evolution 5:1969–77.

Pavlova, A., J. N. Amos, L. Joseph, K. Loynes, J. J. Austin, J. S. Keogh, G. N. Stone, J. A. Nicholls, and P. Sunnucks (2013b). Perched at the mito-nuclear crossroads: Divergent mitochondrial lineages correlate with environment in the face of ongoing nuclear gene flow in an Australian bird. Evolution 67:3412–28.

Payne, B. A. I., I. J. Wilson, P. Yu-Wai-Man, J. Coxhead, D. Deehan, R. Horvath, R. W. Taylor, D. C. Samuels, M. Santibanez-Koref, and P. F. Chinnery (2013). Universal heteroplasmy of human mitochondrial DNA. Human Molecular Genetics 22:384–90.

Pazos, F., and A. Valencia (2008). Protein co-evolution, co-adaptation and interactions. EMBO Journal 27:2648–55.

Pearce, S., C. L. Nezich, and A. Spinazzola (2013). Mitochondrial diseases: Translation matters. Molecular and Cellular Neuroscience 55:1–12.

Pennycuick, C. J. (1975). Mechanics of flight. In Avian Biology, Vol. V (D. S. Farner and J. R. King, Editors). Elsevier, Amsterdam, pp. 1–75.

Pereira, R. J., F. S. Barreto, and R. S. Burton (2014). Ecological novelty by hybridization: Experimental evidence for increased thermal tolerance by transgressive segregation in *Tigriopus californicus*. Evolution 68:204–15.

Pereira, R. J., M. C. Sasaki, and R. S. Burton (2017). Adaptation to a latitudinal thermal gradient within a widespread copepod species: The contributions of genetic divergence and phenotypic plasticity. Proceedings of the Royal Society B: Biological Sciences 284:20170236.

Pérez, V. I., A. Bokov, H. Van Remmen, J. Mele, Q. Ran, Y. Ikeno, and A. Richardson (2009). Is the oxidative stress theory of aging dead? Biochimica et Biophysica Acta—General Subjects 1790:1005–14.

Perlman, S. J., C. N. Hodson, P. T. Hamilton, G. P. Opit, and B. E. Gowen (2015). Maternal transmission, sex ratio distortion, and mitochondria. Proceedings of the National Academy of Sciences 112:10162–8.

Peters, J. L., K. Winker, K. C. Millam, P. Lavretsky, I. Kulikova, R. E. Wilson, Y. N. Zhuravlev, and K. G. McCracken (2014). Mito-nuclear discord in six congeneric lineages of Holarctic ducks (genus *Anas*). Molecular Ecology 23:2961–74.

Peterson, A. T. (1996). Geographic variation in sexual dichromatism in birds. Bulletin of the British Ornithologists Club 116:156–72.

Petit, R. J., and L. Excoffier (2009). Gene flow and species delimitation. Trends in Ecology and Evolution 24:386–93.

Pfenninger, M., H. Lerp, M. Tobler, C. Passow, J. L. Kelley, E. Funke, B. Greshake, U. K. Erkoc, T. Berberich, and M. Plath (2014). Parallel evolution of cox genes in H2S-tolerant fish as key adaptation to a toxic environment. Nature Communications 5:3873.

Pfenninger, M., S. Patel, L. Arias-Rodriguez, B. Feldmeyer, R. Riesch, and M. Plath (2015). Unique evolutionary trajectories in repeated adaptation to hydrogen sulphide-toxic habitats of a neotropical fish (*Poecilia mexicana*). Molecular Ecology 24:5446–59.

Phillips, D., M. J. Reilley, A. M. Aponte, G. Wang, E. Boja, M. Gucek, and R. S. Balaban (2010). Stoichiometry of STAT3 and mitochondrial proteins: Implications for the regulation of oxidative phosphorylation by protein-protein interactions. Journal of Biological Chemistry 285:23532–6.

Picard, M., and B. S. McEwen (2014). Mitochondria impact brain function and cognition. Proceedings of the National Academy of Sciences 111:7–8.

Picard, M., D. C. Wallace, and Y. Burelle (2016). The rise of mitochondria in medicine. Mitochondrion 30:105–16.

Pichaud, N., E. H. Chatelain, J. W. O. Ballard, R. Tanguay, G. Morrow, and P. U. Blier (2010). Thermal sensitivity of mitochondrial metabolism in two distinct mitotypes of *Drosophila simulans*: Evaluation of mitochondrial plasticity. Journal of Experimental Biology 213:1665–75.

Pichaud, N., J. W. O. Ballard, R. M. Tanguay, and P. U. Blier (2011). Thermal sensitivity of mitochondrial functions in permeabilized muscle fibers from two populations of *Drosophila simulans* with divergent mitotypes. Regulatory, Integrative and Comparative Physiology 301:R48–R59.

Pichaud, N., J. W. O. Ballard, R. M. Tanguay, and P. U. Blier (2012). Naturally occurring mitochondrial DNA haplotypes exhibit metabolic differences: Insight into functional properties of mitochondria. Evolution 66:3189–97.

Pichaud, N., M. Messmer, C. C. Correa, and J. W. O. Ballard (2013). Diet influences the intake target and mitochondrial functions of *Drosophila melanogaster* males. Mitochondrion 13:817–22.

Pierron, D., D. E. Wildman, M. Hüttemann, G. C. Markondapatnaikuni, S. Aras, and L. I. Grossman (2012). Cytochrome c oxidase: Evolution of control via nuclear subunit addition. Biochimica et Biophysica Acta—Bioenergetics 1817:590–7.

Pizzari, T. (2009). Sexual selection: Sperm in the fast lane. Current Biology 19:R292–R294.

Plath, M., M. Pfenninger, H. Lerp, R. Riesch, C. Eschenbrenner, P. A. Slattery, D. Bierbach, N. Herrmann, M. Schulte, L. Arias-Rodriguez, J. Rimber Indy, et al. (2013). Genetic differentiation and selection against migrants in evolutionarily replicated extreme environments. Evolution 67:2647–61.

Poole, A. M., and S. Gribaldo (2014). Eukaryotic origins: How and when was the mitochondrion acquired? Cold Spring Harbor Perspectives in Biology 6:a015990.

Popadin, K. Y., S. I. Nikolaev, T. Junier, M. Baranova, and S. E. Antonarakis (2013). Purifying selection in mammalian mitochondrial protein-coding genes is highly effective and congruent with evolution of nuclear genes. Molecular Biology and Evolution 30:347–55.

Pörtner, H. O. (2002). Climate variations and the physiological basis of temperature dependent biogeography: Systemic to molecular hierarchy of thermal tolerance in animals. Comparative Biochemistry and Physiology—A Molecular and Integrative Physiology 132:739–61.

Poulton, J., M. R. Chiaratti, F. V. Meirelles, S. Kennedy, D. Wells, and I. J. Holt (2010). Transmission of mitochondrial DNA diseases and ways to prevent them. PLoS Genetics 6:e1001066.

Pozzi, A., F. Plazzi, L. Milani, F. Ghiselli, and M. Passamonti (2017). SmithRNAs: Could mitochondria "bend" nuclear regulation? Molecular Biology and Evolution 34:1960–73.

Prasai, K. (2017). Regulation of mitochondrial structure and function by protein import: A current review. Pathophysiology 24:107–22.

Prebble, J. (2002). Peter Mitchell and the ox phos wars. Trends in Biochemical Sciences 27:209–12.

Presgraves, D. C. (2010). Speciation genetics: Search for the missing snowball. Current Biology 20:R1073–R1074.

Price, T. (2015). Speciation in Birds. W. H. Freeman, New York.

Pritchard, V. L., and S. Edmands (2013). The genomic trajectory of hybrid swarms: Outcomes of repeated crosses between populations of *Tigriopus californicus*. Evolution 67:774–91.

Projecto-Garcia, J., C. Natarajan, H. Moriyama, R. E. Weber, A. Fago, Z. A. Cheviron, R. Dudley, J. A. McGuire, C. C. Witt, and J. F. Storz (2013). Repeated elevational transitions in hemoglobin function during the evolution of Andean hummingbirds. Proceedings of the National Academy of Sciences 110:20669–74.

Provine, B. (1981). Origins of the genetics of natural populations series. In Dobzhansky's Genetics of Natural Populations (R. C. Lewontin, J. A. Moore, W. B. Provine and B. Wallace, Editors). Columbia University Press, New York, pp. 1–85.

Prum, R. O. (1997). Phylogenetic tests of alternative intersexual selection mechanisms: Trait macroevolution in a polygynous clade (Aves: Pipridae). The American Naturalist 149:668–692.

Prum, R. O. (2010). The Lande-Kirkpatrick mechanism is the null model of evolution by intersexual selection: Implications for meaning, honesty, and design in intersexual signals. Evolution 64:3085–100.

Prum, R. O. (2012). Aesthetic evolution by mate choice: Darwin's really dangerous idea. Philosophical Transactions of the Royal Society B: Biological Sciences 367:2253–65.

Qiu, Q., G. Zhang, T. Ma, W. Qian, J. Wang, Z. Ye, C. Cao, Q. Hu, J. Kim, D. M. Larkin, L. Auvil, et al. (2012). The yak genome and adaptation to life at high altitude. Nature Genetics 44:946–9.

De Queiroz, K. (2007). Species concepts and species delimitation. Systematic Biology 56:879–86.

Quintela, M., M. P. Johansson, B. K. Kristjánsson, R. Barreiro, and A. Laurila (2014). AFLPs and mitochondrial haplotypes reveal local adaptation to extreme thermal environments in a freshwater gastropod. PLoS One 9:e101821.

Qvarnström, A., and R. I. Bailey (2009). Speciation through evolution of sex-linked genes. Heredity 102:4–15.

Van Raamsdonk, J. M., and S. Hekimi (2012). Superoxide dismutase is dispensable for normal animal lifespan. Proceedings of the National Academy of Sciences 109:5785–90.

Radzvilavicius, A. L., Z. Hadjivasiliou, A. Pomiankowski, and N. Lane (2016). Selection for mitochondrial quality drives evolution of the germline. PLOS Biology 14:e2000410.

Radzvilavicius, A. L., H. Kokko, and J. R. Christie (2017). Mitigating mitochondrial genome erosion without recombination. Genetics 207:1079–88.

Rahman, J., and S. Rahman (2018). Mitochondrial medicine in the omics era. The Lancet 391:2560–74.

Raimundo, N., L. Fernandez-Mosquera, and K. F. Yambire (2016). Mitochondrial signaling. In Mitochondria and Cell Death (D. M. Hockenbery, Editor). Springer Science & Business Media, New York, pp. 169–86.

Raj, A., S. A. Rifkin, E. Andersen, and A. Van Oudenaarden (2010). Variability in gene expression underlies incomplete penetrance. Nature 463:913–18.

Rajender, S., P. Rahul, and A. A. Mahdi (2010). Mitochondria, spermatogenesis and male infertility. Mitochondrion 10:419–28.

Rak, M., P. Bénit, D. Chrétien, J. Bouchereau, M. Schiff, R. El-Khoury, A. Tzagoloff, and P. Rustin (2016). Mitochondrial cytochrome c oxidase deficiency. Clinical Science 130:393–407.

Ramesh, M. A., S. B. Malik, and J. M. Logsdon (2005). A phylogenomic inventory of meiotic genes: Evidence for sex in *Giardia* and an early eukaryotic origin of meiosis. Current Biology 15:185–91.

Rand, D. M. (2001). The units of selection on mitochondrial DNA. Annual Review of Ecology and Systematics 32:415–48.

Rand, D. M. (2008). Mitigating mutational meltdown in mammalian mitochondria. PLoS Biology 6:E35.

Rand, D. M. (2017). Fishing for adaptive epistasis using mitonuclear interactions. PLoS Genetics 13:4–7.

Rand, D. M., R. A. Haney, and A. J. Fry (2004). Cytonuclear coevolution: The genomics of cooperation. Trends in Ecology and Evolution 19:645–53.

Ravinet, M., R. Faria, R. K. Butlin, J. Galindo, N. Bierne, M. Rafajlović, M. A. F. Noor, B. Mehlig, and A. M. Westram (2017). Interpreting the genomic landscape of speciation: A road map for finding barriers to gene flow. Journal of Evolutionary Biology 30:1450–77.

Rawson, P. D., D. A. Brazeau, and R. S. Burton (2000). Isolation and characterization of cytochrome c from the marine copepod *Tigriopus californicus*. Gene 248:15–22.

Rawson, P. D., and R. S. Burton (2002). Functional coadaptation between cytochrome c and cytochrome c oxidase within allopatric populations of a marine copepod. Proceedings of the National Academy of Sciences 99:12955–8.

Reeve, H. K., and D. W. Pfennig (2003). Genetic biases for showy males: Are some genetic systems especially conducive to sexual selection? Proceedings of the National Academy of Sciences 100:1089–94.

Reinhardt, K., D. K. Dowling, and E. H. Morrow (2013). Mitochondrial replacement, evolution, and the clinic. Science 341:1345–6.

Reinhold, K. (1998). Sex linkage among genes controlling sexually selected traits. Behavioral Ecology and Sociobiology 44:1–7.

Reynolds, J. D., and M. R. Gross (1990). Costs and benefits of female mate choice: Is there a lek paradox? The American Naturalist 136:230–43.

Rheindt, F. E., and S. V. Edwards (2011). Genetic Introgression: An integral but neglected component of speciation in birds. The Auk 128:620–32.

Ribeiro, Â. M., P. Lloyd, and R. C. K. Bowie (2011). A tight balance between natural selection and gene flow in a Southern African arid-zone endemic bird. Evolution 65:3499–514.

Ricchetti, M. (2018). Replication stress in mitochondria. Mutation Research 808:93–102.

Rice, W. R. (1996). Evolution of the Y sex chromosome in animals. BioScience 46:331–43.

Rice, W. R. (2013). Nothing in genetics makes sense except in light of genomic conflict. Annual Review of Ecology, Evolution, and Systematics 44:217–37.

Rich, P. R., and A. Marechal (2010). The mitochondrial respiratory chain. Essays in Biochemistry 47:1–24.

Richter-Dennerlein, R., S. Dennerlein, and P. Rehling (2015). Integrating mitochondrial translation into the cellular context. Nature Reviews Molecular Cell Biology 16:586–92.

Ricklefs, R. E. (1984). The optimization of growth rate in altricial birds. Ecology 65:1602–16.

Ro, S., H.-Y. Ma, C. Park, N. Ortogero, R. Song, G. W. Hennig, H. Zheng, Y.-M. Lin, L. Moro, J.-T. Hsieh, and W. Yan (2013). The mitochondrial genome encodes abundant small noncoding RNAs. Cell Research 23:759–74.

Rogozin, I. B., Y. I. Wolf, A. V. Sorokin, B. G. Mirkin, and E. V. Koonin (2003). Remarkable interkingdom conservation of intron positions and massive, lineage-specific intron loss and gain in eukaryotic evolution. Current Biology 13:1512–17.

Rogozin, I. B., L. Carmel, M. Csuros, and E. V Koonin (2012). Origin and evolution of spliceosomal introns. Biology Direct 7:11.

Rotig, A. (2011). Human diseases with impaired mitochondrial protein synthesis. Biochimica et Biophysica Acta 1807:1198–205.

Roze, D., and N. H. Barton (2006). The Hill-Robertson effect and the evolution of recombination. Genetics 173:1793–811.

Roze, D., F. Rousse, and Y. Michalakis (2005). Germline bottlenecks, biparental inheritance and selection on mitochondrial variants: A two-level selection model. Genetics 170:1385–99.

Ruby, J. G., M. Smith, and R. Buffenstein (2018). Naked mole-rat mortality rates defy Gompertzian laws by not increasing with age. eLife 7:e31157.

Ruiz-Pesini, E., C. Diez, A. C. Lapena, A. Pérez-Martos, J. Montoya, E. Alvarez, J. Arenas, and M. J. Lopez-Perez (1998). Correlation of sperm motility with mitochondrial enzymatic activities. Clinical Chemistry 44:1616–20.

Ruiz-Pesini, E., D. Mishmar, M. Brandon, V. Procaccio, and D. C. Wallace (2004). Effects of purifying and adaptative selection on regional variation in human mtDNA. Science 303:223–6.

Ruud, J. T. (1954). Vertebrates without erythrocytes and blood pigment. Nature 173:848–850.

Ryan, M. J., and A. S. Rand (1993). Species recognition and sexual selection as a unitary problem in animal communication. Evolution, 47:647–57.

Sabat, P., C. Narváez, I. Peña-Villalobos, C. Contreras, K. Maldonado, J. C. Sanchez-Hernandez., S. D. Newsome, R. Nespolo, and F. Bozinovic (2017). Coping with salt water habitats: Metabolic and oxidative responses to salt intake in the rufous-collared sparrow. Frontiers in Physiology 8:654.

Sackton, T. B., R. A. Haney, and D. M. Rand (2003). Cytonuclear coadaptation in *Drosophila*: Disruption of cytochrome c oxidase activity in backcross genotypes. Evolution 57:2315–25.

Saether, S. A., G.-P. Saetre, T. Borge, C. Wiley, N. Svedin, G. Andersson, T. Veen, J. Haavie, M. R. Servedio, S. Bureš, M. Král, et al. (2007). Sex chromosome-linked species recognition and evolution of reproductive isolation in flycatchers. Science 318:95–7.

Saetre, G.-P., T. Moum, S. Bureš, M. Král, M. Adamjan, and J. Moreno (1997). A sexually selected character displacement in flycatchers reinforces premating isolation. Nature 387:589.

Saetre, G.-P., T. Borge, K. Lindroos, J. Haavie, B. C. Sheldon, C. Primmer, and A.-C. Syvänen (2003). Sex chromosome evolution and speciation in *Ficedula* flycatchers. Proceedings of the Royal Society B: Biological Sciences 270:53–9.

Sagan, L. (1967). On the origin of mitosing cells. Journal of Theoretical Biology 14:225–74.

Saha, B., G. Borovskii, and S. K. Panda (2016). Alternative oxidase and plant stress tolerance. Plant Signaling and Behavior 11:e1256530.

Sahm, A., M. Bens, M. Platzer, and A. Cellerino (2017). Parallel evolution of genes controlling mitonuclear balance in short-lived annual fishes. Aging Cell 16:488–96.

Sanchez-Caballero, L., S. Guerrero-Castillo, and L. Nijtmans (2016). Unraveling the complexity of mitochondrial complex i assembly: A dynamic process. Biochimica et Biophysica Acta—Bioenergetics 1857:980–90.

Saoura, M., C. A. Powell, R. Kopajtich, A. Alahmad, B. Albash, M. Alfadhel, C. L. Alston, and E. Bertini (2018). Mutations in ELAC2 associated with hypertrophic cardiomyopathy impair mitochondrial tRNA 3'-end processing. bioRxiv 2018:378448.

Sapp, J. (1987). Beyond the Gene: Cytoplasmic Inheritance and the Struggle for Authority in Genetics. Oxford University Press, Oxford, UK.

Sato, M., and K. Sato (2013). Maternal inheritance of mitochondrial DNA by diverse mechanisms to eliminate paternal mitochondrial DNA. Biochimica et Biophysica Acta—Molecular Cell Research 1833:1979–84.

Sauna, Z. E., and C. Kimchi-Sarfaty (2011). Understanding the contribution of synonymous mutations to human disease. Nature Reviews Genetics 12:683–91.

Sazanov, L. A. (2015). A giant molecular proton pump: Structure and mechanism of respiratory complex I. Nature Reviews Molecular Cell Biology 16:375–88.

Scarpulla, R. C. (2008). Transcriptional paradigms in mammalian mitochondrial biogenesis and function. Physiological Reviews 88:611–38.

Schatz, G., E. Haslbrunner, and H. Tuppy (1964). Deoxyribonucleic acid associated with yeast mitochondria. Biochemical and Biophysical Research Communications 15:127–32.

Scheffler, K., M. Krohn, T. Dunkelmann, J. Stenzel, B. Miroux, S. Ibrahim, O. von Bohlen et al. (2012). Mitochondrial DNA polymorphisms specifically modify cerebral β-amyloid proteostasis. Acta Neuropathologica 124(2):199–208.

Scheper, G. C., M. S. van der Knaap, and C. G. Proud (2007). Translation matters: Protein synthesis defects in inherited disease. Nature Reviews Genetics 8:711–23.

Schilthuizen, M., M. C. W. G. Giesbers, and L. W. Beukeboom (2011). Haldane's rule in the 21st century. Heredity 107:95–102.

Schindel, D. E., and S. E. Miller (2005). DNA barcoding a useful tool for taxonomists. Nature 435:17.

Schmaljohann, H., B. Bruderer, and F. Liechti (2008). Sustained bird flights occur at temperatures far beyond expected limits. Animal Behaviour 76:1133–8.

Schmidt, T. R., W. Wu, M. Goodman, and L. I. Grossman (2001). Evolution of nuclear- and mitochondrial-encoded subunit interaction in cytochrome c oxidase. Molecular Biology and Evolution 18:563–9.

Schon, I., G. Rossetti, and K. Martens (2009). Darwinulid ostracods: Ancient asexual scandals or scandalous gossip? In Lost Sex: The Evolutionary Biology of Parthenogenesis (I. Schon, K. Martens and P. J. Van Dijk, Editors). Springer, Dordrecht, The Netherlands, pp. 241–57.

Schulte, P. M. (2015). The effects of temperature on aerobic metabolism: Towards a mechanistic understanding of the responses of ectotherms to a changing environment. Journal of Experimental Biology 218:1856–66.

Schurko, A. M., and J. M. Logsdon (2008). Using a meiosis detection toolkit to investigate ancient asexual "scandals" and the evolution of sex. BioEssays 30:579–89.

Schwander, T. (2016). Evolution: The end of an ancient asexual scandal. Current Biology 26:R233–R235.

Scott, G. R., J. G. Richards, and W. K. Milsom (2009). Control of respiration in flight muscle from the high-altitude bar-headed goose and low-altitude birds. American Journal of Physiology—Regulatory, Integrative and Comparative Physiology 297:1066–74.

Scott, G. R., P. M. Schulte, S. Egginton, A. L. M. Scott, J. G. Richards, and W. K. Milsom (2011). Molecular evolution of cytochrome c oxidase underlies high-altitude adaptation in the bar-headed goose. Molecular Biology and Evolution 28:351–63.

Scott, G. R., L. A. Hawkes, P. B. Frappell, P. J. Butler, C. M. Bishop, and W. K. Milsom (2015). How bar-headed geese fly over the Himalayas. Physiology 30:107–15.

Secondi, J., B. Faivre, and S. Bensch (2006). Spreading introgression in the wake of a moving contact zone. Molecular Ecology 15:2463–75.

Secor, S. M., and J. M. Diamond (1995). Adaptive responses to feeding in Burmese pythons: Pay before pumping. The Journal of Experimental Biology 198:1313–25.

Secor, S. M., and J. Diamond (1998). A vertebrate model of extreme physiological regulation. Nature 395:659–62.

Seebacher, F., M. D. Brand, P. L. Else, H. Guderley, A. J. Hulbert, and C. D. Moyes (2010). Plasticity of oxidative metabolism in variable climates: Molecular mechanisms. Physiological and Biochemical Zoology 83:721–32.

Seipel, K., N. Yanze, and V. Schmid (2004). The germ line and somatic stem cell gene Cniwi in the jellyfish *Podocoryne carnea*. International Journal of Developmental Biology 48:1–7.

Seixas, F., P. Boursot, and J. Melo-Ferreira (2018). The genomic impact of historical hybridization with massive mitochondrial DNA introgression. Genome Biology 19:91.

Servedio, M. R., and M. A. Noor (2003). The role of reinforcement in speciation: Theory and data. Annual Review of Ecology, Evolution, and Systematics 34:339–64.

Servedio, M. R., G. S. Van Doorn, M. Kopp, A. M. Frame, and P. Nosil (2011). Magic traits in speciation: "magic" but not rare? Trends in Ecology and Evolution 26:389–97.

Shadel, G. S. (2004). Coupling the mitochondrial transcription machinery to human disease. Trends in Genetics 20:513–19.

Shadel, G. S., and T. L. Horvath (2015). Mitochondrial ROS signaling in organismal homeostasis. Cell 163:560–9.

Sharpley, M. S., C. Marciniak, K. Eckel-Mahan, M. McManus, M. Crimi, K. Waymire, C. S. Lin, S. Masubuchi, N. Friend, M. Koike, D. Chalkia, et al. (2012). Heteroplasmy of mouse mtDNA is genetically unstable and results in altered behavior and cognition. Cell 151:333–43.

Shen, Y.-Y., L. Liang, Z.-H. Zhu, W.-P. Zhou, D. M. Irwin, and Y.-P. Zhang (2010). Adaptive evolution of energy metabolism genes and the origin of flight in bats. Proceedings of the National Academy of Sciences 107:8666–71.

Shoubridge, E. A., and T. Wai (2007). Mitochondrial DNA and the mammalian oocyte. Current Topics in Developmental Biology 77:87–111.

El Shourbagy, S. H., E. C. Spikings, M. Freitas, and J. C. St. John (2006). Mitochondria directly influence fertilisation outcome in the pig. Reproduction 131:233–45.

Sies, H. (2013). Oxidative Stress. Elsevier, New York.

Silva, G., F. P. Lima, P. Martel, and R. Castilho (2014). Thermal adaptation and clinal mitochondrial DNA variation of European anchovy. Proceedings of the Royal Society B: Biological Sciences 281:20141093.

Simčič, T., F. Pajk, M. Jaklic, A. Brancelj, and A. Vrezec (2014). The thermal tolerance of crayfish could be estimated from respiratory electron transport system activity. Journal of Thermal Biology 41:21–30.

Simmons, L. W. (2001). Sperm Competition and Its Evolutionary Consequences in the Insects. Princeton University Press, Princeton, NJ.

Simonson, T. S., Y. Yang, C. D. Huff, H. Yun, G. Qin, D. J. Witherspoon, Z. Bai, F. R. Lorenzo, J. Xing, L. B. Jorde, J. T. Prchal, and R. L. Ge (2010). Genetic evidence for high-altitude adaptation in Tibet. Science 329:72–5.

Sirey, T. M., and C. P. Ponting (2016). Insights into the post-transcriptional regulation of the mitochondrial electron transport chain. Biochemical Society Transactions 44:1491–8.

Sissler, M., L. E. González-Serrano, and E. Westhof (2017). Recent advances in mitochondrial aminoacyl-tRNA synthetases and disease. Trends in Molecular Medicine 23:693–708.

Skold-Chiriac, S., A. Nord, M. Tobler, J.-A. Nilsson, and D. Hasselquist (2015). Body temperature changes during simulated bacterial infection in a songbird: Fever at night and hypothermia during the day. Journal of Experimental Biology 218:2961–9.

Slater, E. C. (2003). Keilin, cytochrome, and the respiratory chain. Journal of Biological Chemistry 278:16455–61.

Slatkin, M. (1985). Gene flow in natural populations. Annual Review of Ecology and Systematics 16:393–430.

Sloan, D. B., A. Nakabachi, S. Richards, J. Qu, S. C. Murali, R. A. Gibbs, and N. A. Moran (2014a). Parallel histories of horizontal gene transfer facilitated extreme reduction of endosymbiont genomes in sap-feeding insects. Molecular Biology and Evolution 31:857–71.

Sloan, D. B., D. A. Triant, M. Wu, and D. R. Taylor (2014b). Cytonuclear interactions and relaxed selection accelerate sequence evolution in organelle ribosomes. Molecular Biology and Evolution 31:673–82.

Sloan, D. B., J. C. Havird, and J. Sharbrough (2017). The on-again, off-again relationship between mitochondrial genomes and species boundaries. Molecular Ecology 26:2212–36.

Sloan, D. B., J. M. Warren, A. M. Williams, Z. Wu, S. E. Abdel-Ghany, A. J. Chicco, and J. C. Havird (2018). Cytonuclear integration and co-evolution. Nature Reviews Genetics 19:635–648.

Van Der Sluis, E. O., H. Bauerschmitt, T. Becker, T. Mielke, J. Frauenfeld, O. Berninghausen, W. Neupert, J. M. Herrmann, and R. Beckmann (2015). Parallel structural evolution of mitochondrial ribosomes and OXPHOS complexes. Genome Biology and Evolution 7:1235–51.

Smith, D. R., and P. J. Keeling (2015). Mitochondrial and plastid genome architecture: Reoccurring themes, but significant differences at the extremes. Proceedings of the National Academy of Sciences 112:201422049.

Smits, P., J. A. M. Smeitink, L. P. van den Heuvel, M. A. Huynen, and T. J. G. Ettema (2007). Reconstructing the evolution of the mitochondrial ribosomal proteome. Nucleic Acids Research 35:4686–703.

Snook, R. R. (2005). Sperm in competition: Not playing by the numbers. Trends in Ecology and Evolution 20:46–53.

Socolich, M., S. W. Lockless, W. P. Russ, H. Lee, K. H. Gardner, and R. Ranganathan (2005). Evolutionary information for specifying a protein fold. Nature 437:512–18.

Solmaz, S. R. N., and C. Hunte (2008). Structure of complex III with bound cytochrome C in reduced state and definition of a minimal core interface for electron transfer. Journal of Biological Chemistry 283:17542–9.

Somero, G. N. (2004). Adaptation of enzymes to temperature: Searching for basic "strategies." Comparative Biochemistry and Physiology—B Biochemistry and Molecular Biology 139:321–33.

Sorenson, M. D., and T. W. Quinn (1998). Numts: A challenge for avian systematics and population biology. The Auk 115:214–21.

Speijer, D. (2015). Birth of the eukaryotes by a set of reactive innovations: New insights force us to relinquish gradual models. BioEssays 37:1268–76.

Speijer, D. (2016). What can we infer about the origin of sex in early eukaryotes? Philosophical Transactions of the Royal Society B: Biological Sciences 371:20150530.

Speijer, D., J. Lukeš, and M. Eliáš (2015). Sex is a ubiquitous, ancient, and inherent attribute of eukaryotic life. Proceedings of the National Academy of Sciences 112:8827–34.

Spiropoulos, J., D. M. Turnbull, and P. F. Chinnery (2002). Can mitochondrial DNA mutations cause sperm dysfunction? Molecular Human Reproduction 8:719–21.

Steinemann, M., and S. Steinemann (1998). Enigma of Y chromosome degeneration: Neo-Y and neo-X chromosomes of *Drosophila miranda* a model for sex chromosome evolution. Genetica 102–103:409–20.

Stewart, J. B., C. Freyer, J. L. Elson, and N.-G. Larsson (2008). Purifying selection of mtDNA and its implications for understanding evolution and mitochondrial disease. Nature Reviews Genetics 9:657–62.

Stewart, J. B., and N.-G. Larsson (2014). Keeping mtDNA in shape between generations. PLoS Genetics 10:e1004670.

Stoeckle, M. Y., and D. S. Thaler (2014). DNA barcoding works in practice but not in neutral theory. PLoS One 9:e100755.

Storchová, R., J. Reif, and M. W. Nachman (2010). Female heterogamety and speciation: Reduced introgression of the z chromosome between two species of nightingales. Evolution 64:456–71.

Storz, J. F. (2007). Hemoglobin function and physiological adaptation to hypoxia in high-altitude mammals. Journal of Mammalogy 88:24–31.

Storz, J. F., G. R. Scott, and Z. A. Cheviron (2010). Phenotypic plasticity and genetic adaptation to high-altitude hypoxia in vertebrates. Journal of Experimental Biology 213:4125–36.

Stroud, D. A., E. E. Surgenor, L. E. Formosa, B. Reljic, A. E. Frazier, M. G. Dibley, L. D. Osellame, T. Stait, T. H. Beilharz, D. R. Thorburn, A. Salim, and M. T. Ryan (2016). Accessory subunits are integral for assembly and function of human mitochondrial complex I. Nature 538:1–17.

Stuart, J. A., L. A. Maddalena, M. Merilovich, and E. L. Robb (2014). A midlife crisis for the mitochondrial free radical theory of aging. Longevity and Healthspan 3:4.

Sun, C., Q.-P. Kong, and Y.-P. Zhang (2007). The role of climate in human mitochondrial DNA evolution: A reappraisal. Genomics 89:338–42.

Sun, N., R. J. Youle, and T. Finkel (2016). The mitochondrial basis of aging. Molecular Cell 61:654–66.

Sung, W., M. S. Ackerman, S. F. Miller, T. G. Doak, and M. Lynch (2012). Drift-barrier hypothesis and mutation-rate evolution. Proceedings of the National Academy of Sciences 109:18488–92.

Sunnucks, P., H. E. Morales, A. M. Lamb, A. Pavlova, and A. Brelsford (2017). Integrative approaches for studying mitochondrial and nuclear genome co-evolution in oxidative phosphorylation. Frontiers in Genetics 8:1–12.

Suzuki, K., K. Mizutani, S. Maruyama, K. Shimono, F. L. Imai, E. Muneyuki, Y. Kakinuma, Y. Ishizuka-Katsura, M. Shirouzu, S. Yokoyama, I. Yamato, and T. Murata (2016). Crystal structures of the ATP-binding and ADP-release dwells of the V1 rotary motor. Nature Communications 7:13235.

Suzuki, T., and T. Suzuki (2014). A complete landscape of post-transcriptional modifications in mammalian mitochondrial tRNAs. Nucleic Acids Research 42:7346–57.

Swaminathan, N. (2008). Why does the brain need so much power? Scientific American April 29:1–4.

Szczepanowska, K., and A. Trifunovic (2017). Origins of mtDNA mutations in ageing. Essays In Biochemistry 61:325–37.

Taanman, J.-W. (1999). The mitochondrial genome: Structure, transcription, translation and replication. Biochimica et Biophysica Acta (BBA)—Bioenergetics 1410:103–23.

Talbot, D. A., C. Duchamp, B. Rey, N. Hanuise, J. L. Rouanet, B. Sibille, and M. D. Brand (2004). Uncoupling protein and ATP/ADP carrier increase mitochondrial proton conductance after cold adaptation of king penguins. Journal of Physiology 558:123–35.

Tangwancharoen, S., and R. S. Burton (2014). Early life stages are not always the most sensitive: Heat stress responses in the copepod *Tigriopus californicus*. Marine Ecology Progress Series 517:75–83.

Tavares, E. S., and A. J. Baker (2008). Single mitochondrial gene barcodes reliably identify sister-species in diverse clades of birds. BMC Evolutionary Biology 8:14.

Teacher, A. G. F., C. Andre, J. Merila, and C. W. Wheat (2012). Whole mitochondrial genome scan for population structure and selection in the Atlantic herring. BMC Evolutionary Biology 12:248.

Tegelström, H. (1987). Transfer of mitochondrial DNA from the northern red-backed vole (*Clethrionomys rutilus*) to the bank vole (*C. glareolus*). Journal of Molecular Evolution 24:218–27.

Templeton, A. R. (2001). Using phylogeographic analyses of gene tress to test species status and processes. Molecular Ecology 10:779–91.

Teshima, K. M., and M. Przeworski (2006). Directional positive selection on an allele of arbitrary dominance. Genetics 172:713–18.

Thangaraj, K., M. B. Joshi, A. G. Reddy, A. A. Rasalkar, and L. Singh (2003). Sperm mitochondrial mutations as a cause of low sperm motility. Journal of Andrology 24:388–92.

Theurey, P., and P. Pizzo (2018). The aging mitochondria. Genes 9:22.

Thibert-Plante, X., and S. Gavrilets (2013). Evolution of mate choice and the so-called magic traits in ecological speciation. Ecology Letters 16:1004–13.

Thompson, L. V. (2006). Oxidative stress, mitochondria and mtDNA-mutator mice. Experimental Gerontology 41:1220–2.

Timmis, J. N., M. A. Ayliffe, C. Y. Huang, and W. Martin (2004). Endosymbiotic gene transfer: Organelle genomes forge eukaryotic chromosomes. Nature Reviews Genetics 5:123–35.

Tobe, S. S., A. C. Kitchener, and A. M. T. Linacre (2010). Reconstructing mammalian phylogenies: A detailed comparison of the cytochrome b and cytochrome oxidase subunit I mitochondrial genes. PLoS One 5:e14156.

Toews, D. P. L., and A. Brelsford (2012). The biogeography of mitochondrial and nuclear discordance in animals. Molecular Ecology 21:3907–30.

Toews, D. P. L., L. Campagna, S. A. Taylor, C. N. Balakrishnan, D. T. Baldassarre, P. E. Deane-Coe, M. G. Harvey, D. M. Hooper, D. E. Irwin, C. D. Judy, N. A. Mason, et al. (2016a). Genomic approaches to understanding population divergence and speciation in birds. The Auk 133:13–30.

Toews, D. P. L., S. A. Taylor, R. Vallender, A. Brelsford, B. G. Butcher, P. W. Messer, and I. J. Lovette (2016b). Plumage genes and little else distinguish the genomes of hybridizing warblers. Current Biology 26:2313–18.

Tomasco, I. H., and E. P. Lessa (2014). Two mitochondrial genes under episodic positive selection in subterranean octodontoid rodents. Gene 534:371–8.

Tońska, K., A. Kodroń, and E. Bartnik (2010). Genotype-phenotype correlations in Leber hereditary optic neuropathy. Biochimica et Biophysica Acta 1797:1119–23.

Tooby, J. (1982). Pathogens, polymorphism, and the evolution of sex. Journal of Theoretical Biology 97:557–76.

Tourmente, M., M. Hirose, S. Ibrahim, D. K. Dowling, D. M. Tompkins, E. R. S. Roldan, and N. J. Gemmell (2017). mtDNA polymorphism and metabolic inhibition affect sperm performance in conplastic mice. Reproduction 154:341–54

Touzet, P., and E. H. Meyer (2014). Cytoplasmic male sterility and mitochondrial metabolism in plants. Mitochondrion 19:166–171.

Trifunovic, A., A. Wredenberg, M. Falkenberg, J. N. Spelbrink, A. T. Rovio, C. E. Bruder, M. Bohlooly-Y, S. Gidlöf, A. Oldfors, R. Wibom, J. Törnell, et al. (2004). Premature ageing in mice expressing defective mitochondrial DNA polymerase. Nature 429:417–23.

Trivers, R. L. L. (1972). Parental investment and sexual selection. In Sexual Selection and the Descent of Man; 1871–1971 (B. Campbell, Editor). Aldine, Chicago, IL, pp. 136–79.

Tsukihara, T., H. Aoyama, E. Yamashita, and T. Tomizaki (1996). The whole structure of the 13-subunit oxidized cytochrome c oxidase at 2.8 angstrom. Science 272:1136.

Tuda, M., N. Wasano, N. Kondo, S.-B. Horng, L.-Y. Chou, and Y. Tateishi (2004). Habitat-related mtDNA polymorphism in the stored-bean pest *Callosobruchus chinensis* (Coleoptera: Bruchidae). Bulletin of Entomological Research 94:75–80.

Tuppen, H. A. L., E. L. Blakely, D. M. Turnbull, and R. W. Taylor (2010). Mitochondrial DNA mutations and human disease. Biochimica et Biophysica Acta 1797:113–28.

Turelli, M., and L. C. Moyle (2007). Asymmetric postmating isolation: Darwin's corollary to Haldane's rule. Genetics 176:1059–88.

Turelli, M., and H. A. Orr (2000). Dominance, epistasis and the genetics of postzygotic isolation. Genetics 154:1663–79.

Turner, J. R. (1967). Why does the genome not congeal? Evolution 21:645–56.

Tuttle, E. M. (2003). Alternative reproductive strategies in the white-throated sparrow: Behavioral and genetic evidence. Behavioral Ecology 14:425–32.

Tuttle, E. M., A. O. Bergland, M. L. Korody, M. S. Brewer, D. J. Newhouse, P. Minx, M. Stager, A. Betuel, Z. A. Cheviron, W. C. Warren, R. A. Gonser, and C. N. Balakrishnan (2016). Divergence and functional degradation of a sex chromosome-like supergene. Current Biology 26:344–50.

Uddin, M., J. C. Opazo, D. E. Wildman, C. C. Sherwood, P. R. Hof, M. Goodman, and L. I. Grossman (2008). Molecular evolution of the cytochrome c oxidase subunit 5A gene in primates. BMC Evolutionary Biology 8:1–12.

Uy, J. A. C., R. G. Moyle, and C. E. Filardi (2009). Plumage and song differences mediate species recognition between incipient flycatcher species of the Solomon Islands. Evolution 63:153–64.

Vaught, R. C., and D. K. Dowling (2018). Maternal inheritance of mitochondria: Implications for male fertility? Reproduction 155:R159–R168.

Vendramin, R., J. Marine, and E. Leucci (2017). Non-coding RNAs: The dark side of nuclear–mitochondrial communication. The EMBO Journal 36:1123–33.

Verberk, W. C. E. P., D. T. Bilton, P. Calosi, and J. I. Spicer (2011). Oxygen supply in aquatic ectotherms: Partial pressure and solubility together explain biodiversity and size patterns. Ecology 92:1565–72.

Verdú, J. R., L. Arellano, and C. Numa (2006). Thermoregulation in endothermic dung beetles (Coleoptera: Scarabaeidae): Effect of body size and ecophysiological constraints in flight. Journal of Insect Physiology 52:854–60.

Verzijden, M. N., R. F. Lachlan, and M. R. Servedio (2005). Female mate-choice behavior and sympatric speciation. Evolution 59:2097–108.

Vinothkumar, K. R., J. Zhu, and J. Hirst (2014). Architecture of mammalian respiratory complex I. Nature 515:80–4.

Vrijenhoek, R. C. (1998). Animal clones and diversity. Are natural clones generalists or specialists? BioScience 48:617–28.

Wade, M. J., and D. M. Drown (2016). Nuclear–mitochondrial epistasis: A gene's eye view of genomic conflict. Ecology and Evolution 6:6460–72.

Wade, M. J., and C. J. Goodnight (2006). Cyto-nuclear epistasis: Two-locus random genetic drift in hermaphroditic and dioecious species. Evolution 60:643–59.

Wai, T., D. Teoli, and E. A. Shoubridge (2008). The mitochondrial DNA genetic bottleneck results from replication of a subpopulation of genomes. Nature Genetics 40:1484–8.

Wai, T., A. Ao, X. Zhang, D. Cyr, D. Dufort, and E. A. Shoubridge (2010). The role of mitochondrial DNA copy number in mammalian fertility. Biology of Reproduction 83:52–62.

Wallace, A. R. (1889). Darwinism. Macmillan, London.

Wallace, B. (1953). On coadaptation in *Drosophila*. The American Naturalist 87:343–58.

Wallace, D. C. (1999). Mitochondrial diseases in man and mouse. Science 283:1482–8.

Wallace, D. C. (2008). Mitochondria as Chi. Genetics 179:727–35.

Wallace, D. C. (2010). Bioenergetics, the origins of complexity, and the ascent of man. Proceedings of the National Academy of Sciences 107:8947–53.

Wallace, D. C. (2016). Genetics: Mitochondrial DNA in evolution and disease. Nature 535:498–500.

Wallace, D. C., and W. Fan (2010). Energetics, epigenetics, mitochondrial genetics. Mitochondrion 10:12–31.

Wallace, D. C., M. D. Brown, and M. T. Lott (1999). Mitochondrial DNA variation in human evolution and disease. Gene 238:211–30.

Wallace, D. C., D. C. Wallace, D. C. Wallace, M. M. Gonzalez-Barroso, F. Bouillaud, D. Ricquier, B. Miroux, G. Sorci, H. A. L. L. Tuppen, E. L. Blakely, D. M. Turnbull, et al. (2015). Mitochondrial-nuclear epistasis impacts fitness and mitochondrial physiology of interpopulation *Caenorhabditis briggsae* hybrids. Molecular Biology and Evolution 6:209–19.

Walsh, J., W. G. Shriver, B. J. Olsen, and A. I. Kovach (2016). Differential introgression and the maintenance of species boundaries in an advanced generation avian hybrid zone. BMC Evolutionary Biology 16:1.

Watmough, N. J., and F. E. Frerman (2010). The electron transfer flavoprotein: Ubiquinone oxidoreductases. Biochimica et Biophysica Acta—Bioenergetics 1797:1910–16.

Weatherhead, P. J., and R. J. Robertson (1979). Offspring quality and the polygyny threshold: "the sexy son hypothesis." The American Naturalist 113:201–8.

Weaver, R. J., P. A. Cobine, and G. E. Hill (2018a). On the bioconversion of dietary carotenoids to astaxanthin in the marine copepod, *Tigriopus californicus*. Journal of Plankton Research 40:142–50.

Weaver, R. J., E. S. A. Santos, A. M. Tucker, A. E. Wilson, and G. E. Hill (2018b). Carotenoid metabolism strengthens the link between feather coloration and individual quality. Nature Communications 9:73.

Webb, W. C., J. M. Marzluff, and K. E. Omland (2011). Random interbreeding between cryptic lineages of the common raven: Evidence for speciation in reverse. Molecular Ecology 20:2390–402.

Weedall, G. D., and N. Hall (2015). Sexual reproduction and genetic exchange in parasitic protists. Parasitology 142:S120–S127.

Wernegreen, J. J. (2012). Endosymbiosis. Current Biology 22:555–61.

Werren, J. H. (1997). Biology of *Wolbachia*. Annual Review of Entomology 124:587–609.

West, S. A., C. M. Lively, and A. F. Read (1999). A pluralist approach to sex and recombination. Journal of Evolutionary Biology 12:1003–12.

Wheeler, Q., and R. Meier (2000). Species Concepts and Phylogenetic Theory: A debate. Columbia University Press, New York.

White, C. R., L. A. Alton, and P. B. Frappell (2012). Metabolic cold adaptation in fishes occurs at the level of whole animal, mitochondria and enzyme. Proceedings of the Royal Society B: Biological Sciences 279:1740–7.

White, C. R., T. M. Blackburn, G. R. Martin, and P. J. Butler (2007). Basal metabolic rate of birds is associated with habitat temperature and precipitation, not primary productivity. Proceedings of the Royal Society B: Biological Sciences 274:287–93.

Whitehead, A. (2009). Comparative mitochondrial genomics within and among species of killifish. BMC Evolutionary Biology 9:11.

Whitnow, G. C. (1986). Regulation of body temperature. In Avian Physiology (P. D. Sturkie, Editor). Springer, Berlin, pp. 221–252.

Whitton, J., C. J. Sears, E. J. Baack, and S. P. Otto (2008). The dynamic nature of apomixis in the angiosperms. International Journal of Plant Sciences 169:169–82.

Widmer, H. R., H. Hoppeler, E. Nevo, C. R. Taylor, and E. R. Weibel (1997). Working underground: Respiratory adaptations in the blind mole rat. Proceedings of the National Academy of Sciences 94:2062–7.

Wiens, J. J. (2004). What is speciation and how should we study it? The American Naturalist 163:914–23.

Willett, C. S. (2010). Potential fitness trade-offs for thermal tolerance in the intertidal copepod *Tigriopus californicus*. Evolution 64:2521–34.

Willett, C. S., and R. S. Burton (2001). Viability of cytochrome c genotypes depends on cytoplasmic backgrounds in *Tigriopus californicus*. Evolution 55:1592–9.

Willett, C. S., and J. T. Ladner (2009). Investigations of fine-scale phylogeography in *Tigriopus californicus* reveal historical patterns of population divergence. BMC Evolutionary Biology 9:139.

Williams, G. C. (1957). Pleiotropy, natural selection, and the evolution of senescence. Evolution 11:398.

Williams, G. C. (1966). Adaptation and Natural Selection: A Critique of Some Current Evolutionary Thought. Princeton University Press, Princeton, NJ.

Williams, G. C. (1975). Sex and Evolution. Princeton University Press, Princeton, NJ.

Williams, T. A. (2014). Evolution: Rooting the eukaryotic tree of life. Current Biology 24:R151–R152.

Williams, T. A., P. G. Foster, C. J. Cox, and T. M. Embley (2013). An archaeal origin of eukaryotes supports only two primary domains of life. Nature 504:231–6.

Wilson, K. H. (1995). Molecular biology as a tool for taxonomy. Clinical Infectious Diseases 20:S117–S121.

Wilson Sayres, M. A., and K. D. Makova (2011). Genome analyses substantiate male mutation bias in many species. BioEssays 33:938–45.

Wink, M., and H. Sauer-Gürth (2002). Mitochondrial-DNA sequences and genomic fingerprinting. British Birds 95:349–55.

Wirtz, P. (1999). Mother species-father species: Unidirectional hybridization in animals with female choice. Animal Behaviour 58:1–12.

Wittig, I., and H. Schägger (2008). Structural organization of mitochondrial ATP synthase. Biochimica et Biophysica Acta 1777:592–8.

Wolf, J. B. W., and H. Ellegren (2017). Making sense of genomic islands of differentiation in light of speciation. Nature Reviews Genetics 18:87–100.

Wolf, J. B. W., J. Lindell, and N. Backström (2010). Speciation genetics: Current status and evolving approaches. Philosophical transactions of the Royal Society B: Biological Sciences 365:1717–33.

Wolf, Y. I., and E. V. Koonin (2013). Genome reduction as the dominant mode of evolution. BioEssays 35:829–37.

Wolff, J. N., and N. J. Gemmell (2013). Mitochondria, maternal inheritance, and asymmetric fitness: Why males die younger. BioEssays 35:93–9.

Wolff, J. N., N. Pichaud, M. F. Camus, G. Cote, P. U. Blier, and D. K. Dowling (2016a). Evolutionary implications of mitochondrial genetic variation: Mitochondrial genetic effects on OXPHOS respiration and mitochondrial quantity change with age and sex in fruit flies. Journal of Evolutionary Biology 29:736–47.

Wolff, J. N., D. M. Tompkins, N. J. Gemmell, and D. K. Dowling (2016b). Mitonuclear interactions, mtDNA-mediated thermal plasticity, and implications for the Trojan female technique for pest control. Scientific Reports 6:30016.

Woodson, J. D., and J. Chory (2008). Coordination of gene expression between organellar and nuclear genomes. Nature Reviews Genetics 9:383–95.

Wright, S. (1942). Genetics and the origin of species. Journal of Heredity 33:283–4.

Wu, C. I., N. A. Johnson, and M. F. Palopoli (1996). Haldane's rule and its legacy: Why are there so many sterile males? Trends in Ecology and Evolution 11:281–4.

Wu, W., M. Goodman, M. I. Lomax, and L. I. Grossman (1997). Molecular evolution of cytochrome c oxidase subunit IV: Evidence for positive selection in simian primates. Journal of Molecular Evolution 44:477–91.

Wu, W., T. R. Schmidt, M. Goodman, and L. I. Grossman (2000). Molecular evolution of cytochrome c oxidase subunit I in primates: Is there coevolution between mitochondrial and nuclear genomes? Molecular Phylogenetics and Evolution 17:294–304.

Wu, Z., J. M. Cuthbert, D. R. Taylor, and D. B. Sloan (2015). The massive mitochondrial genome of the angiosperm *Silene noctiflora* is evolving by gain or loss of entire chromosomes. Proceedings of the National Academy of Sciences 112:10185–91.

Xu, J. (2004). The prevalence and evolution of sex in microorganisms. Genome 47:775–80.

Xu, S., J. Luosang, S. Hua, J. He, A. Ciren, W. Wang, X. Tong, Y. Liang, J. Wang, and X. Zheng (2007). High altitude adaptation and phylogenetic analysis of Tibetan horse based on the mitochondrial genome. Journal of Genetics and Genomics 34:720–9.

Xu, S. Q., Y. Z. Yang, J. Zhou, G. E. Jing, Y. T. Chen, J. Wang, H. M. Yang, J. Wang, J. Yu, X. G. Zheng, and R. L. Ge (2005). A mitochondrial genome sequence of the Tibetan antelope (*Pantholops hodgsonii*). Genomics, Proteomics and Bioinformatics 3:5–17.

Yan, Z., G. Y. Ye, and J. Werren (2018). Evolutionary rate coevolution between mitochondria and mitochondria-associated nuclear-encoded proteins in insects. bioRxiv 288456.

Yang, M., Z. He, S. Shi, C. Wu, C. Hill, S. Diego, L. Jolla, G. P. Wallis, J. W. Arntzen, C. Venditti, A. Meade, et al. (2017). Speciation genomics and a role for the Z chromosome in the early stages of divergence between Mexican ducks and mallards. Evolution 11:1–21.

Yankovskaya, V., R. Horsefield, S. Törnroth, C. Luna-Chavez, H. Miyoshi, C. Léger, B. Byrne, G. Cecchini, and S. Iwata (2003). Architecture of succinate dehydrogenase and reactive oxygen species generation. Science 299:700–4.

Yee, W. K. W., K. L. Sutton, and D. K. Dowling (2013). In vivo male fertility is affected by naturally occurring mitochondrial haplotypes. Current Biology 23:R55–R56.

Yi, X., Y. Liang, E. Huerta-Sanchez, X. Jin, Z. X. P. Cuo, J. E. Pool, N. Xu, H. Jiang, N. Vinckenbosch, T. S. Korneliussen, H. Zheng, et al. (2010). Sequencing of 50 human exomes reveals adaptation to high altitude. Science 329:75–8.

Yu, L., X. Wang, N. Ting, and Y. Zhang (2011). Mitogenomic analysis of Chinese snub-nosed monkeys: Evidence of positive selection in NADH dehydrogenase genes in high-altitude adaptation. Mitochondrion 11:497–503.

Yu, X., U. Gimsa, L. Wester-Rosenlof, E. Kanitz, W. Otten, M. Kunz, and S. M. Ibrahim (2009). Dissecting the effects of mtDNA variations on complex traits using mouse conplastic strains. Genome Research 19:159–65.

Zachos, F. E. (2016). Species Concepts in Biology: Historical Development, Theoretical Foundations and Practical Relevance. Springer International Publishing, New York.

Zeyl, C., B. Andreson, and E. Weninck (2005). Nuclear-mitochondrial epistasis for fitness in *Saccharomyces cerevisiae*. Evolution 59:910–14.

Zhang, F., and R. E. Broughton (2013). Mitochondrial-nuclear interactions: Compensatory evolution or variable functional constraint among vertebrate oxidative phosphorylation genes? Genome Biology and Evolution 5:1781–91.

Zhang, G., C. Cowled, Z. Shi, Z. Huang, K. A. Bishop-Lilly, X. Fang, J. W. Wynne, Z. Xiong, M. L. Baker, W. Zhao, M. Tachedjian, et al. (2013a). Comparative analysis of bat genomes provides insight into the evolution of flight and immunity. Science 339:456–60.

Zhang, Y., F. Humes, G. Almond, A. Kavazis, and W. R. Hood (2017). A mitohormetic response to pro-oxidant exposure in the house mouse. American Journal of Physiology—Regulatory, Integrative and Comparative Physiology 314:R122–R134.

Zhang, Z.-Y., B. Chen, D.-J. Zhao, and L. Kang (2013b). Functional modulation of mitochondrial cytochrome c oxidase underlies adaptation to high-altitude hypoxia in a Tibetan migratory locust. Proceedings of the Royal Society B: Biological Sciences 280:20122758.

Zhao, X., N. Wu, Q. Zhu, U. Gaur, T. Gu, and D. Li (2015). High-altitude adaptation of Tibetan chicken from MT-COI and ATP-6 perspective. Mitochondrial DNA Part A 27:3280–8.

Zhu, C. T., P. Ingelmo, and D. M. Rand (2014). GxGxE for lifespan in *Drosophila*: Mitochondrial, nuclear, and dietary interactions that modify longevity. PLoS Genetics 10:e1004354.

Zhu, J., K. Z. Q. Wang, and C. T. Chu (2013). After the banquet: Mitochondrial biogenesis, mitophagy, and cell survival. Autophagy 9:1663–76.

Zink, R. M., and G. F. Barrowclough (2008). Mitochondrial DNA under siege in avian phylogeography. Molecular Ecology 17:2107–21.

Zink, R. M., and M. C. McKitrick (1995). The debate over species concepts and its implications for ornithology. Auk 112:701–19.

Index

The manufacturer's authorised representative in the EU for product safety is Oxford University Press España S.A. of el Parque Empresarial San Fernando de Henares, Avenida de Castilla, 2 – 28830 Madrid (www.oup.es/en or product.safety@oup.com). OUP España S.A. also acts as importer into Spain of products made by the manufacturer.

www.ingramcontent.com/pod-product-compliance
Ingram Content Group UK Ltd.
Pitfield, Milton Keynes, MK11 3LW, UK
UKHW020426250726
13967UKWH00007B/2829
9780198818267